Baryonic Dark Matter

NATO ASI Series

Advanced Science Institutes Series

A Series presenting the results of activities sponsored by the NATO Science Committee, which aims at the dissemination of advanced scientific and technological knowledge, with a view to strengthening links between scientific communities.

The Series is published by an international board of publishers in conjunction with the NATO Scientific Affairs Division

A Life Sciences	Plenum Publishing Corporation
B Physics	London and New York
C Mathematical and Physical Sciences	Kluwer Academic Publishers Dordrecht, Boston and London
D Behavioural and Social Sciences	
E Applied Sciences	
F Computer and Systems Sciences	Springer-Verlag
G Ecological Sciences	Berlin, Heidelberg, New York, London,
H Cell Biology	Paris and Tokyo

Baryonic Dark Matter

edited by

D. Lynden-Bell

Institute of Astronomy & Clare College,
Cambridge, U.K.

and

G. Gilmore

Institute of Astronomy & King's College,
Cambridge, U.K.

Springer-Science+Business Media, B.V.

Proceedings of the NATO Advanced Study Institute on
Baryonic Dark Matter
Cambridge, U.K.
July 17–28, 1989

Library of Congress Cataloging in Publication Data

Baryonic dark matter : proceedings of the NATO ASI held at the
  Institute of Astronomy, Cambridge, July 17-28, 1989 / edited by D.
  Lynden-Bell & G. Gilmore.
       p.   cm. -- (NATO advanced study institutes series. Series C,
  Mathematical and physical sciences ; v. 306)

    1. Dark matter (Astronomy)--Congresses. 2. Baryons--Congresses.
  3. Nuclear astrophysics--Congresses. I. Lynden-Bell, D.
  II. Gilmore, Gerry, 1951-   . III. Series.
  QB791.3.B37 1990
  523.1'12--dc20
                                                          90-4180

ISBN 978-94-010-6742-3   ISBN 978-94-009-0565-8 (eBook)
DOI 10.1007/978-94-009-0565-8

Published by Kluwer Academic Publishers,
P.O. Box 17, 3300 AA Dordrecht, The Netherlands.

Kluwer Academic Publishers incorporates the publishing programmes of
D. Reidel, Martinus Nijhoff, Dr W. Junk and MTP Press.

Sold and distributed in the U.S.A. and Canada
by Kluwer Academic Publishers,
101 Philip Drive, Norwell, MA 02061, U.S.A.

In all other countries, sold and distributed
by Kluwer Academic Publishers Group,
P.O. Box 322, 3300 AH Dordrecht, The Netherlands.

Printed on acid-free paper

TABLE OF CONTENTS

PREFACE

The visible universe is a small perturbation on the material universe. Zwicky and Sinclair Smith in the 1930s gave evidence of invisible mass in the Coma and Virgo Clusters of Galaxies. Better optical data has only served to confound their critics and the X-ray data confirms that the gravitational potentials are many times larger than those predicted on the basis of the observed stars. Dynamical analyses of individual galaxies have found that significant extra mass is needed to explain their rotational velocities. On much larger scales, tens of megaparsecs, there is suggestive evidence that there is even more mass per unit luminosity.

What is this non-luminous stuff of which the universe is made? How much of it is there? Need there be only one kind of stuff? There are three basic possibilities:– all of it is ordinary (baryonic) matter, all of it is some other kind of (non–baryonic) matter, or some of it is baryonic and some is non–baryonic.

Progress in deciding between these options can be obtained in several ways. One might compile a list of all plausible forms in which baryonic mass might be hiding, and attempt to test each possibility to the level of proof or exclusion in turn. One might attempt to identify a characteristic length scale associated with the dark mass, and thereby constrain its nature. One might deduce the amount, and to some extent constrain the nature of the dark matter from considerations of elegance in cosmological theory. Finally, one can have recourse to the detailed calculation of the creation of the light elements in the Big Bang, since the distribution of these elements is sensitive to the total baryonic mass in the universe.

Each of these approaches is discussed in articles in this book. Reviews are presented of the current status of studies of the distribution of mass in the universe, with particular emphasis on that fraction of the unidentified mass which could be normal baryonic (dark) matter. The consensus result is that a substantial fraction, probably most, of all the baryonic matter which is likely to exist, based on Big Bang element nucleosynthesis models, has yet to be identified. No information is available as to the location or form of this baryonic dark matter, and only lower limits as to its amount are available. There is in fact no dynamical evidence that *any* significant amount of non–baryonic matter exists in the universe. On the other hand, there is irrefutable evidence that some (significant) amount of baryonic matter does exist. It is therefore a conservative hypothesis that considerably more baryonic matter *could* exist than is currently identified from electromagnetic radiation, and it merely remains to find where and how it is hidden from our view.

This volume contains the invited review talks at a 2–week meeting held in Cambridge, UK, in July 1989 to discuss Baryonic Dark Matter. In addition to these invited talks many shorter talks were presented, providing an up–to–date and informative analysis of the distribution of normal matter in the Universe. We thank the participants for a lively meeting, NATO Science Committee for funding it, and Alice Julier for her help in compiling this volume.

D. Lynden-Bell   G. Gilmore

Briggs Mollere Little Biviano White Pesce Moore Kroupa Ingham
Scott George Wszolek Schiller Hudson Edge Peterson Valls-Gabaud

Watt Johnstone Grieger Cancelmo Dejonghe Weinberg Scherrer Canizares Carr
Ashman Salucci von Linde Fruchter Rauch Basu Stebbins Tremaine Thorsteinsson

Raychaudhury Disney Casertano Broeils Sato Turner Secco Ekmekçi Lynden-Bell
Amendt Sackett Gould Sellwood Alimi Lake Cappi Rohlfs Hogan

Evans Saslaw Weidemann Gilmore Sargent Novotny Clarke Tamanaha Silk Julier
Davies Beckman van Albada Rees Fowler Fenton Maurogordato Frenk Bailey

PARTICIPANTS

J.-M. Alimi, *Observatoire de Paris, Meudon, France.*
P. Amendt, *S.I.S.S.A., Strada Costiera 11, Trieste 34014,Italy.*
K.M. Ashman, *Space Telescope Science Institute, Baltimore, MD 21218, U.S.A.*
M. Bailey, *Dept. of Astronomy, The University, Manchester M13 9PL, UK*
B. Basu, *75/3A Hazra Road, Calcutta-700 029, India*
J. Beckman, *Area de Investigacion, Inst. de Astrofis. de Canarias, Tenerife*
A. Biviano, *Dipartimento di Astronomia, Trieste, Italy*
R. Bond, *Canadian Institute for Theoretical Astrophysics, Toronto, Canada*
A. Broeils, *Kapteyn Institute, P.O. Box 800, 9700 AV Groningen, The Netherlands.*
R. Canal, *Dept. de Fisica de l'Atmosfera y Astrofisica, U. de Barcelona, Barcelona, Spain*
D. L. Cancelmo, *5100 W. Camino del Desierto, Tucson, AZ 85745, U.S.A.*
C. Canizares, *Massachusetts Institute of Technology, Cambridge, MA 02139, U.S.A.*
A. Cappi, *Osservatorio Astronomico di Bologna, 40100 Bologna, Italy.*
C. Carignan, *Dept. Physique, Universite de Montreal, Quebec, Canada H3C 3J7*
B.J. Carr, *School of Mathematical Sciences, Queen Mary College, London E1 4NS, UK*
S. Casertano, *Kapteyn Laboratory, Postbus 800, 9700 AV Groningen, The Netherlands*
T.T. Chia, *National University of Singapore, Physics Dept., Singapore 0511*
J. Davies, *Dept. of Physics, University College, Cardiff CF1 3TH, UK*
H. Dejonghe, *Sterrenkundig Observatorium - RUG, B9000 Gent, Belgium*
M. Disney, *Dept. of Physics, University College, Cardiff CF1 3TH, UK*
R. Durrer, *Joseph Henry Labs., Physics Dept., Princeton, NJ 08540, U.S.A.*
A. Edge, *Institute of Astronomy, Madingley Road, Cambridge, UK*
F. Ekmekci, *Ankara University, Dept. of Astronomy & Space Sciences, Ankara, Turkey*
R. Evans, *Dept. of Physics, University of Wales, Cardiff, UK*
A.C. Fabian, *Institute of Astronomy, Madingley Road, Cambridge, UK*
M. Fitchett, *Space Telescope Science Institue, Baltimore MD 21218, U.S.A.*
W. Fowler, *W.K. Kellogg Radiation Lab., CALTECH, Pasadena CA 91125, U.S.A.*
C.S. Frenk, *Durham University, Dept. of Physics, Durham DH1 3LE, UK*
A.S. Fruchter, *Carnegie Institution, Washington D.C. 20015, U.S.A.*
G. Gilmore, *Institute of Astronomy, Madingley Road, Cambridge, UK*
A. Gould, *Institute for Advanced Study, Princeton NJ 08540, U.S.A.*
B. Grieger, *Hamburger Sternwarte, D-2050 Hamburg 80, W. Germany*
S. Hayakawa, *Nagoya University, Furo-Cho, Chikusa-ku, Nagoya, Japan*
C. Hogan, *Steward Observatory, University of Arizona, Tucson, U.S.A.*
M. Hudson, *Institute of Astronomy, Madingley Road, Cambridge, UK*
I. Kovner, *Weizmann Institute, Rehovot 76100, Israel*
K. Kuijken, *Vijfde Dreef 3, B-2080 Kapellen, Belgium*
G. Lake, *Astronomy FM20, University of Washington, Seattle WA 98195, U.S.A.*
D. Lynden-Bell, *Institute of Astronomy, Madingley Road, Cambridge, UK*
A. Lyne, *Dept. of Astronomy, The University, Manchester M13 9PL*
E. Marsch, *Max-Planck-Institut fur Aeronomie, D-3411 Lindau, F.R.G.*
S. Maurogordato, *Observatoire de Meudon - DAEC, Meudon, France*
J.G. Mollere III, *Coll. of General Studies, W. Michigan U., Kalamazoo, U.S.A*

xii

B. Moore, *Durham University, Dept. of Physics, Durham DH1 3LE, UK*

L. Nelson, *Canadian Institute of Theoretical Astrophysics, Toronto, Canada*

B.E.J. Pagel, *Royal Greenwich Observatory, Herstmonceux, Sussex BN27 1RP, UK*

B.A. Peterson, *Mt Stromlo Observatory, Woden A.C.T. 2606, Australia*

S. Pineault, *Laval University, Dept. of Physics, Quebec, Canada, G1K 7P4*

T.J. Ponman, *School of Physics & Space Res., Birmingham U., Birmingham, UK*

M. Rauch, *Windenerstrasse 57, D-5408 Nassau, W. Germany*

S. Raychaudhury, *Institute of Astronomy, Madingley Road, Cambridge, UK*

M.J. Rees, *Institute of Astronomy, Madingley Road, Cambridge, UK*

K. Rohlfs, *Astronom. Institut der Ruhr Universitat, D 4630 Bochum 1, F.R.G.*

P.D. Sackett, *Kapteyn Laboratorium, Postbus 800, 9700 AV Groningen, The Netherlands*

P. Salucci, *S.I.S.S.A., Strada Costiera 11, Trieste 34014, Italy*

W.L.W. Sargent, *Dept. of Astronomy 105-24, CALTECH, Pasadena CA 91125, U.S.A.*

K. Sato, *University of Tokyo, Dept. of Physics, Tokyo 113, Japan*

R. Scherrer, *Dept. of Physics, Ohio State U., Columbus OH 43210, U.S.A.*

P. Schiller, *MPI fur Physik und Astrophysik, D-8046 Garching b. Munchen, F.R.G.*

J. Schneider, *DARC, Observatoire de Meudon, Meudon, France*

D. Schreiber, *Tel Aviv University, 69978 Tel Aviv, Israel*

D. Scott, *Institute of Astronomy, Madingley Road, Cambridge CB3 0HA, UK*

L. Secco, *Dipartimento Astronomia, Univ. Padova, Padova, Italy*

J.A. Sellwood, *Dept. of Astronomy, The University, Manchester M13 9PL, UK*

T. Shanks, *Dept. of Physics, Durham University, Durham DH1 3LE, UK*

J. Silk, *Department of Astronomy, U. of California, Berkeley CA 94720, U.S.A.*

A. Stebbins, *Canadian Institute for Theoretical Astrophysics, Toronto, Canada M5S 1A1*

F. Tamanaha, *601 Campbell Hall, U. of California, Berkeley CA 94720, U.S.A.*

C. Thompson, *130-33 CALTECH, Pasadena CA 91125, U.S.A.*

E.H. Thorsteinsson, *'Die Burse' Zi.703, 2000 Hamburg 54, F.R.G.*

S. Tremaine, *Canadian Institute of Theoretical Astrophysics, Toronto, Canada M5S 1A1*

E.L. Turner, *Princeton University Observatory, Peyton Hall, Princeton, U.S.A.*

D. Valls-Gabaud, *Institute d'Astrophysique, F-75014 Paris, France*

T. van Albada, *Kapteyn Laboratorium, 9700 AV Groningen, The Netherlands*

J. von Linde, *Hamburger Sternwarte, D-2050 Hamburg 80, F.R.G.*

M. Watt, *School of Physics & Space Research, Birmingham U., Birmingham, U.K.*

V. Weidemann, *Institut fur Theoretische Physik und Sternwarte, D-2300 Kiel, F.R.G.*

M. Weinberg, *Institute for Advanced Study, Princeton NJ 08540, U.S.A.*

B. Wszolek, *Jagiellonian University, Astronomical Observatory, Krakow, Poland*

HOW MANY BARYONS ARE THERE?

Craig J. Hogan
Steward Observatory
University of Arizona
Tucson, AZ 85721
USA

ABSTRACT. An overview of baryonic dark matter is presented from a cosmogonical perspective.

According to Eddington, the number of protons and electrons in the entire universe was $\frac{3}{2} \times 136 \times 2^{256}$. Although many now would doubt that he was exactly right, the issue of how much of the mass of the universe is made of such familiar stuff, perhaps in some unfamiliar form, and how much is made of something completely different, for example some new and exotic particle left over from the early universe, is still one that provokes strong opinions and · prejudices based as much on aesthetic taste as on empirical evidence. This brief overview will focus on the question of how much dark baryons contribute to the mean density of the universe, summarizing some of the current prejudices and preferences of the cosmological community, especially as reflected at this workshop. More detailed reviews of specific issues, and more comprehensive references to the literature, appear elsewhere in this volume.

How many baryons are known to exist? Fairly firm lower limits on the mean baryon density are available from several different points of view, reviewed elsewhere in these proceedings. For example, Sargent (Sargent & Steidel 1990) emphasizes that cold diffuse gas can of course be seen through its *absorption* of light, and indeed observation of absorption lines in the spectra of quasi-stellar objects (QSO's) gives one of the firmest lower limits on the density of dark baryons (although since these are seen at substantial redshifts, the same gas might not be dark today). Dozens of QSO's are known at high redshift, and their sightlines traverse a known pathlength through the universe. About one in five passes through a cloud of matter of very high neutral hydrogen column density—high enough that the amount of hydrogen can be accurately measured by the damped wings of the line profile. The physical state of the clouds is somewhat uncertain—they may be gaseous dwarf galaxies or giant gas-rich disks—but their contribution to the mean density is estimated directly from the observations, by dividing the total column density by the total line-of-sight (see Wolfe 1987).

Another lower limit is obtained by just adding up the mass of known stellar populations in galaxies, by taking a theoretical or empirical mass-to-light ratio for stellar populations in various types of galaxies and multiplying by the corresponding luminosity density. There can be procedural differences of detail, but answers agree to within a factor of two. The interesting point is that both of these direct measures of baryonic density always give very

1

D. Lynden-Bell and G. Gilmore (eds.), Baryonic Dark Matter, 1–6.
© 1990 *Kluwer Academic Publishers.*

small values; the absorption clouds at a redshift of 2 contribute only about $\Omega_b \simeq 10^{-2}h^{-1}$, and present-day luminous stellar populations in galaxies (which may be the same material) contribute $\Omega_b \simeq 8 \times 10^{-3}h^{-1}$ (assuming an $M/L = 2$ and a luminosity density $L_B = 3 \times 10^8 h\, L_\odot\, \mathrm{Mpc}^{-3}$). Observationally, more baryons than this need not exist, although there is plenty of indirect evidence [for example in the cluster cooling flows, as discussed here by Fabian (1990)] that not all baryons are accounted for this way.

Here $\Omega = 1$ corresponds to the density required to exactly balance the kinetic energy of expansion and gravitational potential energy, thus giving the universe zero total energy. Relating this to a physical density involves knowing the expansion rate in terms of Hubble's constant $H = 100h\,\mathrm{kms}^{-1}\mathrm{Mpc}^{-1}$, where h probably lies between $1/2$ and 1. The known inventory of baryons is therefore small in this dimensionless sense.

A total density very close to the critical value $\Omega = 1$ is not only predicted by inflation models of the big bang (see eg Turner 1987) but is always expected unless the present epoch is somehow imprinted at the moment of the big bang. The "inflation-agnostic" reason for preferring $\Omega = 1$, and cosmological constant $\Lambda = 0$, is the following. Suppose we rewrite the usual Friedmann equation governing the expansion in the following form:

$$\Omega - 1 = \frac{k}{(aH)^2} + \frac{\Lambda}{3H^2}.$$

Here k and Λ are constants but a, H and aH all evolve with time. We can rule out empirically $aH \ll 1$ or $\Lambda \gg 3H^2$ (the universe is not curvature- or vacuum-dominated by a large factor), so unless the present epoch is imprinted on the initial conditions somehow (allowing $aH \simeq 1$, $\Lambda \simeq 3H^2$), we would expect $aH \gg 1$ and $\Lambda \ll 3H^2$. But this implies that $\Omega = 1 \pm \epsilon$ and $\Lambda = \pm\epsilon$, with $\epsilon \ll 1$.

The reason people continue to defend this prejudice is that if $\Omega = 1 \pm \epsilon$ we can conceive of physical mechanisms, of which inflation provides a concrete example, for getting the expansion going and setting the parameters of the classical period of the expansion. But if not then we will have a generic problem with any mechanism that cannot single out our particular epoch for whatever reason. The only scheme so far proposed which might be able to accommodate an open universe is one based on many parallel universes with different parameters, where the expansion properties of our entire universe are somehow fixed by our presence at a certain epoch (see Barrow and Tipler 1988).

Some current theories of galaxy formation lend support to the idea of a critical density of exotic (i.e. non-baryonic) matter, because without it the gravitational instability picture with scale-free fluctuations and standard recombination history creates excessive anisotropy in the cosmic background radiation on small angular scales (Beckman, 1990). Inert dark matter solves this problem because it can grow density fluctuations gravitationally while being decoupled from the radiation. It should be remembered however that anisotropy can be avoided in other ways as well—for example, by reionizing the universe early enough small-scale fluctuations can be smeared out, in which case the observable anisotropy can be made acceptably small. (Generally speaking this only works if the fluctuation spectrum is not scale-free, so that the large-scale anisotropies, which do survive ionization, can be made small relative to the matter fluctuations on galaxy scales.) It is even possible that the gravitational instability picture is not at all correct, if the growth of fluctuations is influenced appreciably by nongravitational forces, in which case completely different processes dominate the formation of anisotropy.

An indirect but very precise fix on the total baryon density can be obtained by matching the predicted abundances of various elements, which depend on Ω_b, with observations. A complete up-to-date review of this subject is given by Pagel in these proceedings (Pagel 1990). The simplest homogeneous big bang model can simultaneously fit the best current estimate of the primordial abundances of several light elements only for a single density, $\Omega_b h^2 = .01$, and 95% confidence limits on these abundances still permit only a narrow range of density: $.01 \leq \Omega_b h^2 \leq .015$. This range has steadily narrowed as observations have improved; the strictest upper limits now come from very accurate recent determinations of He in gas-rich dwarf galaxies, the lower limits from observations of Li in atmospheres of old stars and interstellar measurements of D and ^{3}He.

If this model is taken literally (and its success in fitting diverse abundances suggests that perhaps we should) then with this new observational data we can draw several interesting conclusions.

First, about 90% of the baryons that exist have not been detected either by absorption or emission, and must exist in some currently dark inert form, the most conservative candidate being some form of burned-out stellar remnants (degenerate dwarfs, neutron stars, or black holes). This idea can be squared with models of chemical enrichment and evolution of galaxies and, with a variable IMF, with models for forming the submillimeter background. Alternatively, it may be in the form of "Jupiters" or "Brown Dwarfs"—bodies too small to burn brightly as main-sequence stars. Such bodies would have no appreciable effect on chemical enrichment. (Another interesting candidate is even smaller degenerate bodies—"hydrogen icebergs"—but it is harder to see how to form these). All of these interesting possibilities are reviewed elsewhere in this volume.

Second, there *must* also be *some* nonbaryonic dark matter, because the mean density of all dark matter is known to contribute at least $\Omega \simeq 0.1$, as measured from the dynamics of various galactic cluster and supercluster systems using the virial theorem or flow models. This conclusion could not be drawn before Pagel's new abundance data were available, and is interesting because literal adherents to the standard nucleosynthesis model are now forced for the first time to accept a nonnegligible contribution of some exotic form of matter, even if they are skeptical about the need for $\Omega = 1$.

Finally, if Ω really is very close to 1 as some would like to think, cosmic matter is almost all (more than 94%) nonbaryonic. These three conclusions form the orthodox view on the subject.

But in spite of its broad success in explaining the pattern of primordial abundances, perhaps one should not believe the standard homogeneous big bang model in literal detail, especially since it now seems to be leading inexorably to the apocalyptic conclusion that much of the dark matter cannot be baryonic. It is often pointed out that indeed some inhomogeneity of matter is required to make galaxies, although the amplitude of these fluctuations can be made small enough to leave the standard picture almost perfectly intact. Nonlinear inhomogeneities could however alter it drastically. To take one extreme example (see Sale and Mathews, 1986), lumps of matter with density contrast $(\delta\rho/\bar{\rho}) \gg 1$ and baryon mass larger than $10^5 \, \mathcal{M}_\odot (\delta\rho/\bar{\rho})^{-1/2}$ would lie above the Jeans mass and collapse at recombination, locking up all the "wrong" abundances in remnants that never mix with observed material. The observed abundances in this case would be indistinguishable from the standard model, but a large quantity of baryons would be locked up in dark remnants.

Perturbations this large are implausible, because they are much larger than the horizon at nucleosynthesis and defy most mechanisms for producing the net baryon number (see below). For lumps on somewhat smaller scales ($10^{-17} \lesssim M \lesssim 10^5 (\delta\rho/\bar{\rho})^{-1/2} \, \mathcal{M}_\odot$) the gas from the lumps would mix with that between them, but only after nucleosynthesis. This tends to mimic a higher-density universe, because some observable matter actually undergoes nucleosynthesis at lower than average entropy—producing the opposite to the effect we are seeking if we want a higher Ω_b to be consistent with the abundances (see Epstein and Petrosian 1975).

On still smaller scales ($10^{-23} \lesssim M \lesssim 10^{-17} \, \mathcal{M}_\odot$) however, a new effect comes into play, emphasized in the contribution here of W. Fowler (Fowler 1990). Because neutrons and protons couple to the plasma differently, the neutrons and protons from a lump of matter disperse at different rates; for lumps in this range of sizes, such differential diffusion would create neutron-rich pockets of matter at nucleosynthesis. This effect leads to modification of the abundance predictions that reproduces the observed abundance pattern over a wider range of baryon density than in the standard model—perhaps even up to the critical density. (Below $\simeq 10^{-23} \, \mathcal{M}_\odot$, inhomogeneities diffuse away in all species before nucleosynthesis, leaving no residual effect.) The attraction of this idea is that no exotic matter would then be required.

Such small-scale departures from homogeneity are more plausible than the larger ones just discussed, and in particular might be generated in a first-order cosmic phase transition from quark to hadronic matter at about 100 MeV; the natural scale for these lumps calculated from simple nucleation theory usually lies in the right range of scales, 10^{-23} to $10^{-17} \, \mathcal{M}_\odot$.

When strong-interaction physics is better understood, such modifications to the simplest big bang might actually become mandatory; present theories are mixtures of calculation and conjecture, particularly concerning the nature of the perturbations. Modeling of these universes, especially the nucleosynthetic processes, is however gradually moving closer to realism. Supercomputers are being applied to compute realistically spatial particle diffusion and nuclear reactions simultaneously; even new laboratory experiments must be deployed in order to measure ordinarily unimportant reaction rates that gain prominence in a neutron-rich environment. New species are emerging as possibly cosmologically prominent; primordial Be and even much heavier elements, such as gold and platinum, may be produced in observably copious quantities. Current indications are that such models may achieve $\Omega_b = 1$ only in a very narrow range of parameter space, but since no calculation to date has included diffusion of all species, including photons, protons, and neutrinos, this conclusion is still preliminary. The excitement now is that abundances might eventually be used to probe not just the mean density of baryons, but also events at a much earlier time—not just the first three minutes, but even the first three microseconds, when the universe turned from quark soup into ordinary hadronic plasma.

Another question revolves around the issue of how many baryons there *ought* to be, and how this compares with the expected density of the exotic dark matter particles. In spite of the proliferation of particle candidates, none of those so far proposed has a mass density which depends on the same theoretical parameters as the baryons. To be sure, the densities of these candidates come out in about the right range–otherwise they would not be candidates– but various parameters must be tuned to make this happen, and there is as far as we know no natural reason to expect a density comparable to baryons.

It is usually thought that the initial baryon number is zero– that is, that the cosmic plasma is initially neutral, with exactly the same numbers of particles and antiparticles.

This is expected to occur as a result of inflation, and is a generic prediction for any model which makes the universe very large by generating vast amounts of entropy, as inflation does. The observed excess of baryons over antibaryons can then be produced by a "Sakharov process" (Sakharov 1967), which has three necessary ingredients: (1) Departure from thermal equilibrium. This is needed to provide a macroscopic arrow of time; it can be provided by the expansion. (2) CP violation in the fundamental theory of matter. This is needed so that the microscopic physics can recognize the arrow of time, and differentiate between b and $\bar{b}$ in some interactions. (3) Baryon number must not be absolutely conserved. This is clearly required to produce a net baryon number out of an initially neutral medium.

This baryogenesis must occur after inflation and before nucleosynthesis. GUT theories put it at $T \simeq 10^{15}$GeV; these theories also predicted proton decay on an observably short timescale $\sim 10^{31}$ y, and inspired the beautiful experiments which detected supernova neutrinos (but no proton decay as yet). Recent proposals have put baryogenesis as low as $T \simeq 100$GeV, perhaps associated with nonperturbative effects in the Standard Model with no extensions at all (Kuzmin et al. 1985; Affleck and Dine 1985; Linde 1985).

The point here is, *no exotic dark matter candidate is produced by a Sakharov process.* Such particles as massive neutrinos, wimps and axions could make contributions to the cosmic mass density, but none of them has a density determined by the same factors that determine the baryon density. The fact that the baryon density is in a very broad sense comparable to the total cosmic density is thus most natural in a scheme where all of the dark matter is baryonic, and other forms of dark matter make a negligible contribution to the total density. Of course, there could be a deeper theory in which the necessary coincidences could be explained to make one of the "ino" candidates dominate over the baryon density by a moderate factor (see Turner and Carr, 1987). Put another way, the whole idea of exotic dark matter would seem more compelling if the theory of this matter could also at the same time predict correctly the abundance of the matter we know exists—the baryonic stuff.

This work was supported by NASA grant NAGW-1703, NSF grant 8714663, and an Alfred P. Sloan Foundation fellowship.

References

Affleck, I., and Dine, M. 1985, *Nucl. Phys.*, **B249**, 361.

Barrow, J., and Tipler, F. 1988, *The Anthropic Cosmological Principle* (Oxford).

Beckman, J. 1990, in *Baryonic Dark Matter* eds D. Lynden-Bell & G. Gilmore (Dordrecht: Kluwer) p265.

Epstein, R., and Petrosian, V. 1975. *Astrophys. J.* **197**, 281.

Fabian, A.C. 1990, in *Baryonic Dark Matter* eds D. Lynden-Bell & G. Gilmore (Dordrecht: Kluwer) p195.

Kuzmin, V. A., Rubakov, V. A., and Shoposhnikov, M. E. 1985, *Phys. Lett.*, **155B**, 36.

Linde, A. D. 1985, *Phys. Lett.*, **160B**, 243.

Pagel, B. 1990, in *Baryonic Dark Matter* eds D. Lynden-Bell & G. Gilmore (Dordrecht: Kluwer) p237.

Sakharov, A. D. 1967, *Pis'ma Zh. Eksp. Teoret. Fiz.*, **5**, 32.

Sale, K. E., and Mathews, G. J. 1986. *Astrophys. J. Lett.* **309**, L1.

Sargent, W.L.W. & Steidel, C. 1990, in *Baryonic Dark Matter* eds D. Lynden-Bell & G. Gilmore (Dordrecht: Kluwer) p223.

Turner, M. S.. and Carr, B. J. 1987, *Mod. Phys. Lett. A.*, **2**, 1.

Turner, M. S. 1987, Fermilab preprint 87/35-A, *Proceedings of the GIFT Seminar* (Singapore: World Scientific) in press

Wolfe, A. M. 1987, in *QSO absorption lines: probing the Universe*, ed. J. C. Blades, D. A. Turnshek, and C. A. Norman (Cambridge: Univ. Press), p. 297.

COMETARY MASSES

M.E. Bailey
Department of Astronomy
The University
Manchester M13 9PL
England

ABSTRACT. The principal uncertainty in relating theories of cometary origin to observation, and theories of cometary dynamics to the formation of the solar system and the evolution of planetary and satellite surfaces, is the question of how accurately to determine cometary masses. Even the intrinsic *brightness* distribution of comets, corrected for observational selection effects, is not generally agreed, while estimates of the masses of comets, mostly based on 'total' absolute magnitudes are themselves subject to large systematic uncertainties. These factors mean that cometary masses are probably uncertain by at least an order of magnitude, making estimates of the total mass of comets in the solar system (or elsewhere) subject to an equally large statistical uncertainty. On the other hand, theoretical arguments suggest that the original mass of comets in the solar system may have ranged up to around 10% of the total mass normally associated with the condensible 'solid' component of the nebula, while similar arguments applied to interstellar material suggest that comets, solid bodies of cometary size, or even partially aggregated 'giant' interstellar grains (of which comets are usually thought to be made), may make up 10% of the mass density associated with dust. Although ordinary comets, as we know them, cannot therefore be the solution to the problem posed by the dark, 'missing mass' in galaxies, galaxy clusters or elsewhere, they may exist in sufficient numbers to have significant astrophysical consequences in a wide range of astronomical environments.

1. Introduction

Comets, when bright enough to be seen with the unaided eye, are rather beautiful objects. They range in appearance from a faint star-like point of light, or a mere smudge on the sky (like the view, for many of us, of Halley's comet at its last apparition), to the rather more spectacular images pictured in most popular books on astronomy — showing a broad, curved dust tail and a straight, highly structured gas tail, both emanating from a round, fuzzy cometary 'head' or coma, the latter frequently having a central inner concentration sometimes described as the cometary nucleus. The source of this activity is indeed the nucleus, though the solid central core of a comet has only recently been visually resolved; first (it seems) by ground-based observations of Comet IRAS-Araki-Alcock (Larson & Johnson 1983), which happened to make an exceptionally close encounter with the earth, and secondly by the Vega and Giotto spacecraft that flew past Halley's comet during March 1986 (*e.g.* Sagdeev *et al.* 1986, Keller *et al.* 1986).

The question of the mass of comets, then, reduces to a consideration of the mass of the central solid nucleus within the observed coma. However, this body cannot easily be observed whilst in the inner planetary system, since its brightness is then often swamped by that of the surrounding coma. Similarly, at large heliocentric distances, where the coma may be faint or completely absent, the reflected sunlight from the nucleus is very weak and

D. Lynden-Bell and G. Gilmore (eds.), Baryonic Dark Matter, 7–35.
© 1990 *Kluwer Academic Publishers.*

detection again becomes extremely difficult. Most estimates of cometary masses, therefore, are forced to rely on indirect arguments. These include, for example, the assumption of an average relationship between the mass and total absolute magnitude of a comet (that is, the apparent magnitude of the nucleus, coma and tail, normalized to a heliocentric and geocentric distance of 1 AU), and dynamical and physical arguments based on the non-gravitational acceleration which affects a comet's motion as a result of outgassing (e.g. Rickman 1989). Alternatively, if the magnitude of the central nucleus (the so-called nuclear magnitude) can indeed be separated from that of the surrounding coma, the dimensions of the nucleus may be obtained in a similar way as that for asteroids. The corresponding mass determination, however, is still plagued by uncertainties, particularly those due to the generally unknown albedo of the surface at the wavelength in question and the effective density of the central body, presumably non-spherical and porous.

For these reasons, despite relatively rapid advances in cometary astronomy overall, the masses of comets remain extremely uncertain. At the present time few, if any, comets have masses that are reliably determined to better than a factor of about four, while the majority of quoted mass estimates are probably uncertain by at least twice this amount. Improving on this situation represents an important problem for the future, and it is important to obtain as many nuclear magnitudes of long-period comets as possible. Comets may have played a crucial rôle in the history of the early solar system, and, particularly in the planetesimal theory of planet formation, may hold the key to understanding the formation and evolution of planet-sized bodies in the protostellar systems and discs now frequently found in and around star-forming regions. Indeed, the impact of comets on the earth has often been suggested as the leading trigger for the major changes in geological epoch that have occurred, producing mass extinctions of species and other 'catastrophic' terrestrial effects — with associated wide implications for the origin and development of life on our own planet. Since a detailed understanding of these phenomena depends crucially on the *masses* of comets, both individually and *in toto*, it is important to piece together some of the evidence relating to this question.

This paper is divided into seven sections. Section 2 briefly outlines the conventional 'Oort cloud' theory of the origin of comets, including the idea, much discussed during the past decade, that the 'observed' Oort cloud (mostly containing comets with semi-major axes a greater than about 2×10^4 AU) may contain a much larger number of comets moving in tightly bound orbits comprising a hypothetical 'dense inner core' within the standard 1950-model. We then review the distribution of long-period comet absolute magnitudes, showing how this distribution together with a reliable mass-magnitude calibration is crucial to the problem of estimating long-period comet masses. Section 4 contrasts these 'observed' masses with theoretical estimates based on particular suggestions for the origin of comets, while the following Section 5 illustrates how the present mass of comets in the solar system is determined according to various models of the Oort cloud. Section 6 then briefly extends this review to include the suggestion that there may be a substantial number of comets in interstellar space, emphasizing the fact that, if comets are frequently formed under normal interstellar conditions (e.g. in dense molecular clouds, or in wind-driven shells around young stars), one might expect a significant population of large, so-called 'giant' grains to be present in the interstellar medium. These grains, with dimensions on the order of $1-10\,\mu$m or larger, should be significantly more concentrated towards the galactic plane than either the observed gaseous component or ordinary interstellar grains. It is possible, therefore, that new stars forming from molecular clouds within such a zone may show signs of an anomalous dust-to-gas ratio in their parent gas. Finally, the principal conclusions from this investigation are summarized in Section 7.

2. Comet nuclei and the Oort cloud

2.1. COMPOSITION

The chemical composition of cometary nuclei has been discussed by many authors during the past few years. The interested reader is referred, for example, to the recent reviews of Greenberg (1988) and Geiss (1988) and references therein. In summary, the central nucleus appears to be a mixture of aggregated interstellar or interplanetary dust and ices, brought together by accretion or gravitational condensation into a single body typically some 10 km across. The nucleus is composed largely of 'solids', mostly the elements carbon, nitrogen, oxygen, hydrogen, sulphur, magnesium, silicon, iron and a few others. Roughly half the mass is made of volatile compounds, often described as 'ices', the bulk of which turns out to be ordinary water ice, H_2O. Water ice typically makes up about 80% or or more of the mass associated with cometary volatiles, the next most commonly occurring molecule being carbon monoxide, CO, with an abundance of around 5–30% that of H_2O by number, the precise figure being variable from one comet to another. There is, in fact, a suspicion that changes in the observed CO abundance correlate with the amount of dust, and that some, therefore, if not most, of the observed CO ultimately originates either through the break-up of complex organic compounds associated with the dust component or in carbon monoxide trapped in the grains.

The proportion of the non-volatile 'dust' component to the more volatile 'gas' or ice component in cometary material also varies amongst different comets, quite apart from being an intrinsically difficult parameter to determine with certainty. Many authors adopt a dust-to-gas ratio by mass in the range 0.5–1.0 in order to derive overall abundances (cf. Delsemme 1982), but even in the case of Halley's comet estimates of this quantity have ranged between 0.2 and 2 (cf. McDonnell et al. 1986, 1989). The total relative abundances of elements within comets are correspondingly uncertain, but, with this proviso, the results generally show that, normalized relative to silicon, the abundances are approximately solar. This implies that cometary material is mostly composed of 'heavy' elements, with the additional implication that comets (at least so far as the word is normally understood, i.e. referring to comet-sized bodies of the kind that have been observed) can never account for the total 'missing' baryonic matter in the universe; that is, if the predictions of 'standard' (or even 'non-standard') theories of big-bang nucleosynthesis are to be believed.

2.2. ORIGIN

So far as the origin of comets is concerned, however, most arguments during the past forty years or so have developed almost independently of theories concerning the detailed chemistry of the central body, and instead have concentrated on purely *dynamical* questions, relating observations of cometary orbits to their theoretical long-term evolution. From this point of view, the observed comets form an arbitrary and very inhomogeneous data set. The observations have mostly accumulated over a timescale measured in centuries or millennia: that is, on a timescale short compared with characteristic astronomical periods over which the cometary flux may be expected to vary, but long compared with the timescale on which observational techniques and selection effects are known to have changed. The result is that present observations provide a rather poor base from which to infer the long-term average properties of the cometary influx through the inner solar system.

Nevertheless, a robust feature of the observations is the relatively large number of long-period comets having orbits of very long periods. When planetary perturbations are taken into account, the 'original' orbits of many of these comets are found to lie close to the parabolic limit, i.e. with reciprocal semi-major axes $1/a \simeq 0$, although when evaluated with sufficient care the majority of these (with just a handful of exceptions) have original orbits that are marginally elliptical and hence bound to the solar system. Following Oort's (1950) pioneering investigation of this feature, the observations are now usually interpreted as providing strong empirical evidence that the solar system is surrounded by a huge, almost spherical, cloud of comets having semi-major axes in the approximate range 2×10^4–2×10^5 AU. This comet swarm, known as the Oort cloud, has been the subject of many theoretical studies during the past forty years; these have had the parallel objectives not only of explaining the origin of comets within a primordial solar system framework, but also of explaining the formation of the Oort cloud by examining how such comets might originally have been scattered into their observed long-period orbits. As a by-product of this primordial hypothesis, it must also be shown that comets can survive more or less in situ within the Oort cloud for the necessary length of time (3–5×10^9 years) since the solar system formed.

The comets moving in these very long-period observed orbits are subjected to the cumulative perturbations of passing stars, molecular clouds, and the galactic tide. These act both randomly and systematically to change the angular momenta of such comets, and provide a mechanism by which apparently 'new' comets are fed from neighbouring orbits in the Oort cloud into ones of sufficiently small perihelia to become observable. All these comets (indeed, all those with perihelia ranging up to about 15 AU) are significantly affected by planetary perturbations around perihelion, and these sufficiently change the cometary energies so as to ensure that virtually none are scattered back to the Oort cloud whence they came. Roughly half are ejected into hyperbolic orbits, while the remainder come back on shorter-period elliptical orbits, subsequently to be ejected or to return in possibly more tightly bound orbits. In this way, the presence of the Oort cloud and the effects of stellar and galactic perturbations on the comets within it 'explain' the observed near-parabolic flux, while successive planetary perturbations acting on those comets scattered into shorter period orbits ultimately provide an explanation for the observed short-period comets, a group defined (by convention) to have periods less than 200 years.

In recent years, however, this simple (and in many ways attractive) explanation for the origin of comets has come increasingly under attack. A detailed discussion of the arguments and references to related work may be found in the reviews by Bailey, Clube & Napier (1986, 1990). In particular, one can identify at least four main 'problems' for

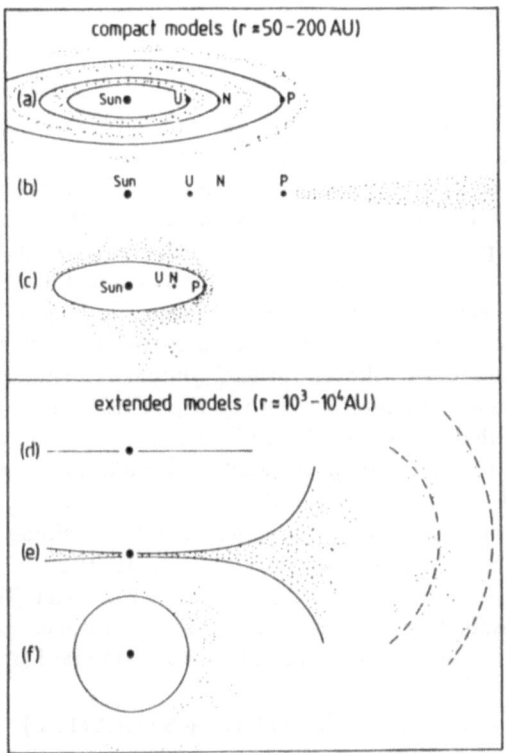

Figure 1: Six proposed variations on the dense inner core hypothesis. Diagram taken from Bailey *et al.* (1990).

the theory: the survival problem, the formation problem, the fading problem, and the short-period comet problem. Each of these has a solution which may be described either in terms of time-dependence of the cometary flux (consistent with the assumptions of a recent disturbance of a primordial Oort cloud or recent capture of comets from interstellar space), or in terms of a natural prediction of a model involving a dense inner core within the Oort cloud (or both). A schematic illustration of the several kinds of inner core that have been proposed is shown in Figure 1 taken from Bailey *et al.* (1990). In particular, one should emphasize that the several models illustrated are not necessarily mutually exclusive: the planetesimal picture, for example, although consistent with a model of type (e), might also include additional components represented by types (a) ± (b). Recent reviews of the inner core of the Oort cloud and of how short-period comets in particular now provide increasingly stringent constraints on viable inner core models have been given by Bailey (1989, 1990), while some of the issues raised by the number and orbital parameters of short-period comets have also been discussed at this meeting by Tremaine (1990) and colleagues (*e.g.* Quinn *et al.* 1989).

3. Comet mass distribution

3.1. ABSOLUTE MAGNITUDES

The apparent brightnesses of comets vary greatly, not only systematically, with changing geocentric distance Δ and heliocentric distance r, but also unpredictably due to intrinsic variations of one sort or another, for example due to variable outgassing or dust production, or to motion through the inhomogeneous solar wind. Leaving these changes aside, however, and assuming that the proportion of a comet's nuclear surface that is active is roughly constant from one comet to the next, it is reasonable to assume that the apparently brighter comets will, in general, correspond to those with larger and more massive nuclei. This leads to the idea that different comets may be compared by normalizing their observed total magnitudes to a standard heliocentric and geocentric distance (taken to be 1 AU), thereby defining a so-called 'absolute' magnitude H_0. This leads in turn to the idea that the cometary mass distribution may be determined from a study of the observed absolute magnitude distribution together with a suitably calibrated mass-magnitude relationship. To a large extent, this is the only way in which the masses of a significant sample of long-period comets may be compared with one another, thereby obtaining an estimate (however uncertain) of the underlying long-period comet mass distribution.

The observed dependence of the 'total' magnitude of a comet (that is, the equivalent stellar magnitude corresponding to the net flux from the nucleus, coma and tail) on heliocentric and geocentric distance is usually fitted to a standard photometric law of the form

$$m = H_0 + 2.5n \log(r/1\,\mathrm{AU}) + 5 \log(\Delta/1\,\mathrm{AU}) \tag{1}$$

where n and H_0 are adjustable parameters to be determined. A somewhat simpler procedure also in frequent use, particularly for predicting cometary magnitudes, has been to set n in this expression equal to a standard value such as $n = 4$. In this case, the derived absolute magnitude is denoted by H_{10}, the subscript indicating that the standard value $n = 4$ (times 2.5) has been used in the derivation of H_0.

Even the casual reader will appreciate the large uncertainties introduced by this kind of procedure. Nevertheless, while there is an undeniable element of wishful thinking involved (at least if one assumes that the result is going to be accurate!), it is still possible to assume that given a large enough data set the individual errors will cancel out, and that the overall *shape* of the inferred absolute magnitude distribution will indeed represent a physically meaningful measurement. By way of illustration, Figure 2 shows a comparison between two differential frequency distributions of long-period comet absolute magnitudes taken from the tables of Everhart (1967) and Hughes (1987a). These data comprise 256 and 519 long-period comets respectively, the former being a restricted (and hopefully more homogeneous) data set corresponding to long-period comets discovered by northern observers during the period 1840–1967, and which were sufficiently bright for a short time to have been discovered (at least in principle) by ordinary visual searches.

The principal differences between these observed distributions (particularly the relative excess of very bright and very faint comets in Hughes's sample) can mostly be attributed to the different selection criteria adopted by the two authors. For example, the brightest

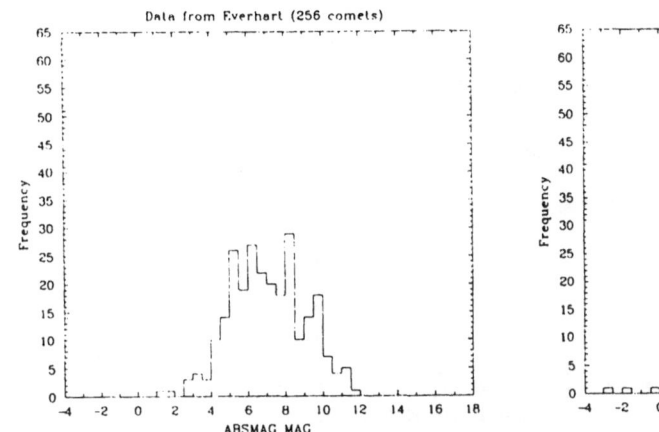
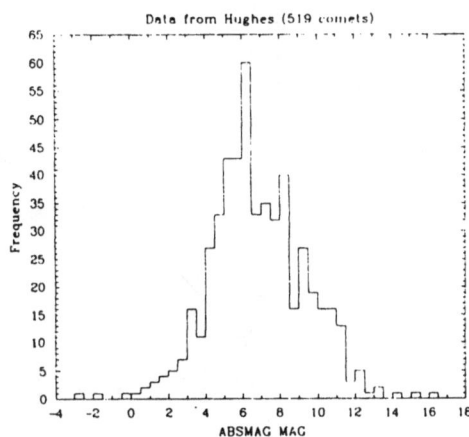

Figure 2: The differential frequency distribution of absolute magnitudes H_{10} taken from the lists of Everhart (1967) and Hughes (1987a).

comets in Hughes's sample often appear to have been discovered long ago (*i.e.* prior to about 1850) or to have had relatively large perihelia (*i.e.* $q \gtrsim 2.5$ AU), thereby making their extrapolated H_{10} values more uncertain; while one also finds that the fainter comets have mostly been discovered during quite recent years. This shows that one is unlikely to obtain an accurate determination of the intrinsic differential brightness distribution by taking the data at face value: Everhart's histogram at bright magnitudes yields an apparent absolute magnitude distribution approximately proportional to $H^{0.35}$, while Hughes's data implies a different distribution roughly proportional to $H^{0.30}$ (*cf.* Hughes & Daniels 1980, Hughes 1987b).

The problem of determining the underlying 'intrinsic' brightness distribution thus requires careful correction of the observed frequency distribution of absolute magnitudes for the complex and probably time-variable selection effects which will have played a significant part in determining the subset of comets that are eventually discovered. A detailed discussion of these factors would go far beyond the scope of the present review, and the interested reader is referred to the papers by Everhart (1967), Kresák (1982), and Hughes (*loc. cit.*) and references therein. Nevertheless, although the main difficulty in passing from the observed absolute magnitude distribution to the intrinsic distribution is indeed that of accurately allowing for the various observational selection effects, a number of other factors appear also to be relevant. First, the 'total' magnitude is itself a relatively poorly defined quantity, and in practice may be subject to large errors. For example, comparisons of absolute magnitudes, even of H_{10} values, quoted by different authors frequently reveal large differences ranging up to a magnitude or more. Secondly, the use of a conventional value $n = 4$ to reduce sometimes very sparse data, and the subsequent extrapolation or interpolation of an inferred light curve to the adopted standard distance may also lead to systematic errors. Thirdly, there is no guarantee that the comets passing perihelion during the first

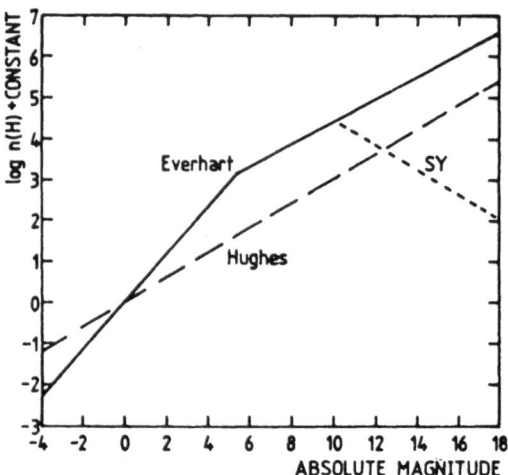

Figure 3: Comparison of the different intrinsic brightness distributions for long-period comets adopted by several authors.

centuries of the present era should have precisely the same brightness distribution as those being discovered at the present epoch; nor, even, is it known whether the present flux of comets is the same. Lastly, it has been emphasized that one should also consider whether there are systematic errors in the measured magnitudes of older comets as determined from photographic plates (Jewitt 1990).

The result, of course, is that the inferred absolute magnitudes of many, if not all, comets in the available historical sample are extremely uncertain, while the derived intrinsic distribution of absolute magnitudes must be regarded as equally unreliable. This has led different authors to adopt one or another extreme approach to the problem: (i) to attempt accurately to correct the observed absolute magnitude distribution for selection effects, assuming that the latter can be reliably quantified (cf. Everhart 1967, Sekanina & Yeomans 1984); (ii) to ignore detailed selection effects completely and assume that the observed distribution is essentially the same as the intrinsic distribution (cf. Hughes 1985, 1987b); or (iii) to throw in the towel completely, and return to the only 'reliable' case we know — Halley's comet — and assume that all comets are essentially equivalent to Comet Halley (cf. Mendis & Marconi 1986, Marochnik et al. 1988)! The latter approach, incidentally, finds some justification from the observed nuclear dimensions of a number of comets (e.g. IRAS-Araki-Alcock 1983 VII, P/Arend-Rigaux, and P/Tempel 2), each of which is surprisingly similar to P/Halley in overall size (Sekanina 1985, Millis et al. 1988, Jewitt & Luu 1989).

The results of applying some of these different approaches to the problem of determining the intrinsic brightness distribution of long-period comets are shown in Figure 3, which highlights the difference between the respective results of Everhart (1967), Sekanina & Yeomans (1984) and Hughes (1987b). In particular, note that Everhart's analysis implies that exceptionally bright comets (e.g. those brighter than $H_0 \simeq -2$) are several powers of ten less likely to be observed than would be predicted on the same basis from Hughes's distribution (assuming both are normalized to the observed flux with absolute magnitudes in the approximate range 3–7), while a similar discrepancy exists between the predictions of Sekanina & Yeomans and those of both Hughes and Everhart for magnitudes fainter than about $H_0 \simeq 12$.

3.2. MASS-MAGNITUDE RELATION

Of course, even if the brightness distribution *is* known, the step to obtaining the mass distribution still relies on determining, and reliably calibrating, a mass-magnitude relation; in particular, on obtaining accurate masses for a number of individual comets. Again, such masses are frequently only approximately determined, and in most cases have been only roughly estimated from indirect determinations of the radius of the nucleus, either involving an assumption as to the albedo and the fraction of the total surface area that is active, or attempting to solve for these additional parameters by a more comprehensive theoretical model. The frequently quoted mass-magnitude relationships (see Bailey & Stagg 1988 for references to earlier work) are often based on a mass calibration that involves relatively few independent mass-magnitude determinations, and, with a few exceptions (e.g. Weissman 1982), simply on an *assumption* as to the appropriate slope of the relationship. Figure 4 is a redetermination of this relationship, based on average values for the radii and absolute magnitudes of 27 long-period comets culled from the literature. The principal sources are noted on the diagram. A least-squares solution to the combined data set yields $\log(m^*) = b_2 + b_1 H$, where $m^* = m/1\,\text{kg}$ and the parameters b_2 and b_1 are given by $b_2 = 18.13 \pm 0.68$ and $b_1 = -0.48 \pm 0.11$. A sufficient approximation to this line has been drawn on the diagram (solid curve) together with dashed lines representing several previous determinations of the best-fitting relationship. In particular, we note that in this diagram the quoted nuclear radii in the literature have been reduced to a common albedo $A = 0.05$ on the assumption that $R \propto A^{-1/2}$, while the corresponding masses have been calculated for a mean internal nuclear density of $500\,\text{kg m}^{-3}$. Given our knowledge of the nuclear structure of long-period comets, however, these assumptions should be regarded as probably no more valid than a rough guess.

If the differential brightness distribution is assumed to be a power law of the form $\log n(H) = A_2 + a_1 H$, it is straightforward to show that, with the above mass-magnitude relation, the corresponding frequency distribution of cometary masses $n(m)\,dm$ is also a power law: $n(m) \propto m^{-s}$, where the mass-distribution index $s = 1 - a_1/b_1$. In this way, assuming $a_1 = 0.59$ or 0.30 (corresponding respectively to Everhart's and Hughes's brightness distributions for $H \lesssim 6$), the cometary mass distribution for the brighter comets is found to have an index s of order 2.2 or 1.6 respectively. Setting $b_1 = -0.4$ or -0.6 would change these values to $(2.5, 2.0)$ and $(1.75, 1.50)$ respectively. Given the uncertainties, therefore, it is not yet securely known whether a few, rare 'giant' comets dominate the total cometary mass, or whether most of the mass has, in effect, already been seen — residing in comets with absolute magnitudes in the range 3–7. The alternative assumption that all comets are essentially like P/Halley would imply an average mass per comet on the order of that of Halley's comet, irrespective of its absolute magnitude, *i.e.* $\bar{m} \approx (2 \pm 1) \times 10^{14}\,\text{kg}$ (*cf.* Fernández 1982, Bailey & Stagg 1988, Rickman 1989).

16

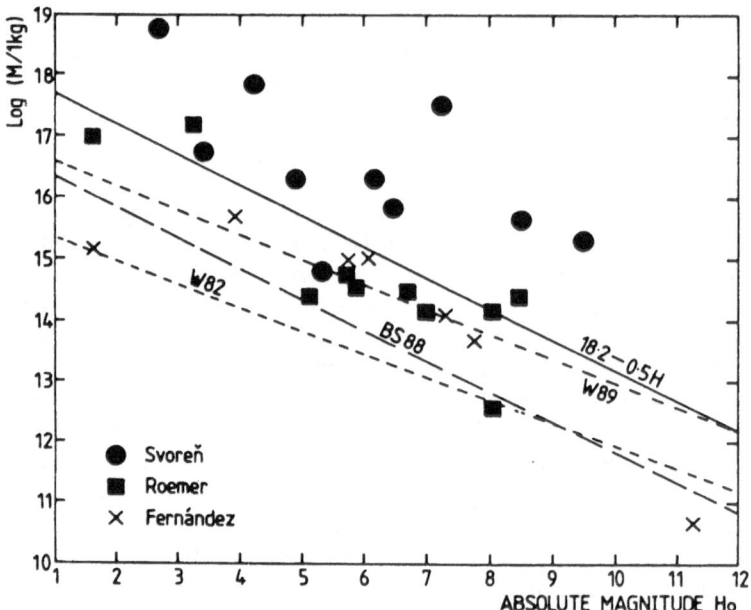

Figure 4: Mass-magnitude relationship for 27 long-period comets. The data are taken mostly from the compilations of Svoreň (1987), Roemer (1966) and Fernández (1982), though in a few cases the precise values adopted are mean values from more than one source. For comparison, the labelled dashed lines W82, W89 and BS88 describe the mass-magnitude relationships adopted by Weissman (1982, 1989) and Bailey & Stagg (1988).

4. Theoretical mass-estimates

In view of the large uncertainties attached to empirical determinations of cometary masses, and to the very wide range in comet masses themselves, it is perhaps not surprising that a survey of *theoretical* mass-estimates gives the superficial impression that observations and theory are in rather good agreement — at least so far as the one is compatible with the other! Any agreement, however, is mostly illusory, being largely attributable to use of a judicious or convenient set of parameters, the observations, despite their uncertainty, continuing to lead theory.

A comprehensive survey of theories of cometary origin has been given by Bailey *et al.* (1990). This shows that, even *since* 1950, a very wide range of ideas and considerations have been advanced by different authors, the total number of distinct proposals ranging up to of order 30! Leaving aside those theories in which comets originate as part of a catastrophic process involving explosions of small bodies or collisions between minor bodies in the outer solar system (*e.g.* Van Flandern 1978, Drobyshevski 1978), or in which they are replenished by accretion or aggregation of the observed dust in the solar system (*e.g.* Mendis & Alfvén 1976, O'Dell 1986), the majority of those remaining can be divided into two main groups. First, comets may be assumed to form in a process some way directly associated with the formation of planets, either as a by-product of ordinary planet formation or during an earlier stage of the same process; and secondly, comet formation can be regarded as totally separate from planet formation, whether occurring on the periphery of a protoplanetary disc, in satellite nebulae or 'fragments' of the sun's original molecular cloud, or in interstellar space itself. The former theories are probably best represented by the ideas of Kuiper (1951), Whipple (1964) and Öpik (1973), and by the so-called 'planetesimal' hypothesis of Safronov (1969), later developed by Greenberg and co-workers (e.g. Greenberg 1985, Greenberg et al. 1978, 1984) and Fernández & Ip (e.g. 1983) among others. The second class of theory represents a much more inhomogeneous group, ranging from the assumption of a primordial origin for comets on the edge of a fairly massive protoplanetary disc (e.g. Cameron 1978, Biermann & Michel 1978), to the ideas that comets form during the initial collapse of the sun's parent cloud (Hills 1982) or in a wind-driven shell of circumstellar material around it (Bailey 1987a), and finally to the idea that comets originate in the interstellar medium more or less in situ (e.g. McCrea 1975, Yabushita 1983, Clube 1985, Napier & Humphries 1986). In the remainder of this section, we consider estimates of cometary masses taken from each of these types of theory.

4.1. DUST AGGREGATION

Following Kuiper (1951) and Öpik (1973), the rate of growth of a comet nucleus as a result of collisions with ambient dust grains may be written in the form

$$\frac{dm_c}{dt} = \xi_c n_f \sigma_c v_{\rm rel} m_f \tag{2}$$

where ξ_c denotes the sticking probability, and n_f and m_f are the number density and mass of the ambient 'field' grains with assumed radii $a_f = (3m_f/4\pi\rho_f)^{1/3}$. The instantaneous mass of the growing comet is denoted by $m_c = \frac{4}{3}\pi\rho_c a_c^3$, where a_c is its radius and ρ_c the internal density, while $\sigma_c = \pi(a_c + a_f)^2$ is the collision cross-section. This expression neglects gravitational focussing of the particle trajectories.

Assuming that the mass-density of large particles (i.e. the 'comets') in the nebula is relatively small compared to that of the small grains, we may write $n_f m_f \simeq \rho_d$, where ρ_d is the local smoothed-out density of dust and ices in the nebula. Hence, since $a_c \gg a_f$ in this approximation, equation (2) can be simplified and integrated to yield

$$a_c \simeq \frac{1}{4}\xi_c(\rho_d/\rho_c)v_{\rm rel}\,t \tag{3}$$

If the relative velocity, $v_{\rm rel}$, is principally determined by epicyclic orbital motions in a disc of semi-thickness H rotating at the Keplerian rate $\Omega = (GM_\odot/r^3)^{1/2}$, we have $v_{\rm rel} \simeq \sqrt{3}v_\perp = \sqrt{3}\Omega H$ and $\rho_d \simeq \Sigma_d/2H$, where Σ_d is the surface density of dust and condensibles in the disc. We thus find (cf. Greenberg et al. 1984)

$$a_c \simeq \frac{\sqrt{3}\xi_c}{8}\frac{\Sigma_d\Omega}{\rho_c}t \tag{4}$$

For example, if comets are assumed to form in the Uranus-Neptune zone of the protoplanetary disc, we may assume $\Sigma_d = M_d/2\pi r\,\Delta r$, where $M_d \simeq 25\,M_\oplus$ (roughly the sum of the masses of Uranus and Neptune), $r \simeq 25\,\mathrm{AU}$ and $\Delta r \simeq 10\,\mathrm{AU}$. This gives $\Sigma_d = 4.25\,\mathrm{kg\,m^{-2}}$, and hence

$$a_c \simeq 9.25\,\xi_c \left(\frac{500\,\mathrm{kg\,m^{-3}}}{\rho_c}\right)\left(\frac{t}{10^8\,\mathrm{yr}}\right) \quad \mathrm{km} \tag{5}$$

This shows that, provided the sticking probability ξ_c is of order unity, the process of direct aggregation of dust in a protoplanetary nebula is probably sufficient to produce bodies of cometary size within a quite reasonable timescale.

However, the precise mass distribution of comets formed in this way and the likely limiting mass for termination of the process cannot yet be specified, since they depend in a complicated way on uncertain details of the system, including the mass-dependence of the product $\sigma_c v_{\mathrm{rel}}$, the effects of gas drag (if any), the presumed lifetime of the dust-disc, the sticking probability of grains, and the mean density of the nucleus. Nevertheless, it does seem that this sort of coagulation may produce a power-law mass distribution of the sort that seems to be observed, with index s in the approximate range 1.4–1.8 (e.g. Safronov & Ruzmaikina 1978). Recent simulations of the collisional aggregation of small dust grains in a protoplanetary nebula (e.g. Weidenschilling et al. 1989, Brooks 1990) also emphasize the probable low density and 'fractal' nature of the aggregates thus formed (cf. data on meteoroids; Verniani 1969), suggesting that the growth of small bodies and comets may be somewhat faster, in practice, than predicted by equation (5) with $\rho_c = 500\,\mathrm{kg\,m^{-3}}$. However, detailed studies of the aerodynamics of fluffy, accreting interplanetary grains have yet to be carried out.

4.2. PLANETESIMALS AND MASSIVE DISCS

An important aspect of the pure dust aggregation theory for forming comets in the proto-planetary disc is that it is not possible, at least according to equation (5), to produce bodies much larger than a few tens of kilometres in radius within a realistic timescale (i.e. $\lesssim 10^8$–$10^9\,\mathrm{yr}$). In particular, the formation of planets by accretion of interplanetary dust grains or their preaggregated cometary products seems to require another process: enhancement in the rate of particle growth due to gravitational focussing or gravitational instability; or the effects of assuming a much higher local surface density of solid material within the disc (or both). Moreover, if the sticking probability were very much less than unity in the above formulae, as argued by Öpik (1973), who adopted $\xi_c = 1/20$, then the formation of bodies even as massive as Halley's comet becomes much harder to explain.

These difficulties with the idea of simple collisional evolution of dust grains to form larger bodies (particularly planets, like Uranus and Neptune) led Safronov and others (e.g. Safronov 1972, Safronov & Ruzmaikina 1978) to suggest that the surface density of condensibles in the protoplanetary nebula might indeed have been very much higher than would be anticipated simply on the basis of the observed planetary masses. This suggestion received additional support from the fact that production of the Oort cloud by gravitational scattering of planetesimals (identified in the theory with comet nuclei) by growing protoplanets was itself probably an inefficient process. The total initial mass of comets (planetesimals) was therefore assumed to be much greater than the total mass of material now in the Oort cloud, leading to a corresponding dust surface density Σ_d rather in excess of that used in equation (5).

Considerations of the interaction between gas and dust during the early evolution of a protoplanetary disc suggest that dust grains will rapidly settle towards the central mid-plane of symmetry, producing a dynamically 'cold' dust layer, possibly comprising dusty aggregates ranging in size up to some millimetres or centimetres across. Goldreich & Ward (1973), following previous authors, emphasized that this central dust layer should be subject to gravitational instability, on a scale depending on the local surface density of dust and the angular velocity of the disc at the point in question. In this way, the dust layer was assumed to fragment into 'blobs' with a size determined by gravitational instability, allowing the former uncertain (and possibly rather slow) initial phase of dust-dust aggregation to be ignored.

Analyses of the gravitational instability in a differentially rotating disc with Keplerian angular velocity $\Omega = (GM_\odot/r^3)^{1/2}$ (e.g. Goldreich & Lynden-Bell 1965a,b; Safronov 1969, Goldreich & Ward 1973, Wetherill 1980, Sekiya 1983, Greenberg et al. 1984) show that, provided the dust velocity dispersion is sufficiently small, the radii R_c of the first gravitationally bound condensations are on the order of $2\pi^2 G\Sigma_d/\Omega^2$, corresponding to a mass $m_c = \pi R_c^2 \Sigma_d \simeq 4\pi^5 G^2 \Sigma_d^3/\Omega^4$, or

$$m_c \simeq \frac{4\pi^5 \Sigma_d^3 r^6}{M_\odot^2} \tag{6}$$

If these condensations are identified with comet nuclei, the mass of the first 'comets' to form may therefore be written in the form

$$m_c \simeq 8.5 \times 10^{20} \, r_{25}^6 \Sigma_{10}^3 \quad \text{kg} \tag{7}$$

where $r_{25} = r/25\,\text{AU}$ and $\Sigma_{10} = \Sigma_d/10\,\text{kg m}^{-2}$. We note parenthetically that this is rather larger than the corresponding expression quoted by Yamamoto & Kozasa (1988), by a factor of about 3.5, the latter authors using the results of Sekiya (1983). However, even assuming an uncertainty of this order of magnitude in m_c, it does not at first sight seem that the planetesimal hypothesis (with $\Sigma_{10} \approx 1$, say, in order to grow planets fast enough) can explain the initial formation of bodies in the protoplanetary disc with masses similar to those of observed comets (cf. Comet Halley, with $m \simeq 2 \times 10^{14}\,\text{kg}$).

There are several ways around this apparent difficulty. First, as emphasized by Goldreich & Ward (1973), the initial unstable regions of the dust disc may not, in fact, collapse all the way down to solid densities, but instead should fragment into many smaller bodies, each with a mass on the order of

$$m_c' \simeq 1.6 \times 10^{14} \, r_{25}^{9/2} \Sigma_{10}^2 \quad \text{kg} \tag{8}$$

(cf. Ward 1976). Here, as in equation (5), we have assumed a density ρ_c for the central solid body of around $500\,\text{kg m}^{-3}$. A second solution to the problem of the excessive initial mass of planetesimals, at least when compared with observed cometary masses, is the suggestion (e.g. Greenberg et al. 1984) that nonhomologous sedimentation of dust grains in the original disc may allow gravitational instability to occur when only a relatively small proportion of the dust, perhaps as low as $\approx 10^{-2} \Sigma_d$, has reached mid-plane. In this case, a much smaller value of Σ_d than that for the whole nebula should be entered into the expression for m_c, producing correspondingly smaller masses, while comet formation might then be expected to occur sequentially, perhaps in as many as 100 distinct phases of local disc gravitational instability. Such an idea is not, of course, restricted to standard planetesimal theories for the evolution of the protosolar nebula (cf. Morfill & Völk 1984, Mizuno et al. 1988). Yet another possibility is the suggestion (e.g. Öpik 1973) that close encounters with the growing planets might themselves lead to fragmentation of the originally weakly bound cometary nuclei, in the same way that large comets are observed to break up whenever they pass within the Roche radius of the sun or a giant planet (e.g. Kresák 1981).

This raises the question whether the first solid bodies or planetesimals to be formed in the disc *should* be identified with observed comet nuclei. For example, the results of Greenberg et al. (1984) and Yamamoto & Kozasa (1988) suggest rather strongly that they should not, while observations of Halley's comet (e.g. Möhlmann & Kührt 1989) are consistent with it being a rather heterogeneous assemblage of smaller particles. Even if solid bodies of a particular size are formed in the protoplanetary disc by gravitational instability or through some other process, collisions between these bodies will inevitably lead to a characteristic (and model-dependent) distribution of sizes for the resulting particles, the final size distribution depending on particular details of the model such as the internal cohesion of the assumed building blocks, their relative velocities, and on whether coalescence or both coalescence and fragmentation are allowed to occur (e.g. Greenberg 1985). Historically, the usual procedure has been to compare the results of theory with observations — in order constrain the theory; however, as we have seen, firm conclusions from observations are themselves so hard to obtain that it is arguable whether this approach could be completely misleading!

Extensions of the planetesimal theory for forming comets via the gravitational instability mechanism have also been described for comet formation in a much more massive nebula, as proposed by Cameron (1962, 1978), or the satellite nebulae advocated by Cameron (1973) and Biermann (1981). Nevertheless, the underlying physics of the dust-aggregation process, at least so far as it is understood, is very similar to that already described (cf. Biermann & Michel 1978), and detailed estimates of cometary masses lead to many of the same difficulties, or apparent difficulties, when compared with observations. In particular, although one can often explain the 'average' comet according to such theories (though, as we have indicated, this may be simply a theoretical selection effect), they generally have greater difficulty in explaining the wide *range* of cometary masses: the apparently largest bodies (with $m \approx 10^{18}$ kg) and the apparently smallest ($m \approx 10^{11}$ kg).

4.3. RADIATION PRESSURE, STELLAR WINDS AND INTERSTELLAR THEORIES

There is no doubt that the planetesimal theory (including its several variants) currently represents the most internally consistent and highly developed picture for the origin of comets. Despite this, and for good reasons (see Bailey et al. 1990), the planetesimal theory has never gained unanimous support, and during the period since 1960 numerous other ideas have been simultaneously investigated. In particular, some of these theories entirely separate the formation of comets from the formation of planets, and appeal to quite different physical processes in explaining the former, whether in a protosolar nebula or beyond. None of these theories, however, yet makes a definite prediction of cometary masses; but, because they do emphasize different physical processes in the formation of solid bodies, it is important to include them in this review. In this section, therefore, we describe the main features of four 'representative' non-planetesimal theories of cometary origin.

The first of these, corresponding to a variant of the usual primordial solar system origin for comet formation, was proposed by Hills (1981, 1982). This author, developing an idea originally elaborated by Spitzer (1941) and Whipple (1946), suggested that the differential radiation pressure acting across a clump of dust in a collapsing protostellar cloud could be strong enough to concentrate the dust enough to form solid bodies during the initial dynamical collapse phase of the nebula. In particular, Hills argued that the formation of 'comets' with final radii on the order of 1 km was not unreasonable, and presented qualitative arguments (e.g. Hills & Sandford II 1982) intended to show that the model might also be capable of explaining a very broad mass distribution (as seems to be required by observations). Differential radiation pressure on dust grains, essentially due to 'shadowing' of one particle by another, may therefore be a means of accelerating the coagulation of dust particles in the interstellar or protostellar medium.

A second proposal for making comets as part of the general process of star formation was advanced by Bailey (1987a). This author showed that the strong stellar wind generated by a newly formed star may sweep the residual, still collapsing, protostellar material into a thin, dense outwardly moving shell. Dust-dust collisions in the shell lead rapidly to the growth of large coagulated dust grains with radii on the order of 100 μm, while such 'giant' grains subsequently decouple from the turbulence associated with the decelerating shell (with assumed radius R_s and velocity V_s) and sediment towards the shell's outer edge. When the dust velocity dispersion v_d falls sufficiently, the giant grains inside a region of characteristic size $\lambda \simeq v_d R_s / V_s$ drift together under self-gravity, the process being analogous to the gravitational instability in a dust disc, but with the initial radial motions of the grains replacing the stabilizing influence of rotation in a protoplanetary nebula. The mass of individual comets formed by such a process is on the order of $v_d^3 R_s / G V_s$, which for the particular models investigated was found to be in the approximate range 10^{14}–10^{16} kg. Grain-grain coagulation, followed by gravitational instability within a sufficiently quiescent dust cloud, is thus a second process by which comets may form within dense regions of the interstellar medium.

On this hypothesis, the comets start their lives moving away from their parent star, and hence the sun's comets (for example) cannot be assumed to produce the sun's Oort cloud. On such a theory, therefore, the Oort cloud must be captured from comets originally formed in the expanding shells around some other star or stars, though probably still formed coevally with the solar system (in order to enhance the likelihood of capture). The alternative assumption that the Oort cloud is captured and is not formed coevally with the solar system, brings us to interstellar theories of cometary origin, particularly the suggestions of McCrea (1975) and Napier & Humphries (1986).

McCrea emphasized that the characteristic Jeans mass for gravitational collapse depends essentially on the velocity dispersion of the material in question. In the case of ordinary interstellar material at high densities and low temperatures, the derived mass is in the range of typical stellar masses: collapse of gas leads to gaseous bodies (stars) of roughly solar mass, as observed. On the other hand, comets are mostly solid material (dust and ices); and the question arises whether the gravitational collapse of dust particles might lead to solid bodies of typical cometary mass. McCrea argued, in fact, that this was the case (provided the dust velocity dispersion corresponded to the Brownian motion of the dust particles involved); and hence suggested that if the dust could be separated from the gas in interstellar space, then comet formation might occur, as with stars, simply by gravitational instability!

In particular, McCrea identified two critical masses within a nebula: the dust Jeans mass, M_D, above which gravitational collapse of dust particles might in principle occur (provided the gas and dust could be decoupled); and an 'evaporation mass', M_E, below which the gas and dust actually are decoupled, and below which collapse might really occur. M_E corresponds to the mass within a length-scale determined by the gas-dust mean free path. Following McCrea (1975), these masses are given approximately by

$$M_D \simeq \frac{1}{3} \left(\frac{kT}{G\rho_c} \right)^{3/2} a_d^{-9/2} (\zeta\rho)^{-1/2} \tag{9}$$

and

$$M_E \simeq 10 \, \rho_c^3 a_d^3 (\zeta\rho)^{-2} \tag{10}$$

where ρ and T are the density and temperature of the gas in which the dust grains are immersed, and ρ_c, $\zeta\rho$ and a_d are the grain material density, the smoothed-out grain density and the mean grain radius respectively. The dust-to-gas ratio by mass in the nebula is ζ.

In this way it was considered that comets of mass M might in principle be directly formed by collapse of the dust component provided that $M_D < M < M_E$. This implies that cometary bodies might form whenever the nebular gas of sufficiently high density is also less than a critical value ρ_M given to order of magnitude by

$$\rho_M \simeq 10 \, \zeta^{-1} \rho_c^3 a_d^5 (G/kT) \tag{11}$$

At this density, comets should form with a single mass of order

$$M_c \simeq 10^{-1} \rho_c^{-3} a_d^{-7} (kT/G)^2 \tag{12}$$

a result illustrated by McCrea (1975, Fig. 2 and Table 1).

For example, if $\rho_c = 500 \, \text{kg m}^{-3}$, $a_d = 0.1 \, \mu\text{m}$, $T = 10 \, \text{K}$ and $\zeta = 10^{-2}$, one finds $M_c \simeq 3 \times 10^{16} \, \text{kg}$ and $\rho_M \simeq 6 \times 10^{-13} \, \text{kg m}^{-3}$, corresponding to a hydrogen number density $n_H \simeq 3 \times 10^8 \, \text{cm}^{-3}$. Although this is much higher than that directly observed in even the densest molecular clouds, it is also much less than must occur at some time during the formation of a typical protostellar disc. The path from one régime to the other must therefore pass through the critical density, at which point a gas-dust separation along the lines proposed by McCrea might occur.

The second interstellar suggestion, proposed by Napier & Humphries (1986; cf. Humphries 1982), is that comets might form in molecular clouds of very much lower density, corresponding to a number density of order $10^3 \, \text{cm}^{-3}$. Their idea, in fact, may be regarded as an extension of that described by Hills (1982); but whereas Hills employed differential radiation pressure to drive the dust grains together, Napier & Humphries used a kind of 'jet reaction' effect. It was suggested that this occurred by a process similar to the photoelectric effect, but where the anisotropic ultraviolet radiation field in interstellar space ejected *molecules* from the icy surfaces of dust grains.

The potential importance of this process was shown to be many times greater than the direct effect of an anisotropic radiation field (as considered, for example by Hills), leading to an effective force on the grains larger by a factor ranging up to 10^2. This greatly extends the range of environments within which grains might be driven together at a fast enough rate, and it seems possible that some kind of accelerated inward drift of dust grains might occur in only moderately dense clouds. For example, in a region of recent star formation containing massive early-type OB stars, the predicted collapse time for the material was reduced to around 10^3 yr, rather less than the corresponding time for the dust to be driven apart by turbulent processes. In this way, by ignoring turbulence and balancing the inward drift of the grains by photodesorption with their tendency to move apart by Brownian motion, Napier & Humphries obtained a rough lower limit to the masses of comets, namely $m_L \simeq 10^{11}$–10^{13} kg. A somewhat different argument allowed them to set an upper limit of order $\approx 10^{22}$ kg to the masses of such comets, based this time on the expectation of thermal instability occurring in very dense small-scale structures in molecular clouds.

In summary, although many of these theories for forming comets in a diffuse gas, whether on the outskirts of the solar nebula, in wind-driven shells or interstellar space may have difficulties of one sort or another, the several ideas described here do illustrate the range of options still available. Thus, comet formation may occur through the action of the radiation field on small dust grains, through turbulent-driven grain-grain coagulation, or even through gravitational instability affecting small regions where the dust may move independently of the gas. However, the efficiency of these comet formation processes, and the range of masses of any comet nuclei thereby produced, remain uncertain.

5. Mass of the Oort cloud

Although the mass of the Oort cloud is a crucial parameter so far as theories of its origin are concerned, recent years have seen a greater range in the several values quoted by different authors than at almost any time since publication of Oort's original paper in 1950. For example, Weissman (1982, 1983, 1985, 1986) has argued for masses increasing from about $1\,M_\oplus$ in 1982 to $25\,M_\oplus$ or more in 1986, while authors such as Mendis & Marconi (1986) and Marochnik et al. (1988) have recently presented arguments for masses ranging up to more than $10^2\,M_\oplus$. On the other hand, Bailey & Stagg (1988) concluded that the mass of the outer, dynamically active cloud could be as low as 2–$3\,M_\oplus$.

There are three main reasons why different authors have arrived at such disparate results for such a fundamental parameter of the theory. First, estimates of the number of comets at great heliocentric distances depend on the observed flux of comets (down to some particular limiting magnitude) together with a correction for the proportion of comets that escape detection. As we have seen, the problem of determining the appropriate correction factor has not been satisfactorily solved. The result is that the adopted influx of nearly parabolic orbits is far from secure, particularly if the nominal value represents an extrapolation to faint magnitudes. Secondly, the numerical conversion from an observed influx to the total number of comets residing at large distance requires the use of a theoretical model of the cloud: for example, whether the cometary velocity distribution is assumed to be isotropic, whether the influx corresponds to the steady-state value, and whether the Oort cloud contains a massive, dense inner core, essentially decoupled from the observed near-parabolic flux. The last possibility may be studied using Monte Carlo simulations that follow the evolution of a population of comets from some assumed initial site of origin for the entire history of the solar system (e.g. Fernández 1982, Duncan et al. 1987), but these too have their limitations. Thirdly, even if such factors were accurately known, the final conversion to a total cometary mass requires one to assume a particular value for the average mass per comet (brighter than some absolute magnitude); and, as we have seen, this too involves large uncertainties. The uncertainties are not reduced, however, as has sometimes happened, by counting the number of comets down to one absolute magnitude, and then converting to masses using the mean mass for a quite different brightness!

In order to illustrate these steps in determining the mass of a given Oort cloud model, we now follow the argument presented by Bailey & Stagg (1988). First, the near-parabolic flux of comets brighter than $H_0 = 7$ is assumed to be $f_{new}(\leq 7) = 0.2$ comets AU^{-1} yr^{-1}. This is estimated on the basis of Everhart's (1967) sample of 256 long-period comets that passed perihelion in an interval of 127 years, together with the assumption that roughly 20% of long-period comets have nearly parabolic orbits and are therefore arriving from the Oort cloud for the first time as representative 'new' comets. The conversion from the observed to the intrinsic influx brighter than a given absolute magnitude depends, of course, on the validity of Everhart's determination of the observational selection effects as a function of both absolute magnitude and perihelion distance. Once the intrinsic flux to $H_0 = 7$ is determined, the equivalent flux (or number of comets) to any other magnitude may be obtained from the differential brightness distribution illustrated in Figure 3. Tables and formulae for the cumulative brightness distribution have been given by Bailey & Stagg (1988).

Next, it is assumed that this near-parabolic flux corresponds to the influx of long-period comets with semi-major axes greater than $a_t = 3.3 \times 10^4$ AU, the semi-major axis beyond which the galactic tide efficiently feeds comets into the inner solar system from the Oort cloud. Any comets in the model with smaller semi-major axes are included by introducing a so-called 'inner edge' to the cloud (at a semi-major axis a_0) and assuming that for $a \geq a_0$ the total number of comets with semi-major axes in the range $(a, a + da)$ is proportional to $a^{\gamma-2}$, where γ is a second parameter defining the model. The smaller is a_0, or γ, the greater is the degree of central concentration of the cloud; typical models are obtained by assuming $a_0 \lesssim 6000$ AU and $\gamma \lesssim 0$.

According to this parameterization (cf. Bailey & Stagg 1988, equation 60), the total number of comets brighter than absolute magnitude $H_0 = 7$ is on the order of

$$N(\leq 7) = 2 \times 10^{10} \frac{(7/2 - \gamma)}{(1 - \gamma)} \left(\frac{a_0}{a_t}\right)^{\gamma-1} \tag{13}$$

the numerical factor corresponding to the product $\frac{1}{2} P(a_t) f_{new}(\leq 7) a_t$, where $P(a_t) \simeq 6$ Myr is the orbital period of a comet with $a = a_t$. (Here we have neglected the weak dependence of the result on the cloud's outer radius.) The total number of comets is thus 5.7×10^{11} for a representative model with $a_0 = 4000\,\mathrm{AU}$ and $\gamma = 0$, such a cloud having about 90% of its mass in bodies with semi-major axes less than $a_t = 3.3 \times 10^4\,\mathrm{AU}$. The number of comets in the assumed 'dynamically active' cloud is on the order 7×10^{10}. Changing either a_0 or γ will obviously change these figures.

Finally, again following Bailey & Stagg (1988), we may adopt Everhart's differential brightness distribution together with the assumption that the mean cometary mass is proportional to $10^{-0.5H}$ (see Figure 4). With these assumptions, the mean mass per comet brighter than $H_0 = 7$ is insensitive to the assumed upper and lower limits of the mass distribution. A value on the order of $2.55 \times 10^{14}\,\mathrm{kg}$ is found for the normalization adopted by Bailey & Stagg (1988), increasing to $5.1 \times 10^{15}\,\mathrm{kg}$ if the rough fit to the mass-magnitude relation shown in Figure 4 is adopted. With the latter normalization, the mass of the dynamically active Oort cloud is almost $60\,M_\oplus$, while the total mass of the model is nearly $500\,M_\oplus$! From the normalization adopted in Figure 4, these values are proportional to $(0.05/A)^{3/2}(\rho_c/500\,\mathrm{kg\,m^{-3}})$, where A is the mean albedo for long-period comets. Such a high mass for the Oort cloud would, of course, virtually rule out a simple planetesimal hypothesis for cometary origin: the original mass of the Oort cloud could hardly have been less than twice its present value, while the initial cometary mass in the protoplanetary disc, allowing for the placement efficiency to the Oort cloud, must have been even larger — far greater, therefore, than the total mass of the observed planets. Whereas comets were formerly regarded as a mere by-product of planet formation, the estimates now being obtained for the total cometary mass in the solar system (though quite uncertain) suggest that comets may once have played a much more significant cosmogonical rôle.

6. Interstellar comets and giant grains

The very large initial mass of comets in the primordial solar nebula indicated by these arguments and also by independent studies of the formation of comets and the evolution of the Oort cloud, suggests that the solar system may have lost a great many comets to interstellar space. An initial cometary mass of order $10^3\,M_\oplus$ corresponds to around 10% the mass of condensibles in the whole solar nebula, suggesting also that such interstellar comets may be a significant sink for elements heavier than helium on a galactic scale (cf. Tinsley & Cameron 1974, Van den Bergh 1982, Vanýsek 1987). Indeed, even were the Oort cloud not primordial (in the solar system sense of having been formed coevally with the sun and planets), it would still be necessary to postulate a large number of interstellar comets, simply in order for capture of the observed Oort cloud to be a reasonably likely event, given the typically very low capture probability per interstellar comet (e.g. Valtonen 1983, Valtonen & Innanen 1982). For these reasons, it is important to explore the consequences of assuming that interstellar comets are, in fact, very numerous, possibly making up 10% or more of the mass usually associated with interstellar dust. Such cometary material would indeed be dark, and baryonic; and in the remainder of this section we briefly consider one possible implication of the presence of such bodies.

6.1. A PROTOCOMETARY DUST DISC?

If comets are a significant, though minor, component of the interstellar medium, it is also likely that the particles of which they are made should be relatively abundant within the interstellar gas. As we have seen, the known chemical composition of comets (and of their inferred decay products) indicates that the solid nuclei are probably formed by the coagulation of interstellar or interplanetary grains, brought together by simple accretion (possibly followed by gravitational instability), or by a process involving the effects of radiation pressure or photodesorptive jet thrusting. The basic building blocks of comets are therefore partially aggregated interstellar or interplanetary dust particles, leading one to suppose that in regions of comet formation 'giant', or much larger than average, interstellar grains should predominate.

The arguments that comets represent a significant component of the interstellar medium therefore also suggest that 'giant' grains — the protocometary building blocks — should be abundant too. Such grains, with sizes much greater than the wavelength of light (i.e. 1–$100\,\mu$m or more; cf. Bailey 1987b), would be difficult to detect, and would probably not contribute significantly to the ordinary interstellar extinction. Indeed, whereas ordinary dust particles tend to be very closely coupled to the interstellar gas, giant grains — relative cannonballs in size — may be partially decoupled from the gas, and may sediment towards the galactic plane to form a dense dust-layer close to $z = 0$.

The density distribution of the large grains normal to the galactic plane may be obtained in a similar way to that used for the gas. That is, if the latter has an isothermal sound speed c_0, its density normal to the plane is given by

$$\rho = \rho_0 \exp(-z^2/2H^2) \tag{14}$$

where self-gravity has been neglected and the gas scale-height, H, is defined by

$$H^2 = c_0^2/4\pi G \rho_{\text{tot}} \tag{15}$$

where $\rho_{\text{tot}} = \rho_* + \rho_{\text{dark}}$ is the total background density of stars and dark matter, assumed to be constant. Observations in the solar neighbourhood yield $c_0 \simeq 5\,\text{km s}^{-1}$ and $H \simeq 50\,\text{pc}$; hence $\rho_{\text{tot}} \simeq 0.185\,M_\odot\,\text{pc}^{-3}$.

In the same way, the density distribution of the dust may be written in the form

$$\rho_d = \rho_{d,0} \exp(-z^2/2h^2) \tag{16}$$

where we now define the dust scale-height, h, in terms of the three-dimensional r.m.s. velocity dispersion v_d of the dust grains, i.e. $h = H v_d/\sqrt{3}c_0$. The corresponding dust column density is $\Sigma_d \simeq \sqrt{2\pi} h \rho_{d,0}$, and if such grains constitute a fraction ζ_d of the overall column density of the interstellar medium, $\rho_{d,0}/\rho_0 \simeq \zeta_d \sqrt{3}c_0/v_d$. In this way, the relative mass-density of giant grains near the galactic plane is enhanced over ζ_d by a factor of order $\sqrt{3}c_0/v_d$.

The remaining question is to estimate v_d, the r.m.s. velocity dispersion of the large grains. Here, there are two principal factors at work: one leading to a systematic *decrease* of dust velocity dispersion with increasing grain size; the other leading to a systematic *increase* of velocity dispersion. The first arises through coupling of the grains to the motions of the gas, and, therefore, to the turbulent velocity field of the interstellar medium. The stronger is this coupling (*i.e.* the smaller are the grains), the closer the dust velocity dispersion approaches the r.m.s. turbulent velocity dispersion of the gas, v_{turb}. On the other hand, the largest grains are essentially decoupled from the gas, tending to move 'ballistically' through the Galaxy under the influence of gravitational forces. As in the case of stars, these lead to a systematic tendency for the velocity dispersion to increase; the effect being mediated by the strength of the gas-dust coupling due to the resistance of motion of the grains through the gas.

In order to estimate the dependence of the dust velocity dispersion on grain size and other parameters of the system, we assume that the net velocity dispersion, $v_d(a)$, can be broken up into two components: $v_{d,turb}$ and $v_{d,grav}$, the relation between the two quantities being given by

$$v_d = \left(v_{d,turb}^2 + v_{d,grav}^2 \right)^{1/2} \tag{17}$$

We evaluate these two contributions by considering each effect separately; *i.e.* by considering the 'turbulent' velocity dispersion in the absence of gravity, and the equilibrium 'gravitational' velocity dispersion in the absence of turbulence. Although this procedure is not expected to be accurate, the general trend of $v_d(a)$ should be approximately described.

Thus, following Völk *et al.* (1980), we first assume that the grain velocity dispersion is dominated by interactions of the grains with turbulent eddies of various sizes, the relevant gas-dust collision cross-section being given solely by the geometric cross-section of the grains. (This ignores, for example, any additional effects due to radiation pressure or grain charge.) Then, following Bailey (1987a), we parameterize the strength of the turbulence by an r.m.s. velocity dispersion $v_{turb} = \kappa c_0$, where κ, for subsonic turbulence, is assumed to be less than unity. In the numerical results below we assume $\kappa = 0.5$. We also assume that the turbulence has a Kolmogoroff spectrum, the power in eddies of wavenumber k being given by $P(k) = C_{turb} k^{-5/3}$ (for $k \geq k_0$), where k_0 is the wavenumber of the largest eddies in the system. We assume these have a wavelength λ of order the scale-height of the gas disc; *i.e.* $k_0 = 2\pi/H$. Normalization of the power spectrum then yields $C_{turb} \simeq \frac{1}{3} k_0^{2/3} v_{turb}^2$.

With these assumptions, the turbulent velocity dispersion of dust grains of radius a is given (Völk *et al.* 1980, Bailey 1987a) approximately by

$$v_d(a) \simeq \kappa c_0 (1 + \tau_s^{1/2})^{-1} \tag{18}$$

where $\tau_s(a)$ is the ratio of the dust-gas slowing down time to the lifetime $t_k(k)$ of the largest turbulent eddies that affect grains of radius a. Following previous authors (*e.g.* Spitzer 1978), we define the dust-gas slowing down time by $t_s = a\rho_{grain}/\rho\bar{c}$, where $\bar{c} = (8/\pi)^{1/2} c_0$ is the mean speed of the gas molecules and ρ_{grain} is the average material density of the grains themselves. The equation of motion of a grain moving with speed v through the gas may then be written (Bailey 1987a) in the form

$$\frac{dv}{dt} = -\frac{v}{t_s} \left(1 + \frac{9}{16} \frac{v^2}{\bar{c}^2} \right)^{1/2} \tag{19}$$

In order to estimate $t_k(k)$, we follow the approximate argument used by Bailey (1987a) to calculate the dust velocity dispersion in a shell of swept-up matter around a newly formed protostar. That is, if the dust has a scale-height h, the wavenumber of the largest eddies that are contained within the dust disc is of order $k = 2\pi/h$, and their lifetime is approximately $(2P(k)k^3)^{-1/2}$ (cf. Völk et al. 1980). In this way (cf. Bailey 1987a, equation 33), we obtain

$$\tau_s(a) = \frac{t_s(a)}{t_k(k)} \simeq \kappa \left(\frac{\pi^3}{3}\right)^{1/2} \frac{a\rho_{\text{grain}}}{\rho H} \left(\frac{H}{h}\right)^{2/3} \tag{20}$$

Combining this with equation (18), and using $h = H v_{d,\text{turb}}/\sqrt{3}c_0$, therefore gives

$$\frac{\tau_s}{(1 + \tau_s^{1/2})^{2/3}} = (3\kappa)^{1/3} \left(\frac{\pi^3}{3}\right)^{1/2} \frac{a\rho_{\text{grain}}}{\rho H} = (3\kappa)^{1/3} \left(\frac{\pi^3}{3}\right)^{1/2} X \tag{21}$$

which may be used to obtain τ_s and hence $v_{d,\text{turb}}$ as a function of both grain and gas-disc parameters. Here, for convenience of notation, we have introduced a gas-dust coupling parameter $X = a\rho_{\text{grain}}/\rho H = 4\sqrt{2}t_s(G\rho_{\text{tot}})^{1/2}$; i.e.

$$X = 0.162 \left(\frac{a}{1\,\mu\text{m}}\right) \left(\frac{\rho_{\text{grain}}}{500\,\text{kg m}^{-3}}\right) \left(\frac{n_H}{1\,\text{cm}^{-3}}\right) \left(\frac{H}{50\,\text{pc}}\right) \tag{22}$$

where the gas density $\rho = n_H \bar{m}$, with $\bar{m} = 2 \times 10^{-27}$ kg.

Setting $\kappa = 0.5$, the critical grain size a_{crit} below which the dust is strongly coupled to the gas may be obtained by setting $\tau_s = 1$, i.e.

$$X_{\text{crit}} = \left(\frac{3}{\pi^3}\right)^{1/2} (3\kappa)^{-1/3} 2^{-2/3} = 0.1712 \tag{23}$$

implying, for the 'typical' parameters adopted in equation (22), $a_{\text{crit}} \simeq 1\,\mu\text{m}$. Grains very much larger than this have $\tau_s \gg 1$ and hence potentially have a much reduced velocity dispersion. Neglecting the extra gravitational effects discussed below, such grains with $X \gg X_{\text{crit}}$ would have $\tau_s \simeq 0.5(X/X_{\text{crit}})^{3/2}$ and would therefore settle to form a dust-disc with scale height asymptotically approaching $h/H \simeq \sqrt{\frac{2}{3}}\kappa(X/X_{\text{crit}})^{-3/4}$. For example, if the conditions are such that $\kappa = 0.5$ and $a_{\text{crit}} = 1\,\mu\text{m}$, then grains with radii in excess of about $5\,\mu\text{m}$ would be expected to occupy a thin dust-disc with scale-height $h \lesssim 0.12\,H$, corresponding to $h \lesssim 6\,\text{pc}$ for a nominal gas scale-height of order $50\,\text{pc}$.

Of course, the predicted decrease in dust velocity dispersion with increasing grain radius cannot be extrapolated to arbitrarily large grains (e.g. comets!). The gravitational effects of stars and molecular clouds must also be considered; and in the remainder of this section we now estimate this contribution to v_d, and finally combine our results, using equation (17), to obtain the overall dependence of v_d on grain size.

The effects of impulsive gravitational perturbations on dust grains due to passing stars and molecular clouds may be estimated in the usual way by squaring the impulse $\Delta v = 2GM/bV_r$ produced by a single perturber of mass M, relative velocity V_r and impact parameter b, and multiplying by the number of such encounters. In this way, neglecting gas drag, we find (for stars) that after a time interval t the stellar contribution to the dust velocity dispersion is given by

$$\Sigma(\Delta v)^2_{\text{stars}} = 8\sqrt{2\pi}\,\frac{G^2 M_*^2 n_*}{\sigma_*}\,\ln(b_{\max}/b_{\min})\,t \tag{24}$$

In deriving this expression, we have assumed that the perturbers (stars) have a Maxwellian velocity distribution with a one-dimensional velocity dispersion σ_*, so that the mean inverse relative velocity $\langle V_r^{-1}\rangle = (1/\sqrt{2}\sigma_*)(\text{erf}(x)/x) \simeq \sqrt{2}/\sqrt{\pi}\sigma_*$, where $x = v_d/\sqrt{2}\sigma_*$ is assumed to be much less than unity.

The effects of molecular clouds may be found in exactly the same way, from which we finally obtain

$$\dot{\Sigma}(\Delta v)^2_{\text{grav}} = \left(\Sigma(\Delta v)^2_{\text{stars}} + \Sigma(\Delta v)^2_{\text{clouds}}\right)/t \simeq 5.3 \times 10^{-8}\quad \text{m}^2\,\text{s}^{-3} \tag{25}$$

In evaluating this expression we have adopted the stellar mass distribution of Bahcall & Soneira (1980), which yields $\langle n_* M_*^2\rangle = 0.03\,M_\odot^2\,\text{pc}^{-3}$, together with $\sigma_* = 30\,\text{km s}^{-1}$ and $\ln(b_{\max}/b_{\min}) \simeq 25$; for the clouds we have assumed $n_c = 5 \times 10^{-8}\,\text{pc}^{-3}$ and $M_c = 2.5 \times 10^5\,M_\odot$, giving $\langle n_c M_c^2\rangle \simeq 3 \times 10^3\,M_\odot^2\,\text{pc}^{-3}$ (cf. Bailey 1986), together with $\sigma_c = 5\,\text{km s}^{-1}$ and $\ln(b_{\max}/b_{\min}) \simeq 7$. Obviously, with these figures, the contribution by molecular cloud perturbations is dominant by several orders of magnitude, indicating (because of their uncertain masses) that the numerical coefficient in equation (25) is probably uncertain by at least a factor of 2.

The equilibrium dust velocity dispersion under gravitational influences may now be estimated by equating the rate of increase of v_d^2 (given by equation (25) with its corresponding rate of decrease due to gas drag, i.e. $dv^2/dt \simeq |2v\,dv/dt|$, where dv/dt is given by equation (19). In this way, assuming that the equilibrium gravitational velocity satisfies $v_{d,\text{grav}}^2 \ll \bar{c}^2$ (this is of most interest in the present application), we obtain the approximate result $v_{d,\text{grav}}^2 = \frac{1}{2}\dot{\Sigma}(\Delta v)^2_{\text{grav}}\,t_s$, which reduces to

$$v_{d,\text{grav}} = 2.2 \left(\frac{\dot{\Sigma}(\Delta v)^2_{\text{grav}}}{5 \times 10^{-8}\,\text{m}^2\,\text{s}^{-3}}\right)^{1/2} \left(\frac{0.185\,M_\odot\,\text{pc}^{-3}}{\rho_{\text{tot}}}\right)^{1/4} X^{1/2}\quad \text{km s}^{-1} \tag{26}$$

These results show that whereas the large-grain turbulent velocity dispersion varies with grain size roughly proportional to $X^{-3/4}$ (for $X \gg X_{\text{crit}}$) the equilibrium velocity dispersion due to gravitational stirring by molecular clouds is proportional to $X^{1/2}$, at least provided $v_{d,\text{grav}}^2 \ll \bar{c}^2$. (If this approximation is reversed, it may be verified that the corresponding expression to (26) indicates $v_{d,\text{grav}} \propto X^{1/3}$.)

Combining these equations, therefore, solving for $v_{d,\text{turb}}$, $v_{d,\text{grav}}$ and v_d in turn, we finally obtain the results illustrated in Figure 5. The dust velocity dispersion has a shallow, broad minimum, indicating that grains with radii on the order of $a_{\text{crit}} \approx 1\,\mu\text{m}$ (depending on the gas density ρ) settle towards the galactic mid-plane with a velocity dispersion on the order of 1–$2\,\text{km s}^{-1}$, corresponding to a dust scale-height h in the range 5–$10\,\text{pc}$. Smaller grains are more strongly coupled to the gas, and have a velocity dispersion close to that, v_{turb}, of the turbulence in the gas, while larger grains are systematically pumped up to

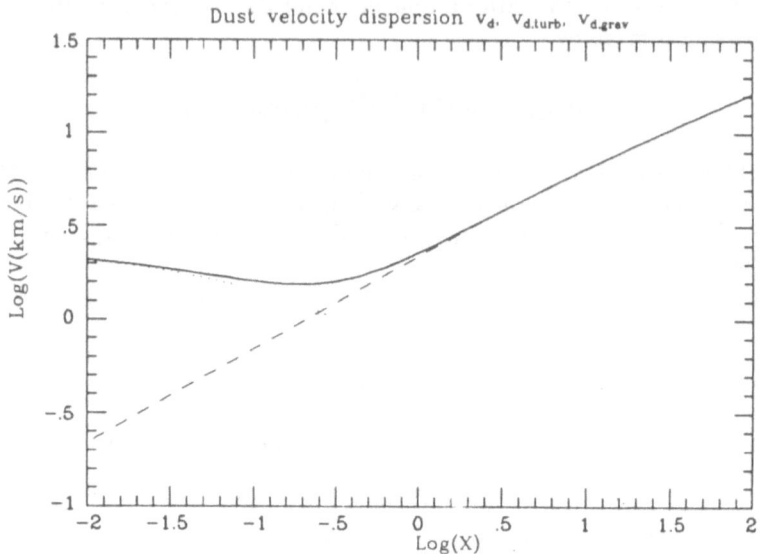

Figure 5: The equilibrium dust velocity dispersion v_d calculated from equation (17) is shown as a solid line together with the separate contributions from turbulent eddies (dotted line) and gravitational stirring (dashed line). $X = a\rho_{grain}/\rho H$, while the critical value of this parameter above which $r_s > 1$ is $X_{crit} = 0.1712$ (i.e. $\log(X_{crit}) = -0.767$). The diagram has been drawn assuming the parameters $\kappa = 0.5$, $c_0 = 5\,\text{km s}^{-1}$, together with the typical gravitational factors indicated by equation (26).

higher velocities by gravitational forces. In this way, protocometary dust grains with radii in the range 1–100 μm, expected to occur in the interstellar medium, should occupy a thin, dense dust-disc with scale-height h on the order of $\lesssim 10$ pc. It would be interesting to explore possible ways of detecting these predicted 'giant' grains, and to speculate (as happened in discussions at this meeting) whether the existence of a dense layer of large dust grains close to $z = 0$ might affect either the earth (e.g. by triggering climatic change or mass-extinctions of species during the earth's 30 Myr periodic passages across the galactic plane), or the chemical composition of new stars formed from gaseous material lying close to the plane.

7. Conclusions

This paper has briefly reviewed the standard Oort cloud theory of cometary origin, concentrating on the physical structure and masses of cometary nuclei. The latter aspect of comets — their masses — emerges as possibly *the* fundamental uncertainty in obtaining firm arguments to discriminate between theories of cometary origin. The masses of individual long-period comets are probably uncertain by at least a factor of four, whilst our detailed knowledge even of the *shape* of the cometary mass-distribution (for example, whether dominated by large, or medium-sized nuclei) has scarcely advanced in more than twenty years. The best hope for improving this situation is to obtain accurate absolute and nuclear magnitudes for as many newly discovered long-period comets as possible; new nuclear magnitudes are especially important, as these still offer the most direct way to estimate the radii of distant cometary nuclei, though other techniques (for example based on radar measurements, or non-gravitational forces) also provide valuable information.

A redetermination of the mass-absolute-magnitude relationship, based on data for 27 long-period comets, indicates a correlation of the form $\log m^* \simeq 18.2 - 0.5H_0$, where $m^* = m/1\,\mathrm{kg}$ and the normalization assumes an average albedo $A = 0.05$ and density $\rho_c = 500\,\mathrm{kg}$ m^{-3}. With Everhart's (1967) cometary brightness distribution, the mean mass of comets brighter than $H_0 = 7$ is found to be $5 \times 10^{15}\,\mathrm{kg}$, twenty times larger than assumed by Bailey & Stagg (1988). These masses could be reduced somewhat if the quoted nuclear magnitudes were systematically too bright, or if the assumed albedo and density were changed. Nevertheless, on the basis of this new mass-magnitude relationship, the mass of the outer, dynamically active Oort cloud is $\simeq 60\,M_\oplus$, while the total mass of a representative Oort cloud with a 'conservative' inner core is around $500\,M_\oplus$. These high estimates for the total mass of comets in the solar system have important implications for theories of cometary origin; apart from anything else, the dark cometary matter originally in the solar system would almost certainly have outweighed the total mass of planets.

The assumption of a very large cometary mass in the solar system suggests that there may be an equally large relative abundance of comets in interstellar space. Similarly, the basic building blocks of comets — large, or partially aggregated 'giant' interstellar grains — should also be common. Such grains, with radii in the approximate range 1–$100\,\mu\mathrm{m}$, are predicted to sediment towards the galactic plane to form a thin, dense sub-disc in the Galaxy with scale-height h on the order of 5–$10\,\mathrm{pc}$. Comets and their partially aggregated precometary dust grains may thus represent an important component of the interstellar medium that remains to be discovered. The total mass of condensed material in this form, which is partially decoupled from the gas, has potentially important implications for a wide range of studies (cf. Bailey 1988). The total mass, however, may be only $\simeq 10\%$ of that normally associated with interstellar dust.

ACKNOWLEDGMENTS. I thank colleagues, particularly Dr. G.J. Hahn, for discussions. This work was supported by the SERC, while my attendance at the meeting was supported by funds from the Local Organizing Committee.

References

Bahcall, J.N. & Soneira, R.M., 1980. *The Universe at faint magnitudes. I. Models for the Galaxy and the predicted star counts*, Astrophys. J. Suppl. Ser., **44**, 73–110.

Bailey, M.E., 1986. *The mean energy transfer rate to comets in the Oort cloud and implications for cometary origins*, Mon. Not. Roy. Astron. Soc., **218**, 1–30.

Bailey, M.E., 1987a. *The formation of comets in wind-driven shells around protostars*, Icarus, **69**, 70–82.

Bailey, M.E., 1987b. *Giant grains around protostars*, Q. Jl. R. Astron. Soc., **28**, 242–247.

Bailey, M.E., 1988. *Comets in star-forming regions*, Dust in the Universe, eds. Bailey, M.E. & Williams, D.A., 113–120. Cambridge University Press.

Bailey, M.E., 1989. *Cometary dynamics — the inner core of the Oort cloud*, Catastrophes and Evolution: Astronomical Foundation, ed. Clube, S.V.M., in press. Cambridge University Press.

Bailey, M.E., 1990. *Short-period comets: probes of the inner core*, Asteroids, Comets, Meteors III, eds. Rickman, H., Lagerkvist, C.-L., *et al.*, in press. Uppsala Observatory.

Bailey, M.E., Clube, S.V.M. & Napier, W.M., 1986. *The origin of comets*, Vistas Astron., **29**, 53–112.

Bailey, M.E., Clube, S.V.M. & Napier, W.M., 1990. *The Origin of Comets*. Pergamon Press, Oxford.

Bailey, M.E. & Stagg, C.R., 1988. *Cratering constraints on the inner Oort cloud: steady-state models*, Mon. Not. Roy. Astron. Soc., **235**, 1–32.

Biermann, L., 1981. *The smaller bodies of the solar system*, Philos. Trans. R. Soc. London, Ser. A., **303**, 351–352.

Biermann, L. & Michel, K.W., 1978. *The origin of cometary nuclei in the presolar nebula*, Moon & Planets, **18**, 447–464.

Brooks, A.M., 1990. *Aggregation of grains in the protosolar nebula*, Asteroids, Comets, Meteors III, eds. Rickman, H., Lagerkvist, C.-L., *et al.*, in press. Uppsala Observatory.

Cameron, A.G.W., 1962. *The formation of the sun and planets*, Icarus, **1**, 13–69.

Cameron, A.G.W., 1973. *Accumulation processes in the primitive solar nebula*, Icarus, **18**, 407–450.

Cameron, A.G.W., 1978. *Physics of the primitive solar accretion disk*, Moon & Planets, **18**, 5–40.

Clube, S.V.M., 1985. *Molecular clouds: comet factories?* Dynamics of Comets: Their Origin and Evolution, eds. Carusi, A. & Valsecchi, G.B., IAU Coll. No. 83, 19–30. Reidel, Dordrecht, The Netherlands.

Delsemme, A.H., 1982. *Chemical composition of cometary nuclei*, Comets, ed. Wilkening, L., IAU Coll. No. 61, 85–130. University of Arizona Press, Tucson.

Drobyshevski, E.M., 1978. *The origin of the solar system: implications for transneptunian planets and the nature of the long-period comets*, Moon & Planets, **18**, 145–194.

Duncan, M., Quinn, T. & Tremaine, S.D., 1987. *The formation and extent of the solar system comet cloud*, Astron. J., **94**, 1330–1338.

Everhart, E., 1967. *Intrinsic distributions of cometary perihelia and magnitudes*, Astron. J., **72**, 1002–1011.

Fernández, J.A., 1982. *Dynamical aspects of the origin of comets*, Astron. J., **87**, 1318–1332.

Fernández, J.A. & Ip, W.-H., 1983. *On the time-evolution of the planetary influx in the region of the terrestrial planets*, Icarus, **54**, 377–387.

Geiss, J., 1988. *Composition in Halley's comet: clues to origin and history of cometary matter*, Cosmic Chemistry, ed. Klare, G., 1–27. (Reviews in Modern Astronomy, Volume 1.) Springer-Verlag, Berlin.

Goldreich, P. & Lynden-Bell, D., 1965a. *I. Gravitational instability of uniformly rotating disks*, Mon. Not. Roy. Astron. Soc., **130**, 97–124.

Goldreich, P. & Lynden-Bell, D., 1965b. *II. Spiral arms as sheared gravitational instabilities*, Mon. Not. Roy. Astron. Soc., **130**, 125–158.

Goldreich, P. & Ward, W.R., 1973. *The formation of planetesimals*, Astrophys. J., **183**, 1051–1061.

Greenberg, J.M., 1988. *The interstellar dust model of comets: post Halley*, Dust in the Universe, eds. Bailey, M.E. & Williams, D.A., 121–143. Cambridge University Press.

Greenberg, R., 1985. *The origin of comets among the accreting outer planets*, Dynamics of Comets: Their Origin and Evolution, eds. Carusi, A. & Valsecchi, G.B., IAU Coll. No. 83, 3–10. Reidel, Dordrecht, The Netherlands.

Greenberg, R., Hartmann, W.K., Chapman, C.R. & Wacker, J.F., 1978. *The accretion of planets from planetesimals*, Protostars & Planets, ed. Gehrels, T., IAU Coll. No. 52, 599–622. University of Arizona Press, Tucson.

Greenberg, R., Weidenschilling, S.J., Chapman, C.R. & Davis, D.R., 1984. *From icy planetesimals to outer planets and comets*, Icarus, **59**, 87–113.

Hills, J.G., 1981. *Comet showers and the steady-state infall of comets from the Oort cloud*, Astron. J., **86**, 1730–1740.

Hills, J.G., 1982. *The formation of comets by radiation pressure in the outer protosun*, Astron. J., **87**, 906–910.

Hills, J.G. & Sandford II, M.T., 1982. *The formation of comets by radiation pressure in the outer protosun. II. Dependence on the radiation-grain coupling*, Astron. J., **88**, 1519–1521.

Hughes, D.W., 1985. *The transition between long period comets, short period comets and meteoroid streams*, Dynamics of Comets: Their Origin and Evolution, eds. Carusi, A. & Valsecchi, G.B., IAU Coll. No. 83, 129–142. Reidel, Dordrecht, The Netherlands.

Hughes, D.W., 1987a. *Cometary magnitude distribution: the tabulated data*, Symposium on the Diversity and Similarity of Comets, eds. Rolfe, E.J. & Battrick, B., ESA SP-278, 43–48. ESA Publications, ESTEC, Noordwijk, The Netherlands.

Hughes, D.W., 1987b. *On the distribution of cometary magnitudes*, Mon. Not. Roy. Astron. Soc., **226**, 309–316.

Hughes, D.W. & Daniels, P.A., 1980. *The magnitude distribution of comets*, Mon. Not. Roy. Astron. Soc., **191**, 511–520.

Humphries, C.M., 1982. *Photodesorptive jet thrusting — a mechanism for efficient formation of kilometre-sized bodies in molecular clouds?* Proc. Workshop on Interstellar Comets, eds. Clube, S.V.M. & McInnes, B., Occ. Rep. Roy. Obs. Edinb., No. 9, 33–36.

Jewitt, D., 1990. Paper presented at Uppsala meeting *Asteroids Comets Meteors III*. In press.

Jewitt, D. & Luu, J.X., 1989. *A CCD portrait of Comet P/Tempel 2*, Astron. J., **97**, 1766–1790.

Keller, H.U., Arpigny, C., Barbieri, C., Bonnet, R.M., Cazes, S., Coradini, M., Cosmovici, C.B., Delamere, W.A., Huebner, W.F., Hughes, D.W., Jamar, C., Malaise, D., Reitsema, H.J., Schmidt, H.U., Schmidt, W.K.H., Seige, P., Whipple, F.L. & Wilhelm, K., 1986. *First Halley Multicolour Camera imaging results from Giotto*, Nature, **321**, 320–326.

Kresák, L., 1981. *Evolutionary aspects of the splits of cometary nuclei*, Bull. Astron. Inst. Czechosl., **32**, 19–40.

Kresák, L., 1982. *Comet discoveries, statistics and observational selection*, Comets, ed. Wilkening, L., 56–82. University of Arizona Press, Tucson.

Kuiper, G.P., 1951. *On the origin of the solar system*, Astrophysics, ed. Hynek, J.A., 357–424. McGraw-Hill, New York.

Larson, S.M. & Johnson, R.R., 1983. *Comet IRAS-Araki-Alcock (1983d)*, IAU Circ. No. 3811.

Marochnik, L.S., Mukhin, L.M. & Sagdeev, R.Z., 1988. *Estimates of mass and angular momentum in the Oort cloud*, Science, **242**, 547–550.

McCrea, W.H., 1975. *Solar system as space probe*, Observatory, **95**, 239–255.

McDonnell, J.A.M., Alexander, W.M., Burton, W.M., Bussoletti, E., Clark, D.H., Grard, R.J.L., Grün, E., Hanner, M.S., Hughes, D.W., Igenbergs, E., Kuczera, H., Lindblad, B.A., Mandeville, J.-C., Minafra, A., Schwehm, G.H., Sekanina, Z., Wallis, M.K., Zarnecki, J.C., Chakaveh, S.C., Evans, G.C., Evans, S.T., Firth, J.G., Littler, A.N., Massonne, L., Olearczyk, R.E., Pankiewicz, G.S., Stevenson, T.J. & Turner, R.R., 1986. *Dust density and mass distribution near comet Halley from Giotto observations*, Nature, **321**, 338–341.

McDonnell, J.A.M., Green, S.F., Nappo, S., Pankiewicz, G.S., Perry, C.H. & Zarnecki, J.C., 1989. *Dust mass distributions: the perspective from Giotto's measurements at P/Halley*, preprint. (Paper presented at IAU Coll. No. 116: *Comets in the Post-Halley Era*, submitted to Icarus.)

Mendis, D.A. & Alfvén, H., 1976. *On the origin of comets*, The Study of Comets: Part 2, eds. Donn, B., Mumma, M., Jackson, W., A'Hearn, M. & Harrington, R., IAU Coll. No. 25, 638–659. NASA SP-393, Washington D.C.

Mendis, D.A. & Marconi, M.L., 1986. *A note on the total mass of comets in the solar system*, Earth, Moon & Planets, **36**, 187–191.

Millis, R.L., A'Hearn, M.F. & Campins, H., 1988. *An investigation of the nucleus and coma of Comet P/Arend-Rigaux*, Astrophys. J., **324**, 1194–1209.

Mizuno, H., Markiewicz, W.J. & Völk, H.J., 1988. *Grain growth in turbulent protoplanetary accretion disks*, Astron. Astrophys., **195**, 183–192.

Möhlmann, D. & Kührt, E., 1989. *Comet nucleus models*, Adv. Space Res., **9** (No. 3), (3)17–(3)23. Pergamon Press.

Morfill, G. & Völk, H.J., 1984. *Transport of dust and vapor and chemical fractionation in the early protosolar cloud*, Astrophys. J., **287**, 371–395.

Napier, W.M. & Humphries, C.M., 1986. *Interstellar planetesimals – II. Radiative instability in dense molecular clouds*, Mon. Not. Roy. Astron. Soc., **221**, 105–117.

O'Dell, C.R., 1986. *A possible comet and asteroid link in the formation of comets*, Icarus, **67**, 71–79.

Oort, J.H., 1950. *The structure of the cloud of comets surrounding the solar system and a hypothesis concerning its origin*, Bull. Astron. Inst. Neth., **11**, 91–110.

Öpik, E.J., 1973. *Comets and the formation of planets*, Astrophys. Space Sci., **21**, 307–398.

34

Quinn, T., Tremaine, S. & Duncan, M., 1989. *Planetary perturbations and the origin of short-period comets*, CITA preprint.

Rickman, H., 1989. *The nucleus of Comet Halley: surface structure, mean density, gas and dust production*, Adv. Space Res., **9** (No. 3), (3)59–(3)71. Pergamon Press.

Roemer, E., 1966. *The dimensions of cometary nuclei*, Mem. Soc. Roy. des Sciences de Liège (Ser. 5), **12**, 23–28.

Safronov, V.S., 1969. *Evolution of the Protoplanetary Cloud and Formation of the Earth and the Planets*, Translated by Israel program for scientific translations, Jerusalem 1972.

Safronov, V.S., 1972. *Ejection of bodies from the solar system in the course of the accumulation of the giant planets and the formation of the cometary cloud*, The Motion, Evolution of Orbits, and Origin of Comets, eds. Chebotarev, G.A., Kazimirchak-Polonskaya, E.I. & Marsden. B.G., IAU Symp. No. 45, 329–334. Reidel, Dordrecht, The Netherlands.

Safronov, V.S. & Ruzmaikina, T.V., 1978. *On angular momentum transfer and accumulation of solid bodies in the solar nebula*, Protostars & Planets, ed. Gehrels, T., IAU Coll. No. 52, 545–564. University of Arizona Press, Tucson.

Sagdeev, R.Z., Szabó, F., Avanesov, G.A., Cruvellier, P., Szabó, L., Szegő, K., Abergel, A., Balazs, A., Barinov, I.V., Bertaux, B.-L., Blamont, J., Detaille, M., Demarelis, E., Dul'nev, G.N., Endrőczy, G., Gardos, M., Kanyo, M., Kostenko, V.I., Krasikov, V.A., Nguyen-Trong, T., Nyitrai, Z., Reny, I., Rusznyak, P., Shamis, V.A., Smith, B., Sukhanov, K.G., Szabó, F., Szalai, S., Tarnopolsky, V.I., Toth, I., Tsukanova, G., Valníček, B.I., Varhalmi, L., Zaiko, Yu.K., Zatsepin, S.I., Ziman, Ya.L., Zsenei, M. & Zhukov, B.S., 1986. *Television observations of comet Halley from Vega spacecraft*, Nature, **321**, 262–266.

Sekanina, Z., 1985. *Precession model for the nucleus of periodic comet Giacobini-Zinner*, Astron. J., **90**, 827–845.

Sekanina, Z. & Yeomans, D.K., 1984. *Close encounters and collisions of comets with the earth*, Astron. J., **89**, 154–161.

Sekiya, M., 1983. *Gravitational instabilities in a dust-gas layer and formation of planetesimals in the solar nebula*, Prog. Theor. Phys., **69**, 1116–1130.

Spitzer, L., 1941. *The dynamics of the interstellar medium II. Radiation pressure*, Astrophys. J., **94**, 232–244.

Spitzer, L., 1978. *Physical Processes in the Interstellar Medium*. Wiley, New York.

Svoreň, J., 1987. *Consequences of the size determination of P/Halley by space probes on the scale of sizes of cometary nuclei*, Symposium on the Diversity and Similarity of Comets, eds. Rolfe, E.J. & Battrick, B., ESA SP-278, 707–712. ESA Publications, ESTEC, Noordwijk, The Netherlands.

Tinsley, B.M. & Cameron, A.G.W., 1976. *Possible influence of comets on the chemical evolution of the Galaxy*, Astrophys. Space Sci., **31**, 31–35.

Tremaine, S., 1990 *Dark Matter in the Solar System*, Baryonic Dark Matter, eds D. Lynden-Bell & G. Gilmore, Kluwer, Dordrecht, p 37.

Valtonen, M.J., 1983. *On the capture of comets into the solar system*, Observatory, **103**, 1–4.

Valtonen, M.J. & Innanen, K.A., 1982. *The capture of interstellar comets*, Astrophys. J., **255**, 307–315.

Van den Bergh, S., 1982. *Giant molecular clouds and the solar system comets*, J. R. Astron. Soc. Canada, **76**, 303–308.

Van Flandern, T.C., 1978. *A former asteroidal planet as the origin of comets*, Icarus, **36**, 51–74.

Vanýsek, V., 1987. *A note on comets and the chemical evolution of the Galaxy*, Symposium on the Diversity and Similarity of Comets, eds. Rolfe, E.J. & Battrick, B., ESA SP-278, 745–746. ESA Publications, ESTEC, Noordwijk, The Netherlands.

Verniani, F., 1969. *Structure and fragmentation of meteoroids*, Space Sci. Rev., **10**, 230–261.

Völk, H.J., Jones, F.C., Morfill, G.E. & Röser, S., 1980. *Collisions between grains in a turbulent gas*, Astron. Astrophys., **85**, 316–325.

Ward, W.R., 1976. *The formation of the solar system*, Frontiers of Astronomy, ed. Avrett, E.H., 1–39. Harvard University Press.

Weidenschilling, S.J., Donn, B. & Meakin, P., 1989. *The physics of planetesimal formation*, Formation and Evolution of Planetary Systems, eds. Weaver, H., Paresce, F. & Danly, L., in press. Cambridge University Press.

Weissman, P.R., 1982. *Terrestrial impact rates for long and short-period comets*, Geol. Soc. Amer. Spec. Pap., **190**, 15–24.

Weissman, P.R., 1983. *The mass of the Oort cloud*, Astron. Astrophys., **118**, 90–94.

Weissman, P.R., 19ยั5. *Dynamical evolution of the Oort cloud*, Dynamics of Comets: Their Origin and Evolution, eds. Carusi, A. & Valsecchi, G.B., IAU Coll. No. 83, 87–96. Reidel, Dordrecht, The Netherlands.

Weissman, P.R., 1986. *The mass of the Oort cloud: a post-Halley reassessment*, Bull. Amer. Astron. Soc., **18**, 799.

Weissman, P.R., 1989. *The cometary impactor flux at the earth*, Proceedings of the Conference on Global Catastrophes in Earth History, Snowbird, October 1988. In press.

Wetherill, G.W., 1980. *Formation of the terrestrial planets*, Annu. Rev. Astron. Astrophys., **18**, 77–113.

Whipple, F.L., 1946. *Concentrations of the interstellar medium*, Astrophys. J., **104**, 1–11.

Whipple, F.L., 1964. *The evidence for a comet belt beyond Neptune*, Proc. Natl. Acad. Sci. (USA), **51**, 565–594.

Yabushita, S., 1983. *On the formation of cometary nuclei in dense globules*, Astrophys. Space Sci., **89**, 159–161.

Yamamoto, T. & Kozasa, T., 1988. *The cometary nucleus as an aggregate of planetesimals*, Icarus, **75**, 540–551.

DARK MATTER IN THE SOLAR SYSTEM

Scott Tremaine
Canadian Institute for Theoretical Astrophysics
University of Toronto
60 St. George St.
Toronto M5S 1A1
Canada

ABSTRACT. There is little direct dynamical evidence for dark matter in the solar system. However, fairly general models for the formation of the planetary system predict that dark matter in the form of fossil planetesimals is likely to be present in two locations: an extended spherical cloud with semi-major axes between 10^3 AU and 4×10^4 AU, and a flat disk in the region of the outer planets or just beyond them. The outer part of the extended cloud is probably the source of new comets (the Oort cloud), and the disk is probably the source of the Jupiter-family comets. The mass in the extended cloud may exceed the mass in the cores of the giant planets; thus most of the metals in the solar system outside the Sun may be in the form of dark matter. I also review the evidence for a tenth planet, Planet X, and for the companion star Nemesis. Neither is very probable, since the observations that they are invoked to explain (residuals in the orbits of Uranus and Neptune, and periodicities in the cratering and extinction records) are only marginally significant. In addition, the proposed objects have unusual properties: Planet X must be at least ten times more distant and massive than any other planet, and Nemesis must have a very improbable orbit. Finally, I summarize dynamical constraints that strongly suggest that the Sun has no companion more massive than about $0.03 \mathcal{M}_\odot$.

Introduction

Most of us are mainly interested in baryonic dark matter because of the role it plays in galactic structure and cosmology. Thus I begin by describing why dark matter in the solar system is relevant to this broader context.

First, of course, all of the dynamical arguments for dark matter in galaxies are based on the assumption that Newton's laws of motion and law of gravity are correct, and the solar system provides by far the most accurate tests of these laws on large scales.

A second reason is that some forms of baryonic dark matter may be most easily detected in the solar system. For example, if brown dwarfs comprise most of the dark matter in the Galactic disk, and if they form binaries at the same rate as main sequence stars, then there may very likely be one or more brown dwarfs orbiting the Sun.

It seems certain that dark matter in galaxies is intimately related to galaxy formation;

37

D. Lynden-Bell and G. Gilmore (eds.), Baryonic Dark Matter, 37–65.

similarly, I shall argue that the nature and distribution of dark matter in the solar system provides both insights and puzzles that bear on the formation of the solar system.

Finally, recall that there have been two major dynamical puzzles in the solar system since 1800: the unexplained residuals in the orbit of Uranus that led to the discovery of Neptune, and the anomalous precession of Mercury's perihelion that was explained by general relativity. One puzzle was explained by a new but commonplace object, and the other by radical new physical laws. These two classes of explanation are also seen in one of the main issues addressed by this meeting: will dark matter in galaxies be explained by commonplace objects such as brown dwarfs, or by novel physical laws and exotic new particles?

1. Dark Matter in the Planetary System

Dark matter located between the planets can be detected by its gravitational influence on planetary orbits. The best limits come from Kepler's law, which states that in the absence of dark matter the period P and semi-major axis a of a planet of negligible mass are related by

$$\frac{4\pi^2 a^3}{P^2} = G\,\mathcal{M}_\odot. \tag{1}$$

The perturbing effects of the planetary masses are straightforward to account for and will not be discussed here.

Distances are measured in terms of the astronomical unit (AU), which is defined by the relation

$$G\,\mathcal{M}_\odot = (0.01720209895)^2 \frac{\text{AU}^3}{(\text{day})^2}. \tag{2}$$

The constant in the brackets is approximately 2π divided by the number of days in a year, so that the Earth's semi-major axis $a_\oplus$ is very nearly 1 AU.

Direct range measurements by radar and spacecraft tracking give both the value of 1 AU and the semi-major axes to the other planets in centimeters. The standard IAU (1977) value for the astronomical unit is 1 AU = $1.49597870 \times 10^{13}$ cm.

If there is dark matter in the system, the semi-major axes a_r for other planets that are deduced from range measurements will differ from the semi-major axes a_p deduced from the orbital periods and Kepler's law (1). If for simplicity I assume that the dark matter is distributed spherically, with mass $\Delta M(r)$ interior to radius r, then the apparent mass of the Sun determined from equation (2) and the value of the astronomical unit will be $\mathcal{M}_\odot' = \mathcal{M}_\odot + \Delta M(a_\oplus)$. Then the semi-major axis to another planet as deduced from ranging, and the semi-major axis as deduced from Kepler's law, will be given by

$$\frac{4\pi^2 a_r^3}{P^2} = G[\mathcal{M}_\odot + \Delta M(a_r)], \qquad \frac{4\pi^2 a_p^3}{P^2} = G\,\mathcal{M}_\odot', \tag{3}$$

which yields for $\Delta M \ll \mathcal{M}_\odot$

$$\frac{a_r}{a_p} \equiv 1 + \tfrac{1}{3}\epsilon, \qquad \text{where} \qquad \epsilon \simeq \frac{\Delta M(a_r) - \Delta M(a_\oplus)}{\mathcal{M}_\odot}. \tag{4}$$

The values of ϵ determined from least-squares fits to solar system ephemerides are all consistent with zero (Talmadge *et al.* 1988, Anderson *et al.* 1989) and the 1σ upper limits

Table 1 Limits on dark mass in the planetary system

planet	method	distance (AU)	$\epsilon(1\sigma)$
Mercury	radar	0.4	$< 3 \times 10^{-8}$
	Mariner 10		
Venus	radar	0.7	$< 3 \times 10^{-8}$
Mars	Mariner 9	1.9	$< 1 \times 10^{-9}$
	Viking		
Jupiter	Voyager	5.2	$< 2 \times 10^{-7}$
Saturn	Voyager	9.5	–
Uranus	Voyager	19.2	$< 3 \times 10^{-6}$
Neptune	Voyager	30.0	–

NOTES: Data from Talmadge *et al.* (1988) and Anderson *et al.* (1989). Dashes indicate that data exist but are not yet analyzed and published.

are shown in Table 1. These provide approximate limits on the dark mass in solar masses contained between the planets, although of course any dark mass contained inside Mercury's orbit would not be detected.

Limits on dark mass can also be derived by comparing observed perihelion precession rates with those predicted from mutual planetary perturbations and general relativistic effects. This test is sensitive to the gradient $d\Delta M(a)/dr$ at the planetary orbit rather than the difference $\Delta M(a) - \Delta M(a_\oplus)$. For $\Delta M(r) \propto r$ the limits (from the orbits of Mercury and Mars) are somewhat less sensitive than those from ranging (Talmadge *et al.* 1988).

The main known source of dark matter in the planetary system is the asteroid belt. The total mass of the belt is quite uncertain, due to the unknown contribution from faint asteroids, but is at least $2 \times 10^{-9} \mathcal{M}_\odot$ (Hughes 1982). The masses of the largest few asteroids can be directly determined from their perturbations on the orbit of Mars (Standish and Hellings 1989).

2. Residuals in the Orbits of Uranus and Neptune

Residuals in the orbit of Uranus of ~ 100" led to the prediction of Neptune by LeVerrier and Adams in 1845. Neptune was discovered in 1846 within 1° of LeVerrier's predicted position.

Remaining residuals in Uranus's orbit of a few arc seconds led to predictions of a tenth planet by Lowell in 1915 and Pickering in 1928. Lowell initiated a systematic survey for a tenth planet which led to the discovery of Pluto by Tombaugh in 1930, within 6° longitude of the predicted positions (see Hoyt 1980 for a history).

The Pluto mass needed to remove the residuals is between $0.5M_\oplus$ and $5M_\oplus$ (see Duncombe and Seidelmann 1980 for a review of Pluto mass estimates; $M_\oplus = 1$ Earth mass$= 3.04035 \times 10^{-6} \mathcal{M}_\odot$). However, discovery of Pluto's satellite Charon (Christy and Harrington 1978) showed that the mass of the Pluto-Charon system was only $0.002M_\oplus$, far too small to have any detectable effect on the orbits of Uranus and Neptune. Since Pluto cannot account for the residuals, the discovery of Pluto close to the predicted position must have been accidental; thus, if the residuals are real, it is likely that some undiscovered dark mass is responsible.

The best available theory for the orbits of Uranus and Neptune is the JPL ephemeris DE 200, which is based on fitting observations to 1978. The residuals between DE 200 and the observations show three distinct anomalies (Seidelmann and Harrington 1988):

(i) Observations of the right ascension of both Uranus and Neptune made since 1978 already deviate systematically from the DE 200 predictions, by 0.5" after only 10 years.

(ii) Uranus observations prior to 1900 cannot be fit in right ascension. It is possible that systematic errors in the early observations are the source of this discrepancy.

(iii) Prediscovery observations of Neptune by Lalande in 1795 and Galileo in 1613 cannot be fit. However, the deviations are only four times the estimated error for the Lalande observations, and the scale of Galileo's drawing showing Neptune is uncertain (Kowal and Drake 1980).

While the residuals are larger than expected from the known observational errors, it is probably prudent to regard them as setting upper limits on, rather than providing firm evidence for, dark matter in the solar system.

Constraints on dark matter are also available from two other types of probe at similar distances (20 to 40 AU).

The Pioneer 10 spacecraft shows no evidence of unmodelled acceleration to a level $|\Delta\ddot{r}| \lesssim 5 \times 10^{-9}$ cm sec^{-2} out to about 35 AU (Anderson and Standish 1986). If I set $|\Delta\ddot{r}| = G\Delta M/r^2$, $r = 40$ AU, then I obtain a crude limit on dark mass

$$\Delta M \lesssim 1.3 \times 10^{-5} \, \mathcal{M}_\odot = 4 M_\oplus. \tag{5}$$

In the near future this limit will be improved substantially by the addition of data from Pioneer 11 and Voyager 1 and 2, and the distance will increase to ~ 50 AU.

Comets are excellent probes of distant dark mass, both because their aphelion distances can exceed those of any planet, and because their eccentricity and inclination are large so that precession is easier to detect. Limits on the anomalous precession of the orbit of Halley's comet since 1835 yield (Hamid, Marsden and Whipple 1968, Yeomans 1986)

$$\Delta M \lesssim 0.3 M_\oplus \quad \text{at} \quad 40 \, \text{AU}. \tag{6}$$

There are reasons to treat the limits (5) and (6) with caution. They are obtained from the largest extra acceleration that can be added to a fixed solar system model without introducing unacceptable errors in the trajectory of the spacecraft or comet. However, the mass causing this acceleration would also perturb the planet orbits and change the best-fit values for the planetary masses and orbital elements. Thus the models are not self-consistent. A proper approach would require a simultaneous fit to both planetary data and the data from the spacecraft or comet. The limits (5) and (6) could be substantially too low if there is a large covariance between the dark mass and one or more of the other free parameters.

Another concern is that the solution for the orbit of Halley's comet contains free parameters that model the nongravitational acceleration due to mass loss (Yeomans 1986). These parameters may mask acceleration due to dark mass.

Conclusion. The residuals in the orbits of Uranus and Neptune are poorly understood, and it is not clear whether they are large enough to require the presence of dark mass. However, if dark mass is the major cause of the residuals, then the needed amount is at most about $5M_\oplus$ located at about $40\,\mathrm{AU}$ (since this is the largest Pluto mass that dynamicists have invoked to explain the residuals). The Pioneer spacecraft gives a similar limit; Halley's comet gives a substantially lower but less rigorous limit.

Since the perturbing effects of a distant mass M at radius r scale as M/r^3, the limit given by residuals in the orbits of the outer planets can be written as

$$\Delta M(r) \lesssim 5M_\oplus \left(\frac{r}{40\,\mathrm{AU}}\right)^3. \tag{7}$$

2.1 PLANET X

The most popular explanation for residuals in the Uranus and Neptune orbits is an undiscovered Planet X that orbits beyond Pluto.

The search that led to the discovery of Pluto still provides the best limits on the brightness of any undiscovered planets (Tombaugh 1961). Tombaugh covered 75% of the sky, mostly to a limiting magnitude $V = 17$, almost two magnitudes fainter than Pluto. Unless Planet X is on a highly inclined orbit and happened to be near the ecliptic pole at the time of the search, then it must be substantially fainter than Pluto.

The apparent brightness of a distant planet is proportional to pR^2/r^4, where p is the planet's albedo, R is its radius and $r \gg 1\,\mathrm{AU}$ is its distance from the Sun. Pluto's geometric albedo is high (0.6; see Tholen *et al.* 1987), and the albedo of Planet X may be lower; since both the albedo and the exact limiting magnitude of Tombaugh's search are uncertain, I will simply assume that Tombaugh's failure to detect Planet X implies that its ratio R^2/r^4 must be smaller than Pluto's. Thus

$$\frac{R_x^2}{r_x^4} < \frac{R_p^2}{r_p^4} \quad \text{or} \quad \frac{M_x}{r_x^6} < \frac{M_p}{r_p^6}, \tag{8}$$

where Pluto's distance and mass are $r_p = 40\,\mathrm{AU}$ and $M_p = 0.002M_\oplus$, and I have assumed that Planet X and Pluto have the same density. If Planet X is to explain the residuals in the orbits of Uranus and Neptune, its mass must be close to the upper limit (7), so I shall write

$$M_x = \eta \times 5M_\oplus \left(\frac{r_x}{40\,\mathrm{AU}}\right)^3, \tag{9}$$

where η cannot be much smaller than unity.

Combining equations (8) and (9), I find

$$r_x \gtrsim 500\eta^{1/3}\,\mathrm{AU}, \qquad M_x \gtrsim 10^4\eta^2 M_\oplus. \tag{10}$$

Thus, even for η as low as 0.3, Planet X must be at least a factor of ten more distant than all the known giant planets and more massive than any of them ($r_x \simeq 300\,\mathrm{AU}$, $M_x \simeq 10^3 M_\oplus = 0.003\,\mathcal{M}_\odot$; Jupiter's mass is only $314M_\oplus$). In fact, the required mass is so large that the object should probably be regarded as a degenerate dwarf star rather than a planet. A brown dwarf with this mass and distance would probably be visible in the IRAS Point Source Catalog (Chester 1986; the luminosity and effective temperature are given by

Stevenson 1986), but only at high Galactic latitude. More distant brown dwarfs satisfying equation (9) would be even more massive and even brighter in the infrared; they would also violate other dynamical constraints (§5).

There have been several attempts to determine the location of Planet X from the residuals. Since it is probably very distant, its orbital motion is negligible, so all that can be determined is its position (to within a reflection through the Sun) and the ratio M_x/r_x^3. At present there is no consensus on a predicted location (Harrington 1988, Gomes 1989; see Seidelmann and Harrington 1988 for a comparative review).

My own opinion is that the residuals in the orbits of Uranus and Neptune are not due to a tenth planet. Either the observational errors have been underestimated, or the accuracy of the theoretical models has been overestimated, or some unmodelled dynamical effect is responsible.

Our understanding of the Uranus and Neptune residuals will improve in the future. The Voyager encounters have now provided much more accurate masses for both planets as well as direct ranges. More accurate angular positions of the outer planets will become available through ring occultations and Hubble Space Telescope observations of their satellites. There will also be general improvements to solar system ephemerides from millisecond pulsars, VLA observations of satellites, and spacecraft such as Magellan, Galileo, and Cassini (Standish 1986).

3. The Outer Regions of the Solar System

Before discussing specific models for dark matter beyond the planetary system, it is worthwhile to review some general aspects of dynamics in the outer solar system.

3.1 THE ROCHE SURFACE

Tidal forces from the Galaxy establish an outer boundary to the solar system (Antonov and Latyshev 1972). If I approximate the Galaxy as axisymmetric and assume that the Sun travels on a circular orbit in the Galactic plane, then a test particle orbiting in the combined field of the Sun and Galaxy conserves its Jacobi integral (e.g. Binney and Tremaine 1987)

$$E_J = \tfrac{1}{2}\mathbf{v}^2 - \frac{G\mathcal{M}_\odot}{r} - \tfrac{1}{2}|\mathbf{\Omega} \times (\mathbf{R}_0 + \mathbf{r})|^2 + \Phi_G(\mathbf{R}_0 + \mathbf{r}) \equiv \tfrac{1}{2}\mathbf{v}^2 + W(\mathbf{r}). \tag{11}$$

Here $\mathbf{R}_0$ and $\mathbf{r}$ are the positions of the Sun relative to the Galactic center and the test particle relative to the Sun, Φ_G is the potential due to the Galaxy, and $\mathbf{\Omega}$ is the angular velocity of the Sun in its Galactic orbit. The velocity $\mathbf{v}$ is measured in a frame rotating with the Sun at $\mathbf{\Omega}$. I shall write $\mathbf{r} = (x, y, z)$ where the x-axis points away from the Galactic center and the z-axis to the Galactic pole. I also write $\Phi_G = \Phi_G(R, z)$ where R measures distance from the Galactic center in cylindrical coordinates. Then

$$\mathbf{\Omega} = \Omega(R_0)\mathbf{e}_z, \qquad \text{where} \qquad \Omega^2(R) = \frac{1}{R}\frac{\partial\Phi_G(R,0)}{\partial R}. \tag{12}$$

Since $\mathbf{v}^2$ is non-negative, the region accessible to a particle with a given Jacobi integral is bounded by the inequality $W(\mathbf{r}) \leq E_J$. Expanding $W(\mathbf{r})$ to second order in the small quantity r/R_0 and eliminating $\Omega(R_0)$ using equation (12) yields

$$W(\mathbf{r}) = -\frac{G\mathcal{M}_\odot}{r} + \frac{1}{2}\left[\frac{\partial^2\Phi_G}{\partial R^2} - \frac{1}{R}\frac{\partial\Phi_G}{\partial R}\right]_{(R_0,0)} x^2 + \frac{1}{2}\frac{\partial^2\Phi_G}{\partial z^2}\bigg|_{(R_0,0)} z^2 + \text{constant}. \tag{13}$$

The quantity in square brackets is just $Rd\Omega^2(R)/dR$; when evaluated at R_0 this is equal to $-4A(A - B)$ where A and B are the usual Oort constants (e.g. Binney and Tremaine 1987). Also, $\partial^2\Phi_G/\partial z^2$ can be eliminated using Poisson's equation $\nabla^2\Phi_G = 4\pi G\rho(R, z)$, where ρ is the local density of Galactic material. Thus, equation (13) becomes (Antonov and Latyshev 1972, Heisler and Tremaine 1986)

$$W = -\frac{G\mathcal{M}_\odot}{r} - 2A(A - B)x^2 + [2\pi G\rho_0 + A^2 - B^2]z^2 + \text{constant}, \tag{14}$$

where $\rho_0 = \rho(R_0, 0)$.

The last closed surface $W = $ constant surrounding the Sun is known as the Roche surface, $W \equiv W_R$. The Roche surface is a natural measure of the boundary of the solar system in that (i) a particle at rest that is inside the Roche surface will always remain inside; (ii) any particle outside the Roche surface is not prevented from escaping by the Jacobi integral or any other analytic integral of motion. Numerical calculations (Hénon 1970) confirm that the Roche surface provides a qualitative measure of the boundary of a mass subject to tidal forces. Hénon found that the fractional volume of phase space at a given Jacobi integral E_J that was occupied by escape orbits increased rapidly once $E_J > W_R$, although a few retrograde orbits can remain bound at arbitrarily large distances.

The Roche surface intersects the coordinate axes at $\pm x_L$, $\pm y_L$, $\pm z_L$, where

$$x_L = \left[\frac{G\mathcal{M}_\odot}{4A(A - B)}\right]^{1/3}, \quad y_L = \tfrac{2}{3}x_L,$$
$$\frac{G\mathcal{M}_\odot}{z_L} - [2\pi G\rho_0 + A^2 - B^2]z_L^2 = \tfrac{3}{2}(G\mathcal{M}_\odot)^{2/3}[4A(A - B)]^{1/3}. \tag{15}$$

The Roche surface has roughly the shape of a triaxial ellipsoid, except for cusps at the points $(\pm x_L, 0, 0)$, which are saddle points of W (the collinear Lagrange points).

For $\rho_0 = 0.15\,\mathcal{M}_\odot\,\text{pc}^{-3}$ (the mean from Bahcall 1984 and Kuijken and Gilmore 1989), $A = 14.4\,\text{km s}^{-1}\text{kpc}^{-1}$, $B = -12.0\,\text{km s}^{-1}\text{kpc}^{-1}$ (Kerr and Lynden-Bell 1986), the Roche surface crosses the coordinate axes at

$$x_L = 1.41\,\text{pc}, \qquad y_L = 0.94\,\text{pc}, \qquad z_L = 0.67\,\text{pc}.$$

Thus the Roche surface is about $(2 - 3) \times 10^5$ AU from the Sun. Numerical calculations of the stability of orbits in the Galactic tidal field are roughly consistent with this result, yielding a maximum aphelion distance of about 2×10^5 AU (Smoluchowski and Torbett 1984).

3.2 EVAPORATION

Large orbits are unlikely to survive because they are perturbed onto escape orbits by stochastic gravitational forces from passing stars and giant molecular clouds (GMCs).

The half-life of bound test particles of semi-major axis a in a background of passing stars has often been discussed in the context of survival of binary star systems. Heggie (1975) estimated that half of an ensemble of particles bound to the Sun would escape in a time

$$t_{1/2} = 6 \times 10^{10}\,\text{yr}\left(\frac{2 \times 10^4\,\text{AU}}{a}\right). \tag{16}$$

Heggie considered only "catastrophic" encounters, that is, encounters strong enough to disrupt the system completely. In fact, weaker "diffusive" encounters that lead to a gradual random walk of the test particle to less tightly bound orbits are more important than catastrophic encounters. Including both types of encounter, Bahcall, Hut and Tremaine (1985) found

$$t_{1/2} = 4.5 \times 10^9 \, \text{yr} \left(\frac{2 \times 10^4 \, \text{AU}}{a} \right), \tag{17}$$

roughly a factor of ten shorter than Heggie's estimate.

These calculations neglect the effects of the Galactic tidal field, which increases the escape rate. Weinberg, Shapiro and Wasserman (1987) account crudely for the tidal field by assuming that the binary system is disrupted when its semi-major axis exceeds 1 pc and find

$$t_{1/2} = 4 \times 10^9 \, \text{yr} \left(\frac{2 \times 10^4 \, \text{AU}}{a} \right)^{1.4} \qquad \text{for } 1.5 \times 10^4 \, \text{AU} \lesssim a \lesssim 8 \times 10^4 \, \text{AU}. \tag{18}$$

The disruptive effects of GMCs are much harder to estimate, both because the parameters of the clouds are poorly known and because the encounters are relatively rare (see Hut and Tremaine 1985 for a discussion of the uncertainties). Using a plausible set of GMC parameters, Weinberg *et al.* (1987) found that perturbations from GMCs and stars led to a half-life for a binary with $a = 2 \times 10^4$ AU that was about a factor of two shorter than the half-life due to perturbations from stars alone.

These results suggest that the half-life of dark matter bound to the Sun is less than the age of the solar system if its initial semi-major axis exceeds about $(1 - 2) \times 10^4$ AU. Once the initial semi-major axis of the dark matter exceeds 5×10^4 AU, the probability that it will still be bound to the Sun is substantially less than 10%. Thus perturbations from stars and GMCs establish an effective outer boundary to the distribution of dark matter that is several times smaller than the Roche distance of $(2 - 3) \times 10^5$ AU.

3.3 THE OUTER BOUNDARY OF THE ECLIPTIC

The orbits of most of the planets lie within a few degrees of a common plane known as the ecliptic, reflecting the formation of the planets from a flat disk of dust and gas. However, at larger distances, the torque exerted by Galactic tides over the lifetime of the solar system is large enough to destroy any disk structure that was originally present.

The strongest component of the Galactic tide is described by the potential (Morris and Muller 1986, Heisler and Tremaine 1986)

$$\Phi_G = 2\pi G \rho_0 z^2 \tag{19}$$

(cf. eq. 14; I use the fact that $A^2, B^2 \ll 2\pi G \rho_0$). This potential exerts a torque $\mathbf{N} = -\mathbf{r} \times \nabla \Phi_G$ which causes the angular momentum vector $\mathbf{L}$ of a particle orbit to precess around the Galactic pole. The Galactic pole is tipped by an angle $i = 60.2°$ from the ecliptic pole. Thus a flat ecliptic disk of collisionless particles on circular orbits is converted into a fattened spheroidal distribution whose symmetry axis is the Galactic pole; the thickness of the distribution can be measured by $\langle z^2 \rangle / (\langle x^2 \rangle + \langle y^2 \rangle) = \sin^2 i / (2 - \sin^2 i) = 0.604$, not far from the value 0.5 expected for a spherical distribution. The time scale for fattening-up or isotropizing the disk is $t_{\text{iso}} \approx |\mathbf{L}| / |\mathbf{N}|$; thus disk structure can only survive out to a radius

r_{iso} where t_{iso} equals the age of the solar system, $t_{\text{ss}} = 5 \times 10^9 \, \text{yr}$. To order of magnitude I find

$$r_{\text{iso}} \approx \frac{\mathcal{M}_\odot^{1/3}}{(2\pi\rho_0 t_{\text{ss}})^{2/3} G^{1/3}} = 5 \times 10^3 \, \text{AU}. \tag{20}$$

At larger distances the weaker components of the Galactic tidal field, as well as perturbations from passing stars and GMCs, will isotropize the dark matter distribution even further.

4. A Minimal Theory of Solar System Dark Matter

It is useful to ask whether theories of formation of the solar system naturally give rise to dark matter, and, if so, what its nature and distribution would be.

We believe that the Sun was originally surrounded by a gas disk from which the planets formed. If most of the disk mass in non-volatile material was incorporated into the cores of the giant planets, then the surface density in non-volatiles can be estimated by smearing each giant planet core into an annulus reaching halfway to the next planet (see Figure 1). The resulting distribution can be fit to a power law

$$\Sigma(r) \simeq 1.4 \, \text{g cm}^{-2} \left(\frac{10 \, \text{AU}}{r} \right)^2. \tag{21}$$

As the gas disk cools, the non-volatile material condenses into grains that settle into a thin disk. Once enough material is gathered in this disk, it becomes gravitationally unstable. The most unstable wavelength is (Goldreich and Ward 1973, Binney and Tremaine 1987, eq. (6-54); see Safronov 1960 for an early discussion)

$$\lambda_u = \frac{2\pi^2 \Sigma r^3}{\mathcal{M}_\odot}, \tag{22}$$

so the dominant mass of the condensations ("planetesimals") may be written as

$$m_u = \Sigma(f\lambda_u)^2 = 3 \times 10^{20} \, \text{g} \left(\frac{f}{0.5} \right)^2 \left(\frac{\Sigma}{1 \, \text{g cm}^{-2}} \right)^3 \left(\frac{r}{10 \, \text{AU}} \right)^6, \tag{23}$$

where f is a dimensionless number less than unity. A natural choice is $f = 0.5$, since the overdense region is half a wavelength in size. However, f may be much less than unity, if instabilities develop before all the non-volatile material has settled. Notice also that m_u is only weakly dependent on distance from the planet so long as the radial density distribution is close to that of equation (21).

The orbits and sizes of the planetesimals evolve through a variety of processes including collisions, fragmentation, accretion, gas drag and gravitational scattering. Many are eventually incorporated into the cores of the giant planets. Residual planetesimals that have survived to the present time are a possible source of dark matter in the solar system, and in the rest of this section I will discuss dynamical constraints on the distribution of residual material of this kind.

Orbits in the solar system can be grouped into two classes: regular and chaotic (see Lichtenberg and Lieberman 1983). Regular orbits are characterized by linear divergence of

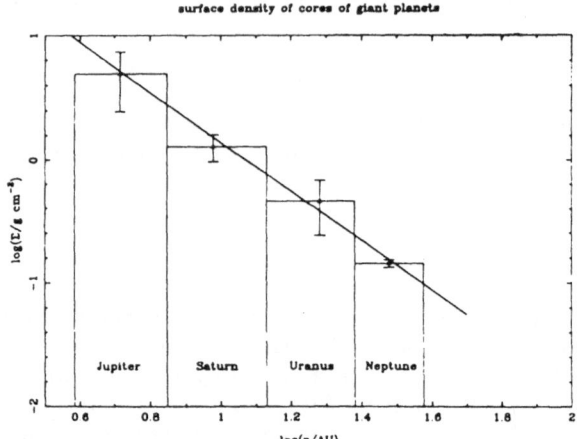

surface density of cores of giant planets

Figure 1. Estimated surface density distribution of non-volatile material in the protoplanetary disk. The masses contained in the rock-ice cores of the giant planets (Stevenson 1982) have been spread uniformly over annuli reaching halfway (logarithmically) to the adjacent planets. The solid line is the power-law fit given in equation (21).

nearby trajectories and discrete power spectra (i.e. they are quasiperiodic). Examples of objects on regular or nearly regular orbits include the planets, their satellites, and most of the asteroids. The evolution of regular orbits can be predicted to high accuracy using the standard methods of celestial mechanics. Chaotic orbits are characterized by exponential divergence of adjacent trajectories and power spectra with a continuous component. The evolution of chaotic orbits is very sensitive to the initial conditions and hence is difficult to predict. For example, planet-crossing orbits are generally chaotic because they can be strongly affected by a chance close encounter with the planet. Pluto's orbit is nearly regular, even though it crosses Neptune's, because the 3:2 Neptune-Pluto resonance prohibits a close encounter. In practical terms, a solar system orbit can only be said to be regular over a given time scale, since it may exhibit weak chaos on a much longer time scale. For example, Sussman and Wisdom (1988) argue that the orbit of Pluto is actually weakly chaotic, even though its orbital elements appear to vary quasiperiodically over time scales as long as several hundred million years.

4.1 PLANETESIMALS ON REGULAR ORBITS

Planetesimals can survive if their orbits are regular or nearly regular over time scales comparable to the age of the solar system, $t_{ss} = 5 \times 10^9$ yr. Unfortunately, the behavior of orbits over these time scales is not well understood. As an example, consider the evolution of a planetesimal in a circular orbit in the ecliptic plane, subjected to perturbations from the planets. What will the orbit look like after a time t_{ss} as a function of the initial radius? Is the orbit stable, in the sense that it remains nearly circular? Will its eccentricity grow, so that it crosses the orbit of an adjacent planet and eventually is ejected by a close encounter?

Or will it, like Pluto, evolve into a planet-crossing orbit that is stabilized by a low-order resonance?

We have only fragmentary answers to these basic questions. Direct numerical integrations show that most near-circular orbits between Jupiter and Saturn are ejected in $\lesssim 10^7$ yr (Franklin, Lecar and Soper 1989). On the other hand, the existence of the asteroid belt strongly suggests that most near-circular orbits with semi-major axes between about 2.2 AU and 3.2 AU are stable over an interval t_{ss}—with the exception of orbits near some low-order resonances with Jupiter (Wisdom 1982, 1983). Investigations using an approximate area-preserving map instead of the accurate equations of motion suggest that most near-circular orbits between the planets are stable over a time t_{ss}; however, ejection occurs for orbits in narrow bands around each planet, for most orbits between Jupiter and Saturn, and for some orbits between Uranus and Neptune (Duncan, Quinn and Tremaine 1989a).

Bands of semi-major axis in which near-circular orbits are stable may host substantial populations of planetesimals. However, no solar system objects other than the planets and their satellites have been found so far on stable orbits, except for the asteroids between Mars and Jupiter; nor is there any dynamically detectable extra mass between the planets (§1). This lack of debris in the planetary system is a remarkable fact that so far has eluded explanation. Perhaps some aspect of the formation process swept the planetesimals out of the system, or perhaps accurate orbit integrations would show that most near-circular orbits are weakly chaotic, so that almost all planetesimals are ejected on 10^9 yr time scales.

An interesting speculation is that the estimate (21) for the density of the protoplanetary disk may be valid to distances much larger than Neptune's semi-major axis of 30 AU. If so, and if I assume that the formation of Neptune depleted the protoplanetary disk only out to about 35 AU, then there should be a residual mass $\Delta M(r) \approx 30 M_\oplus \log(r/r_0)$ inside radius $r > r_0 = 35$ AU, probably in the form of planetesimals. This hypothetical disk was discussed by Kuiper (1951) and is sometimes called the Kuiper belt. It is amusing that the mass $\Delta M(r)$ derived in this way is just below the dynamically determined upper limit (7).

4.1.1 The Kuiper Belt as a Source for Jupiter-family Comets. Indirect evidence for the Kuiper belt is provided by the Jupiter-family comets (comets with orbital period $P < 20$ yr). These have a much flatter distribution of orbits than comets with longer period: their median inclination relative to the ecliptic is only 10° whereas comets with much longer periods have a roughly isotropic distribution of inclinations (Marsden 1983). Also, the Jupiter-family comets are much more numerous than comets with longer period: there are 104 known Jupiter-family comets, compared with only 17 in the period range 20 yr $< P < 200$ yr (this may however partly be a selection effect, since there are more chances to discover a comet with a shorter period). Numerical simulations show that the inclination distribution and orbital period distribution can only be reproduced if the Jupiter-family comets originate on low-inclination planet-crossing orbits with perihelia in the outer planetary system (Duncan *et al.* 1988, 1989b). Roughly $10-20\%$ of these objects are then scattered by the giant planets to perihelia $q \lesssim 2$ AU where they become visible. It is likely that these comets come from the Kuiper belt, and that they are either the original planetesimals that condensed out of the protoplanetary disk, or possibly fragments of planetesimals broken up by collisions (see Greenberg *et al.* 1984 for a discussion). Thus there are two distinct sources of comets: those with periods $P \lesssim 20$ yr come from the Kuiper belt (Edgeworth 1949, Kuiper 1951,

Whipple 1972), while those with $P \gtrsim 20\,\mathrm{yr}$ come from a separate isotropic source, the Oort cloud (see below).

The lifetimes of planet-crossing orbits are generally short compared with the age of the solar system. Thus Jupiter-family comets must spend most of their lives on nearly regular orbits in the Kuiper belt, and evolve onto chaotic planet-crossing orbits at a slow, steady rate. The mechanism of this process is still uncertain. One possibility is that weak instabilities gradually convert regular near-circular orbits onto chaotic planet-crossing ones; simple dynamical models suggest that such instabilities are likely to be present for near-circular orbits between Uranus and Neptune (Duncan *et al.* 1989a).

4.1.2 Chiron. The object Chiron was discovered by C. Kowal during the Palomar Solar System Survey (Kowal, Liller and Marsden 1979, Kowal 1989). Chiron is unresolved with magnitude $m_{\mathrm{pg}} = 18$, suggesting a radius $150\,\mathrm{km}(p/0.05)^{-1/2}$, where p is its geometric albedo. Chiron is the only known minor planet with perihelion well beyond Jupiter's orbit. It orbits between Saturn and Uranus ($a = 13.7\,\mathrm{AU}$, $e = 0.38$, $i = 6.9°$) on a chaotic trajectory that is expected to lead to ejection by Saturn or a close encounter with Jupiter in $10^5 - 10^6\,\mathrm{yr}$ (Oikawa and Everhart 1979). The short lifetime of Chiron's present orbit suggests that it evolved from a more stable orbit in the recent past. It is probable that Chiron spent most of its life on a nearly regular orbit in the Kuiper belt, and that it is simply a large comet following the same evolutionary path that produces the Jupiter-family comets.

4.1.3 Detection of Objects in the Kuiper Belt. The detection of belt objects would offer considerable insight into the formation of the solar system.

To obtain a simple "straw man" model for comparison with the observational limits, I shall assume that the total belt mass is $\Delta M = 1 M_{\oplus}$, that the density of belt objects is $\rho = 1\,\mathrm{g\,cm^{-3}}$, that the total angular extent of the belt normal to the ecliptic is $2\Delta\theta = 0.2$ radians, and that the belt material is concentrated at a distance $r = 40\,\mathrm{AU}$ from the Sun. I assume that the number of objects in the belt with radii between R and $R + dR$ is $n(R)dR$, where

$$
\begin{aligned}
n(R) &= \frac{K}{R_0}\left(\frac{R_0}{R}\right)^{b_1+1}, \quad R < R_0, \\
&= \frac{K}{R_0}\left(\frac{R_0}{R}\right)^{b_2+1}, \quad R > R_0.
\end{aligned}
\tag{24}
$$

I set $b_1 = 2$ since this reproduces the size distribution of cometary nuclei (Shoemaker and Wolfe 1982). This index is also consistent with (i) the inferred size distribution of the objects responsible for cratering the Galilean satellites (Shoemaker and Wolfe 1982); (ii) experimental measurements of the size distribution resulting from fragmentation (Hartmann 1969); (iii) the observed size distribution of the particles in Saturn's rings (Cuzzi *et al.* 1984). I shall also take $R_0 = 10\,\mathrm{km}$, since this is close to the maximum observed size of cometary nuclei.

The power-law index b_2 at the high-mass end is quite uncertain. I shall consider two values, $b_2 = 3.5$, roughly the largest value occurring in sources such as gravel, crushed rock, crater fragments, etc. (Hartmann 1969), and $b_2 = 7$, which occurs in a numerical model of planet growth by accretion of planetesimals (Greenberg *et al.* 1984, Figure 12).

One approach is to search optically for slow-moving objects, much as Tombaugh searched for Pluto. The apparent magnitude of a belt object will depend on its optical geometric albedo p, which I take to be 0.05 since albedos in the outer solar system are usually low. The expected number of objects per unit solid angle brighter than visual magnitude V is then

$$
\begin{aligned}
&0.04 \text{ deg}^{-2} h \left(\frac{R_0}{10\,\text{km}}\right)^{0.5} \left(\frac{p}{0.05}\right)^{1.75} \left(\frac{40\,\text{AU}}{r}\right)^{7} 10^{-0.7(20-V)} \quad \text{for } b_2 = 3.5, \\
&7 \times 10^{-8} \text{ deg}^{-2} h \left(\frac{R_0}{10\,\text{km}}\right)^{4.5} \left(\frac{p}{0.05}\right)^{3.5} \left(\frac{40\,\text{AU}}{r}\right)^{14} 10^{-1.4(20-V)} \quad \text{for } b_2 = 7,
\end{aligned}
\tag{25}
$$

where $h = (\Delta M / 1 M_\oplus)(1\,\text{g cm}^{-3}/\rho)(0.1\,\text{rad}/\Delta\theta)$.

A number of searches for slow-moving objects in the ecliptic have been carried out, and the limits from these searches are plotted in Figure 2, along with the estimates from equation (25). Tombaugh (1961) carried out a visual search covering 1530 $(\text{deg})^2$ to an approximate limiting magnitude $V = 17.5$. Luu and Jewitt (1988) searched 300 $(\text{deg})^2$ to $V = 20$ using Schmidt plates, and 0.34 $(\text{deg})^2$ using a CCD detector to $V = 24.5$. Kowal's (1989) Solar System Survey covered 6400 $(\text{deg})^2$ to $V = 20$ using Schmidt plates. Some of these magnitude limits may be over-optimistic since faint trailed objects are harder to detect than point sources (e.g. Gehrels 1981). Asteroid surveys also provide useful limits, although it is not certain that asteroid observers would always notice the short trails that an object at 40 AU would produce in a typical exposure. Major surveys include the McDonald Survey (Kuiper et al. 1958) of 14,400 $(\text{deg})^2$ to $V = 16$; the Palomar-Leiden Survey (van Houten et al. 1970) of 216 $(\text{deg})^2$ to $V = 19.5$; and the Kiso Schmidt Survey (Ishida et al. 1984) of 1944 $(\text{deg})^2$ to $V = 18.4$. Additional searches for slow-moving objects are presently being undertaken by M. Duncan and H. Levison using a CCD and by R. Webster, A. Żytkow, and me using Schmidt plates. None of these searches has yielded any slow-moving objects except for Tombaugh's discovery of Pluto and Kowal's discovery of Chiron.

Figure 2 shows that theoretical predictions of the density of Kuiper belt objects are so uncertain that they provide very little guidance. However, it is encouraging that one of our estimates predicts that Kuiper belt objects should already have been discovered. The discovery of even a few belt objects would be so informative that it is worthwhile to strive to improve the observational limits.

Since the optical albedo of belt objects is likely to be low, it is tempting to look for the belt in the infrared. The thermal emission from a black body at 40 AU peaks (i.e. $\nu B_\nu(T)$ is maximized) at $\lambda = 80\mu$. The IRAS satellite detected emission from the ecliptic plane at 60μ and 100μ, but it is difficult to remove the effects of interplanetary dust at smaller radii to measure the contribution from the Kuiper belt (see Jackson and Killen 1988 for a discussion). The expected contribution is also difficult to estimate, since the thermal emission is dominated by very small particles in a belt with $b_1 \geq 2$.

Planned space-based infrared telescopes such as SIRTF could detect individual large bodies in the belt. However, the advantage of higher emissivity is more than offset by the limited aperture, limited telescope time, and limited number of pixels in infrared array detectors.

Belt objects will occasionally occult stars (Bailey 1976, McClintock 1985). The size of the first Fresnel zone in the V band is $\sqrt{\lambda r} = 1.8\,\text{km}(r/40\,\text{AU})^{1/2}$, so that objects smaller than a few kilometers will not produce sharp shadows. For simplicity, let me then

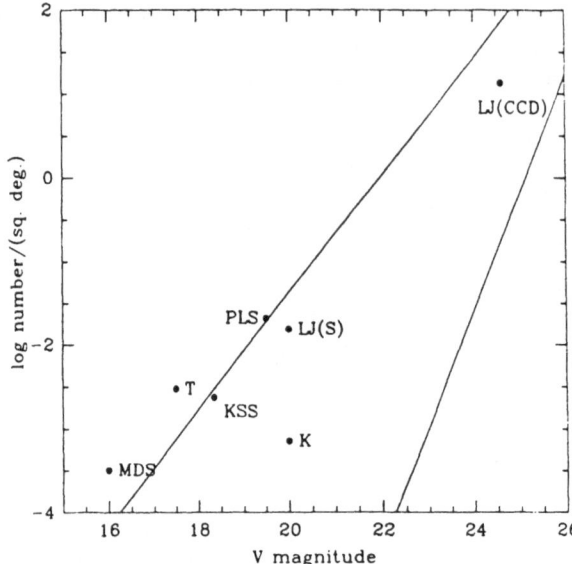

Figure 2. Limits on the number density per square degree of Kuiper belt objects brighter than visual magnitude V. The data points are 99% confidence upper limits. The papers referred to by the symbols are: MDS=Kuiper *et al.* (1958); T=Tombaugh (1961); PLS=van Houten *et al.* (1970); KSS=Ishida *et al.* (1984); LJ=Luu and Jewitt (1988); K=Kowal (1989). The upper and lower lines represent estimates from the top and bottom lines of equation (25).

concentrate on objects with radii $\geq R_0 = 10\,\mathrm{km}$, for which geometrical optics should be a good approximation. Early-type stars are better candidates for occultations than late-type stars, since they have smaller angular diameters for a given apparent magnitude. The angular diameter of the star must be substantially smaller than the angular diameter of the belt object (7×10^{-4} arcsec for a 10 km body at 40 AU); for example, A0 stars satisfy this constraint if their apparent magnitude is substantially fainter than $V = 2$. At opposition, the apparent angular speed of the belt objects mainly reflects the Earth's orbital speed $v_\oplus = 30\,\mathrm{km\ s^{-1}}$; thus the duration of an equatorial occultation is $\Delta t = 2R/v_\oplus = 0.7\,\mathrm{s}(R/10\,\mathrm{km})$. For the assumed size distribution of belt objects (eq. 25), the mean interval between occultations by belt objects larger than R_0 is

$$1.1 \times 10^5\ \mathrm{hr} \left(\frac{1 M_\oplus}{\Delta M}\right) \left(\frac{\rho}{1\,\mathrm{g\,cm^{-3}}}\right) \left(\frac{R_0}{10\,\mathrm{km}}\right)^2 \left(\frac{r}{40\,\mathrm{AU}}\right)^2 \left(\frac{\Delta\theta}{0.1\,\mathrm{rad}}\right), \qquad (26)$$

for both $b_2 = 3.5$ and $b_2 = 7$. The detection rate can be greatly improved by monitoring a field rich in blue stars (e.g. a nearby open cluster) with a CCD, and also by searching for partial occultations by bodies whose size is comparable to the size of the Fresnel zone. Thus the expected occultation rate could be as high as one every few hundred hours. With two telescopes separated by about 1 km, the signature of an occultation would be unmistakable, and the size distribution of the occulting bodies could also be estimated. Thus detection of belt objects by occultations appears to be technically feasible.

4.2 PLANETESIMALS ON CHAOTIC ORBITS

Planetesimals on chaotic orbits can be lost through collision with the Sun or a planet or through escape from the solar system. I will focus here on some interesting aspects of

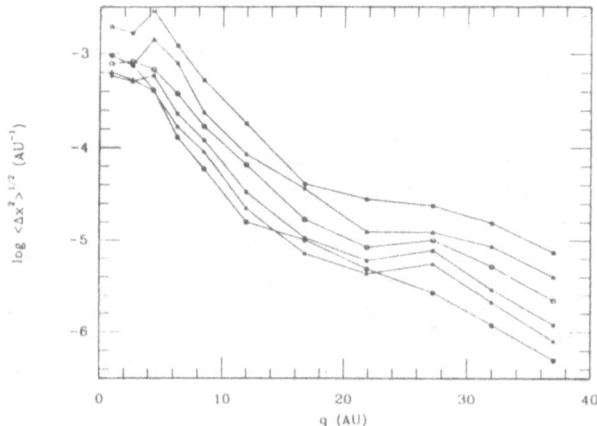

Figure 3. The rms energy change per perihelion passage due to planetary perturbations, as a function of perihelion distance. The light squares, light triangles, light circles, dark squares, dark triangles and dark circles correspond to inclination ranges relative to the ecliptic of $0° - 30°$, $30° - 60°, \ldots, 150° - 180°$. Each point is determined from 3000 passages of parabolic orbits with random argument of perihelion. From Duncan *et al.* (1987), following a figure in Fernández (1981).

the escape process, partly because they have direct observational consequences and partly because the escape process is relatively well-understood. The following arguments are mostly taken from Duncan *et al.* (1987).

In most cases, escape does not result from a single close encounter with a planet, but rather occurs through a gradual random walk or diffusion to less and less tightly bound orbits (this is also true for escape of stars from star clusters and is a consequence of the long range of the gravitational force; see Binney and Tremaine 1987, §8.4.1). During this random walk the perihelion of the planetesimal remains fairly constant, so that the orbit becomes more and more eccentric as it becomes less tightly bound. The planetesimal receives an energy kick each time it passes through the planetary system near perihelion; the orbit is chaotic because each kick depends on the detailed configuration of the planets at that time and hence is effectively random.

For convenience I shall parameterize the energy using the variable $x \equiv 1/a$, where a is the semi-major axis (the usual energy is $E = -\frac{1}{2}G\,\mathcal{M}_\odot m/a$). Let $\langle(\Delta x)^2\rangle^{1/2} \equiv D_x$ be the rms change in x per perihelion passage (the "diffusion coefficient"); D_x is a function of the perihelion distance q and inclination i (see Figure 3) but is almost independent of a since all high-eccentricity orbits are nearly parabolic near perihelion. The diffusion coefficient decreases by a factor of 100 between $q \lesssim 5\,\mathrm{AU}$ and $q \simeq 20\,\mathrm{AU}$, reflecting the fact that Jupiter dominates the perturbations of orbits with $q \lesssim 5\,\mathrm{AU}$ while the smaller planets Uranus and Neptune dominate for $q \gtrsim 20\,\mathrm{AU}$.

The characteristic time scale t_d for evolution of the semi-major axis is related to the square of the rms energy change per orbit, as is usual for a random walk:

$$t_d = P\frac{x^2}{D_x^2} = 1 \times 10^6 \, \text{yr} \left(\frac{10^4 \, \text{AU}}{a}\right)^{1/2} \left(\frac{10^{-4} \, \text{AU}^{-1}}{D_x}\right)^2, \tag{27}$$

where P is the orbital period (eq. 1). The diffusion time is generally much less than the age of the solar system, and thus almost all planetesimals on chaotic planet-crossing orbits should reach escape energy long before the present epoch.

However, diffusion toward zero energy does not lead inevitably to escape. The torque from the Galactic potential (19) changes the angular momentum L and thus the perihelion q of the planetesimal orbits. Once the perihelion exceeds $q_{max} \simeq 35 \, \text{AU}$ (i.e. once there are no longer close encounters with the planets) planetary perturbations become ineffective, energy diffusion stops, and the orbital energy is frozen. Since $L = \sqrt{2G\mathcal{M}_\odot q}$ for highly eccentric orbits and $dL/dt \approx 4\pi G\rho_0 a^2$, the characteristic time for the perihelion to reach q_{max} is $t(q_{max})$, where

$$t(q) = \frac{(2G\mathcal{M}_\odot q)^{1/2}}{4\pi G\rho_0 a^2} = 6 \times 10^7 \, \text{yr} \left(\frac{10^4 \, \text{AU}}{a}\right)^2. \tag{28}$$

The energy is frozen in at the semi-major axis a_f where $t(q_{max})$ first becomes smaller than the diffusion time t_d,

$$a_f = 1.5 \times 10^5 \, \text{AU} \left(\frac{D_x}{10^{-4} \, \text{AU}^{-1}}\right)^{4/3}. \tag{29}$$

Many planetesimals escape on the next orbit once the energy $x \lesssim D_x$. Thus most planetesimals on highly eccentric orbits will eventually escape if $x_f = a_f^{-1} \lesssim D_x$; otherwise Galactic tides will remove most of them from planet-crossing orbits before they escape. The criterion that most orbits do not escape is therefore

$$D_x \lesssim 3 \times 10^{-5} \, \text{AU}^{-1}, \tag{30}$$

which holds for $q \gtrsim 15 \, \text{AU}$ (Figure 3).

These results show that planetesimals on chaotic, highly eccentric orbits with perihelia in the Jupiter-Saturn region mostly escape, in a time much less than the age of the solar system. However, most planetesimals on orbits with initial perihelia in the Uranus-Neptune zone will remain bound to the solar system, on orbits with semi-major axes $\approx (5 - 10) \times 10^3 \, \text{AU}$ (based on equation 29, with the diffusion coefficient $D_x = 10^{-5} \, \text{AU}^{-1}$, which Figure 3 shows is typical for orbits with random inclinations and $20 \, \text{AU} \lesssim q \lesssim 30 \, \text{AU}$).

Thus it is likely that the solar system is surrounded by a cloud of planetesimals at semi-major axes $a \approx (5 - 10) \times 10^3 \, \text{AU}$. The cloud is slowly disrupted by encounters with passing stars and GMCs (eq. 18) and by the removal of planetesimals that happen to re-enter the planetary system, but neither of these processes will disrupt the cloud by the present time.

These crude arguments are confirmed by Monte Carlo simulations of the evolution of eccentric orbits subject to planetary perturbations, the Galactic tide, and perturbations from passing stars (Duncan *et al.* 1987). In particular, the simulations confirm that:

(i) The survival of planetesimals on very eccentric orbits depends strongly on their initial perihelion distance q. Less than 6% of planetesimals with $q < 10 \, \text{AU}$ remain bound to the solar system after a time $t_{ss} = 5 \times 10^9$ yr, but $30 - 40\%$ of planetesimals with perihelia in the Uranus-Neptune region ($20\,\mathrm{AU} < q < 30\,\mathrm{AU}$) are still bound after this time. The survival probability may be decreased by the uncertain effects of GMCs.

(ii) The surviving planetesimals are distributed in an extended cloud that surrounds the Sun, with a median semi-major axis 5×10^3 AU. More than 90% of the planetesimals in the cloud have semi-major axes between 1×10^3 AU and 4×10^4 AU.

(iii) The planetesimal cloud is roughly spherical for semi-major axes $\gtrsim 5 \times 10^3$ AU and is flattened towards the ecliptic at smaller semi-major axes, consistent with equation (20). The planetesimals are uniformly distributed over each energy hypersurface in phase space (i.e. the eccentricity distribution is uniform in e^2).

The formation of this planetesimal cloud is a natural consequence of any model for the formation of the solar system in which the giant planet cores are formed by accreting planetesimals. The distribution of orbits of planetesimals in the cloud is determined by the interplay between the Galactic tide and perturbations from the giant planets. Although the orbital distribution is straightforward to predict, at present we cannot reliably predict the total number or total mass of planetesimals in the cloud, since both the distribution of planetesimal masses and the fraction of planetesimals that evolve to chaotic planet-crossing orbits are unknown.

4.3 COMETS

The most striking feature of the distribution of comet orbits is a sharp peak in the distribution of the energy $x = 1/a$ near zero energy. The peak is centered near $x = x_c \equiv 5 \times 10^{-5}\,\mathrm{AU}^{-1}$, with a width $\pm 5 \times 10^{-5}\,\mathrm{AU}^{-1}$ (see Fernández 1985 and Oort 1986 for reviews and Marsden, Sekanina and Everhart 1978, Everhart and Marsden 1983 for the data; here a refers to the "original" semi-major axis that the comet had before entering the planetary system). The existence of this peak led Oort (1950) to propose that the solar system was surrounded by a spherical cloud of comets with typical semi-major axis $x_c^{-1} = 2 \times 10^4$ AU. Oort pointed out that any comets that we see have sufficiently small perihelion ($q \lesssim 2\,\mathrm{AU}$) that they receive an rms energy impulse $\langle(\Delta x)^2\rangle^{1/2} \simeq 10^{-3}\,\mathrm{AU}^{-1}$ (see Figure 3) as they pass through the planetary system; since this is much larger than x_c the comets will not return to the Oort cloud but will either escape (if $\Delta x < 0$) or return on a much more tightly bound orbit (if $\Delta x > 0$). Thus we require a flux of fresh comets from the cloud to resupply the peak; this is provided by stellar perturbations and the Galactic tide, which continually change the perihelion distances of comets in the cloud.

There is every reason to believe that Oort's comet cloud is the same as the planetesimal cloud that was derived on dynamical grounds in the previous subsection.

If this belief is correct, we must explain why the typical semi-major axis of comets in the observed peak is $x_c^{-1} = 2 \times 10^4$ AU, whereas the median semi-major axis in the planetesimal cloud was predicted to have the smaller value 5×10^3 AU. A comet is only visible if its present perihelion $q \lesssim 2\,\mathrm{AU}$; however, for any perihelion $\lesssim 15\,\mathrm{AU}$ the diffusion coefficient is so large (Figure 3) that diffusion in energy is more rapid than the rate of change of perihelion (t_d from equation 25 is shorter than $t(q = 15\,\mathrm{AU})$ from equation 28). Thus objects from the planetesimal cloud diffuse in energy at roughly constant perihelion once $q \lesssim 15\,\mathrm{AU}$. Comets can therefore only be detected near energy x_c if their *present* perihelion is $\lesssim 2\,\mathrm{AU}$ but their *last* perihelion was $\gtrsim 15\,\mathrm{AU}$; this requires that $t(q = 15\,\mathrm{AU})$ from equation (28) is less than the orbital period P (eq. 1). In other words, planetesimals only reach $q \lesssim 2\,\mathrm{AU}$ with the semi-major axes that they had in the cloud if that semi-major axis exceeds (Heisler and Tremaine 1986, Morris and Muller 1986)

$$a_0 = \left[\frac{(15\,\mathrm{AU})\,\mathcal{M}_\odot{}^2}{2^5\pi^4\rho_0^2}\right]^{1/7} = 2.9 \times 10^4\,\mathrm{AU}, \qquad (31)$$

which is in adequate agreement with the location of the observed peak of $2 \times 10^4\,\mathrm{AU}$.

There are several interesting consequences of the identification of the Oort comet cloud with the theoretical planetesimal cloud:

i The theoretical arguments predict the radial distribution of comets, which is not directly accessible to observation since the only detectable comets are those from the outermost part of the cloud, $a \gtrsim 2 \times 10^4\,\mathrm{AU}$. Hills (1981) already noted the coincidence of a_0 with the observed semi-major axes of comets from the Oort cloud and suggested that most of the cloud could be hidden at smaller semi-major axes; however, Hills estimated that the hidden inner cloud ($a < 2 \times 10^4\,\mathrm{AU}$) could contain several hundred times as many comets as the observable cloud ($a > 2 \times 10^4\,\mathrm{AU}$), whereas the numerical simulations of the formation of the planetesimal cloud yield a smaller ratio, about five (Duncan *et al.* 1987).

ii The result implies that comets formed in the protoplanetary disk, either through gravitational instability (eq. 23) or by fragmentation of larger bodies. Moreover, the rate of discovery of new comets from the Oort cloud permits us to estimate the mass of the planetesimal cloud. For this we need three independent numbers: the rate at which new comets brighter than some limiting flux pass through a perihelion $< 1\,\mathrm{AU}$ (4 per year brighter than magnitude $H = 10$ following Everhart 1967 and Weissman 1983); the number of new comets per year passing through a perihelion $< 1\,\mathrm{AU}$ per comet in the cloud (1.1×10^{-12} from simulations by Heisler 1989, assuming that we are not now in a comet shower [see below]); and the total mass in comets per comet brighter than $H = 10$ ($1.2 \times 10^{17}\,\mathrm{g}$ following Weissman 1986). Thus we arrive at a total cloud mass of $70M_\oplus$, with an uncertainty of at least a factor of three. This is almost the same as the total mass in the giant planet cores ($\simeq 75M_\oplus$ from Stevenson 1982); in other words, of order half of the mass in metals outside the Sun is likely to be dark.

iii The flux of comets reaching the Earth will not be constant in time, since a close or slow encounter with a passing star will shake the cloud strongly enough that comets from the inner part of the cloud may be thrown onto orbits with perihelion $< 1\,\mathrm{AU}$. For $1 - 2\,\mathrm{Myr}$ after such an encounter, the flux of comets reaching the Earth may increase by a factor of 20 or so (Figure 4). These comet "showers" (Hills 1981) occur every 50 Myr or so. It has been suggested that multiple comet impacts on the Earth during such a shower may cause substantial environmental stress and lead to mass extinctions, including the extinction of the dinosaurs at the Cretaceous-Tertiary boundary 65 Myr ago (Hut *et al.* 1987; see van den Bergh 1989 for a general review).

iv The size and mass of the comet cloud depends strongly on the orbits and masses of the planets, and the strength of the Galactic tide. Thus the comet clouds surrounding other stars are probably very different from the Oort cloud.

5. Has the Sun a Companion Star?

Many stars are members of binary systems, and it is possible that the Sun has a distant

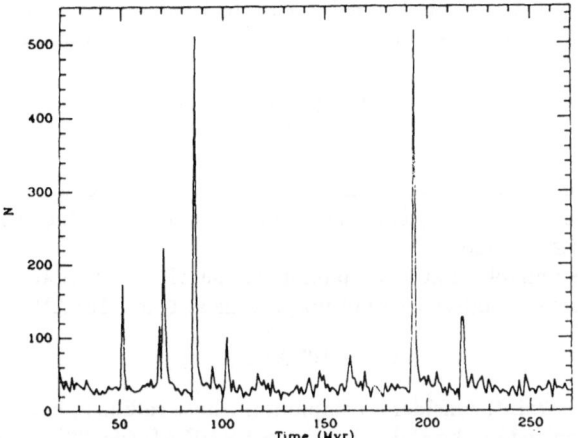

Figure 4. A Monte Carlo simulation of the flux of new comets over a 250 Myr interval. The simulation includes the effects of perturbations from the Galactic tide and passing stars, but not GMCs. The graph shows the total number of comets passing through perihelion with $q < 2\,\mathrm{AU}$ in each 1 Myr interval, from an initial population of 1.44×10^7 comets. The initial distribution of semi-major axes of comets in the Oort cloud is taken from Duncan *et al.* (1987), and the simulation uses a modified version of the code described by Heisler, Tremaine and Alcock (1987). From Heisler (1989).

undiscovered companion, perhaps a neutron star, brown dwarf, or black hole. Optical and infrared searches have not yet revealed any candidate companions (Perlmutter *et al.* 1986, Chester 1986). A hypothetical distant companion called "Nemesis" has been invoked to explain possible periodicities in the record of mass extinctions and impact craters on Earth (see below). Here I review the dynamical limits on the mass M_x and distance r_x of a possible companion. Many of the limits are crude and the discussion here is only intended to produce order-of-magnitude estimates rather than reliable dynamical bounds.

All of the limits derived below are shown on Figure 5.

Constraints from Celestial Mechanics. Residuals in the orbits of the outer planets give the limit (7), which can be rewritten as

$$M_x \lesssim 0.2 \, \mathcal{M}_\odot \left(\frac{r_x}{10^3 \, \mathrm{AU}} \right)^3 . \tag{32}$$

The *Voyager* range to Uranus (Table 1) implies that the mass of any companion inside the orbit of Uranus is

$$M_x \lesssim 3 \times 10^{-6} \, \mathcal{M}_\odot, \qquad r_x \lesssim 20 \, \mathrm{AU}. \tag{33}$$

The binary pulsar PSR 1913+16 provides another limit. The orbital period of the system changes at a rate $\dot{P}/P = -8.6077 \times 10^{-17} \, \mathrm{s}^{-1}$, which is 1.010 ± 0.011 times the rate predicted by general relativity (Taylor and Weisberg 1989). The Sun would accelerate toward a companion at a rate $a = GM_x/r_x^2$, which would change $\dot{P}$ by an amount $\delta \dot{P} = -Pa \cos \phi / c$, where ϕ is the angle between the binary pulsar and the solar companion as seen from the Sun. I replace $|\cos\phi|$ by 0.5, which should be a typical value, and assume that general relativity is correct so that $|\delta\dot{P}/P| < 0.02$ from the timing data, to obtain the limit

$$M_x \lesssim 0.17\,\mathcal{M}_\odot \left(\frac{r_x}{10^3\,\text{AU}}\right)^2. \tag{34}$$

Constraints from Orbital History. These limits are based on the probable evolution of the companion orbit and assume that the companion has been present since the origin of the solar system $5 \times 10^9\,\text{yr}$ ($\equiv t_{ss}$) ago.

As shown in §3.2, it is unlikely that a companion can survive perturbations from passing stars and GMCs for a time t_{ss} unless its semi-major axis is $\lesssim 5 \times 10^4\,\text{AU}$. Thus I assume

$$r_x \lesssim 5 \times 10^4\,\text{AU}. \tag{35}$$

To obtain another constraint, consider a companion whose radius r_x exceeds the distance at which orbits are isotropized by the Galactic tide, $\approx 5 \times 10^3\,\text{AU}$ (eq. 20). Its orbital angular momentum vector $\mathbf{L}_x$ will not generally be aligned with that of the planetary system $\mathbf{L}_p$. Moreover, $|\mathbf{L}_x|$ will exceed $|\mathbf{L}_p|$, at least for $M_x \gtrsim 10^{-4}\,\mathcal{M}_\odot$, and hence $\mathbf{L}_p$ will precess around $\mathbf{L}_x$ due to the torque exerted on the planets by the companion. The characteristic time in which the direction of $\mathbf{L}_p$ will change by one radian is

$$t_x \approx \frac{4r_x^3}{GM_x}\frac{\sum_i m_i(G\mathcal{M}_\odot r_i)^{1/2}}{\sum_i m_i r_i^2} = 2 \times 10^{10}\,\text{yr}\left(\frac{r_x}{10^4\,\text{AU}}\right)^3\left(\frac{\mathcal{M}_\odot}{M_x}\right), \tag{36}$$

where the factor 4 accounts crudely for projection effects at a typical orientation of $\mathbf{L}_x$ relative to $\mathbf{L}_p$, and the sum is over the orbital radii r_i and masses m_i of the giant planets.

Mutual torques between the planets are strong enough that their orbits will precess together so long as $t_x \gtrsim 1 \times 10^6\,\text{yr}$. Thus it is not surprising that the planetary orbits lie close to a common plane. However, the solar quadrupole moment is so small that its spin angular momentum is almost unaffected by planetary torques. (The precession time for the solar spin due to planetary torques is $10^{10} - 10^{11}\,\text{yr}$.) Thus the fact that the solar spin axis lies only 7° from the ecliptic strongly suggests that the orientation of the ecliptic has not changed by more than about 0.1 radian since the formation of the solar system. This implies that $t_{ss} \lesssim 0.1 t_x$, or,

$$M_x \lesssim 0.4\,\mathcal{M}_\odot \left(\frac{r_x}{10^4\,\text{AU}}\right)^3 \qquad \text{for } r_x \gtrsim 5 \times 10^3\,\text{AU}. \tag{37}$$

Another constraint (Hills 1985) is that no companion more massive than about $M_x = 0.02\,\mathcal{M}_\odot$ could have passed within about $30\,\text{AU}$ of the Sun without imparting excessive eccentricities and inclinations to the planetary orbits. The probability that a companion has had no perihelion passage with $q \lesssim 30\,\text{AU}$ is $\exp(-\psi)$, where

$$\psi(q) = \frac{2q}{a}\frac{t_{ss}}{\max[P, t(q)]}. \tag{38}$$

The factor $2q/a$ is the fractional area of the constant energy surface in phase space that is occupied by orbits with perihelion $< q$, and appears because the Galactic tide and stellar perturbations cause the companion orbit to fill the energy surface uniformly. The second factor involves the orbital period P (eq. 1) and the time $t(q)$ required for the perihelion to change by q (eq. 28). If $t(q) < P$ then consecutive perihelion passages will have perihelion distances that differ by $\gtrsim q$, so the number of independent chances the companion has to hit the target area with perihelion $< q$ is just t_{ss}/P; on the other hand, if $t(q) > P$ there will be of order $t(q)/P$ successive perihelion passages with perihelion distance $< q$, so the number of independent chances is reduced to $t_{ss}/t(q)$.

Equation (38) yields

$$\psi = \min\left[30\left(\frac{10^4\,\text{AU}}{a}\right)^{5/2}, 0.5\left(\frac{a}{10^4\,\text{AU}}\right)\right]. \tag{39}$$

Thus the chance that the companion has passed within 30 AU is large ($\psi > 1$) for $2 \times 10^4\,\text{AU} \lesssim a \lesssim 4 \times 10^4\,\text{AU}$, and I arrive at the constraint (Hills 1985)

$$M_x \lesssim 0.02\,\mathcal{M}_\odot \qquad \text{for} \qquad 2 \times 10^4\,\text{AU} \lesssim r_x \lesssim 4 \times 10^4\,\text{AU}. \tag{40}$$

Constraints from Comets. The model for the formation and evolution of the comet cloud that I described in §4.2 fits the cometary orbit distribution quite well. Any companion must be small enough that this agreement is preserved.

For example, the observed distribution of inverse semi-major axes in the Oort cloud peaks within $x_c = 5 \times 10^{-4}\,\text{AU}^{-1}$ of zero energy (energy measured in units of $-G\mathcal{M}_\odot/2$), in good agreement with the dynamical prediction (31). Any companion with $r_x \lesssim x_c^{-1}$ contributes $-GM_x/r_x$ to the measured energy of the comets and hence destroys this agreement unless (Kirk 1978)

$$M_x \lesssim \tfrac{1}{2}\mathcal{M}_\odot r_x x_c = 0.025\,\mathcal{M}_\odot\left(\frac{r_x}{10^3\,\text{AU}}\right). \tag{41}$$

I have argued that new comets with $a \lesssim 10^4\,\text{AU}$ are not seen because the time $t(q = 15\,\text{AU})$ required for their perihelia to change by 15 AU exceeds the orbital period P (eq. 31). A companion star exerts an additional torque on the comet orbit given approximately by $N_c = fGM_x \min(a^2/r_x^3, r_x^2/a^3)$ where I take $f \simeq 0.3$ to account for projection effects. If the companion is massive enough, this torque will exceed the torque from Galactic tides, and the time $t(q)$ in equation (28) must be replaced by $t_c(q) = (2G\mathcal{M}_\odot q)^{1/2}/N_c$. Then the condition that the comets with $a = (5-10) \times 10^3\,\text{AU}$ do not contribute to the flux of new comets is that $t_c(q = 15\,\text{AU}) > P$ in this range of semi-major axis. The resulting constraints on the companion mass are

$$
\begin{aligned}
M_x &\lesssim 0.04\,\mathcal{M}_\odot\left(\frac{5 \times 10^3\,\text{AU}}{r_x}\right)^2 & r_x &\lesssim 5 \times 10^3\,\text{AU}, \\
&\lesssim 0.03\,\mathcal{M}_\odot\left(\frac{1 \times 10^4\,\text{AU}}{r_x}\right)^{1/2} & 5 \times 10^3\,\text{AU} &\lesssim r_x \lesssim 10^4\,\text{AU}, \\
&\lesssim 0.03\,\mathcal{M}_\odot\left(\frac{r_x}{1 \times 10^4\,\text{AU}}\right)^3 & 10^4\,\text{AU} &\lesssim r_x.
\end{aligned}
\tag{42}
$$

The combination of all these limits (Figure 5) shows that we do not expect the Sun to have any companion of mass $M_x \gtrsim 0.1\,\mathcal{M}_\odot$ and that for most of the possible range of

58

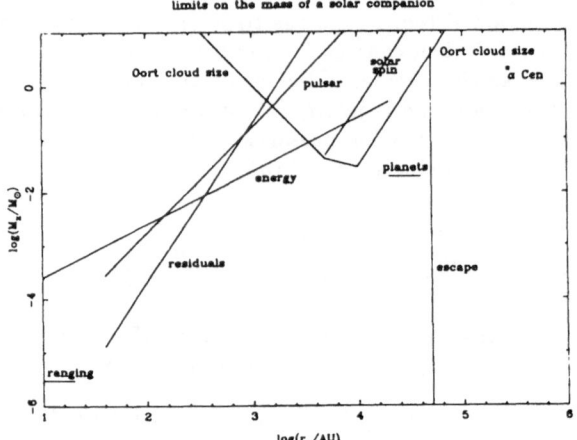

Figure 5. Upper limits on the mass M_x of a solar companion star at radius r_x. The limits come from residuals in the orbits of the outer planets (eq. 32, label "residuals"), ranging to Uranus (eq. 33, label "ranging"), timing the binary pulsar (eq. 34, label "pulsar"), evaporation by passing stars (eq. 35, label "escape"), the alignment of the solar equator with the ecliptic (eq. 37, label "solar spin"), the small eccentricities of the planets (eq. 40, label "planets"), the typical semi-major axes of Oort cloud comets (eq. 41, label "energy"), and the non-observance of new comets with $a \lesssim 10^4$ AU (eq. 42, label "Oort cloud size"). The nearest star, α Cen, is marked for comparison.

distances the limit is closer to $0.03 \mathcal{M}_\odot$. Thus it is unlikely that the companion is a neutron star (minimum mass $\simeq 0.09 \mathcal{M}_\odot$, Shapiro and Teukolsky 1983), or a hydrogen-burning main sequence star (minimum mass $\simeq 0.08 \mathcal{M}_\odot$, D'Antona and Mazzitelli 1985). The only remaining possibilities are black holes, brown dwarfs ($M \gtrsim 0.001 \mathcal{M}_\odot$, gravitational forces balanced by degeneracy pressure) or planets ($M \lesssim 0.001 \mathcal{M}_\odot$, Coulomb attraction balanced by degeneracy pressure).

5.1 NEMESIS

Raup and Sepkoski (1984) pointed out that the record of mass extinctions over the past 250 Myr showed a possible periodicity at a period of 26 Myr. This result led Alvarez and Muller (1984) to suggest that a similar period, 28.4 Myr, was present in the record of major impact craters. Both these periodicities may be due to a single astronomical phenomenon: periodic showers of comets. In a strong shower some comets would strike the Earth, leaving large craters and causing substantial environmental stress that may lead to widespread extinctions.

It is difficult to find a mechanism that produces comet showers with this period. The most interesting suggestion is that the Sun has a companion star, "Nemesis", with a semi-major axis of 9.2×10^4 AU (so that the orbital period $P = 28$ Myr), and that Nemesis plunges through the comet cloud at perihelion and thereby triggers a shower (Davis, Hut and Muller 1984, Whitmire and Jackson 1984; see Shoemaker and Wolfe 1986 or Tremaine 1986 for reviews of this and other suggestions).

Unfortunately Nemesis is very vulnerable to perturbations from passing stars. Its semi-major axis and period can change significantly within the duration of the cratering and extinction records (250 Myr), which would erase any detectable periodicity. The probability that the period wanders by less than about 15% during this time (about the maximum allowable) is only $p_1 \simeq 0.14$ (Weinberg et $al.$ 1987). In addition, the half-life of a binary with this semi-major axis is only about $t_{1/2} = 5 \times 10^8$ yr (Weinberg et $al.$ 1987, without GMCs; with GMCs included the lifetime is even shorter); thus the a $priori$ probability that Nemesis was discovered at this special time (i.e. just before it escapes) is $p_2 = t_{1/2}/t_{ss} = 0.1$. The joint probability $p_1 p_2 = 0.014$ is small enough that the Nemesis hypothesis is very unlikely. Weissman (1985) and Shoemaker and Wolfe (1986) reach similar conclusions.

A more prosaic explanation is that the statistical evidence for periodicity is misleading. Many authors have commented on this issue (e.g. Kitchell and Pena 1984, Hoffman 1985, Shoemaker and Wolfe 1986, Grieve et $al.$ 1987, Heisler and Tremaine 1989) so here I will mention only two points, relating to the cratering and extinction record respectively.

(i) The statistical significance of the periodicity in the cratering record is based on the null hypothesis that craters are independent events described by a uniform Poisson process. Figure 4 shows that this is not so: a substantial fraction of the total comet flux arrives in brief, intense bursts or showers. The resulting distribution of crater ages may show correlations which could lend spurious extra significance to periodic models.

(ii) The original list of 12 extinction events in Raup and Sepkoski (1984) was revised to a list of 8 events by Sepkoski and Raup (1986). Both the original and revised list were claimed to be periodic at the 99% confidence level. Some problems with the statistical analysis in these two papers were pointed out by Tremaine (1986); after correcting these problems Raup and Sepkoski (1986) concluded that their list of 8 events did not show significant periodicity. However, they then revised the timings of two of the events to obtain a new list of extinction events which was once again significant, this time at the 99.9% level. I cannot comment on the paleontological plausibility of these revisions, but any revisions occurring after the original periodicity hypothesis was introduced may be unintentionally and subjectively biased. This is a well-known problem in psychology experiments, where considerable effort is expended to minimize or eliminate such bias by using blind observers; unfortunately, similar precautions are difficult to apply in this case.

6. Summary

There is little or no direct dynamical evidence for substantial quantities of dark matter in the solar system. However, the observational and theoretical constraints on the amount and distribution of dark matter lead to several interesting conclusions:

(i) The upper limit on dark mass in the region of the terrestrial planets is $\lesssim 3 \times 10^{-8}\,\mathcal{M}_\odot$; within the orbit of Uranus the limit is $\lesssim 3 \times 10^{-6}\,\mathcal{M}_\odot$. On the other hand, we believe that the planets were assembled from billions of planetesimals that condensed out of the gaseous protoplanetary disk. Then why is the planetary system so clean? Was planet formation so efficient that virtually every planetesimal was incorporated into a planet or ejected from the system? Did something sweep the residual planetesimals out of the system? Or are most orbits in the planetary system weakly chaotic, so that residual planetesimals are ejected on $\lesssim 10^9$ yr time scales?

(ii) The orbits of Uranus and Neptune show unexplained residuals from the best available theoretical models, which may be due to dark matter. Spacecraft and comet trajectories show no unexplained anomalies at a comparable level of accuracy, so the residuals may arise simply from an underestimate of the systematic observational errors. If the residuals are due to an undiscovered Planet X, it must be at least ten times more distant and massive than any known planet, and is more properly called a brown dwarf rather than a planet.

(iii) The outer boundary of the solar system is set by the tidal field of the Galaxy at about 2×10^5 AU from the Sun. However, orbits with semi-major axes exceeding about 5×10^4 AU are unlikely to survive perturbations from passing stars and GMCs for the lifetime of the solar system. Orbits exceeding 5×10^3 AU in size will typically have a random orientation and will not lie in the ecliptic.

(iv) Planetesimals that are perturbed onto Uranus- and Neptune-crossing orbits will naturally evolve into a cloud surrounding the Sun at distances of 10^3 AU to 4×10^4 AU. Objects from the outer part of the cloud can be identified with "new" comets. Thus the existence and properties of the Oort comet cloud follow naturally from fairly general models for the formation of the planetary system.

(v) The mass of the planetesimal cloud may be $100 M_\oplus$ or more; thus the dark mass in the cloud may exceed the total mass in the cores of the giant planets.

(vi) The Jupiter-family comets probably arise from a disk-like source in the outer planetary system. Objects in this source (the Kuiper belt) could be detected by systematic occultation surveys, and may be visible if the mass function decays slowly at the high-mass end. Chiron may be a member of the Kuiper belt that has wandered into the region of the giant planets.

(vii) The hypothetical solar companion Nemesis probably does not exist since its orbit is improbable and the evidence for periodicity in the cratering and extinction records is weak.

(viii) It is unlikely that the Sun has a companion with mass $\gtrsim 0.03\, \mathcal{M}_\odot$.

References

Alvarez, W., and Muller, R. A. (1984) 'Evidence from crater ages for periodic impacts on the Earth'. *Nature* **308**, 718-720.

Anderson, J. D., and Standish, E. M. (1986) 'Dynamical evidence for planet X', in R. Smoluchowski, J. N. Bahcall, and M. S. Matthews (eds.), The Galaxy and the Solar System, University of Arizona Press, Tucson, 286-296.

Anderson, J. D., Lau, E. L., Taylor, A. H., Dicus, D. A., Teplitz, D. C., and Teplitz, V. L. (1989) 'Bounds on dark matter in solar orbit'. *Astrophys. J.* **342**, 539-544.

Antonov, V. A., and Latyshev, I. N. (1972) 'Determination of the form of the Oort cometary cloud as the Hill surface in the Galactic field', in G. A. Chebotarev, E. I. Kazimirchak-Polonskaya, and B. G. Marsden (eds.), The Motion, Evolution of Orbits, and Origin of Comets, Reidel, Dordrecht, 341-345.

Bahcall, J. N. (1984) 'K giants and the total amount of matter near the Sun'. *Astrophys. J.* **287**, 926-944.

Bahcall, J. N., Hut, P., and Tremaine, S. (1985) 'Maximum mass of objects that constitute unseen disk material'. *Astrophys. J.* **290**, 15-20.

Bailey, M. E. (1976) 'Can "invisible" objects be observed in the solar system?' *Nature* **259**, 290-291.

Binney, J. J., and Tremaine, S. (1987) Galactic Dynamics, Princeton University Press, Princeton.

Chester, T. (1986) 'A statistical analysis and overview of the IRAS point source catalog', in F. P. Israel (ed.), Light on Dark Matter, Reidel, Dordrecht, 3-22.

Christy, J. W., and Harrington, R. S. (1978) 'The satellite of Pluto'. *Astron. J.* **83**, 1005-1008.

Cuzzi, J. N., Lissauer, J. J., Esposito, L. W., Holberg, J. B., Marouf, E. A., Tyler, G. L., and Boischot, A. (1984) 'Saturn's rings: properties and processes', in R. Greenberg and A. Brahic (eds.), Planetary Rings, University of Arizona Press, Tucson, 73-199.

D'Antona, F., and Mazzitelli, I. (1985) 'Evolution of very low mass stars and brown dwarfs. I. The minimum main-sequence mass and luminosity'. *Astrophys. J.* **296**, 502-513.

Davis, M., Hut, P., and Muller, R. A. (1984) 'Extinction of species by periodic comet showers'. *Nature* **308**, 715-717.

Duncan, M., Quinn, T., and Tremaine, S. (1987) 'The formation and extent of the solar system comet cloud'. *Astron. J.* **94**, 1330-1338.

Duncan, M., Quinn, T., and Tremaine, S. (1988) 'The origin of short-period comets'. *Astrophys. J. Lett.* **328**, L69-L73.

Duncan, M., Quinn, T., and Tremaine, S. (1989a) 'The long-term evolution of orbits in the solar system: a mapping approach'. *Icarus* **82**, (in press).

Duncan, M., Quinn, T., and Tremaine, S. (1989b) 'Planetary perturbations and the origin of short-period comets', submitted to *Astrophys. J.*

Duncombe, R. L., and Seidelmann, P. K. (1980) 'A history of the determination of Pluto's mass'. *Icarus* **44**, 12-18.

Edgeworth, K. E. (1949) 'The origin and evolution of the solar system'. *Mon. Not. Roy. Astron. Soc.* **109**, 600-609.

Everhart, E. (1967) 'Intrinsic distributions of cometary perihelia and magnitudes'. *Astron. J.* **72**, 1002-1011.

Everhart, E., and Marsden, B. G. (1983) 'New original and future comet orbits'. *Astron. J.* **88**, 135-137.

Fernández, J. A. (1981) 'New and evolved comets in the solar system'. *Astron. Astrophys.* **96**, 26-35.

Fernández, J. A. (1985) 'The formation and dynamical survival of the comet cloud', in A. Carusi and G. B. Valsecchi (eds.), Dynamics of Comets: Their Origin and Evolution, Reidel, Dordrecht, 45-70.

Franklin, F., Lecar, M., and Soper, P. (1989) 'On the original distribution of the asteroid II. Do stable orbits exist between Jupiter and Saturn?' *Icarus* **79**, 223-227.

Gehrels, T. (1981) 'Faint comet searching'. *Icarus* **47**, 518-522.

Goldreich, P., and Ward, W. R. (1973) 'The formation of planetesimals'. *Astrophys.* . **183**, 1051-1061.

Gomes, R. S. (1989) 'On the problem of the search for Planet X based on its perturbatior on the outer planets'. *Icarus* **80**, 334-343.

Greenberg, R., Weidenschilling, S. J., Chapman, C. R., and Davis, D. R. (1984) 'From ic planetesimals to outer planets and comets'. *Icarus* **59**, 87-113.

Grieve, R. A. F., Sharpton, V. L., Goodacre, A. K., and Rupert, J. D. (1987) 'Detectin a periodic signal in the terrestrial cratering record', in G. Ryder (ed.), Proceeding of the 18th Lunar and Planetary Science Conference, Cambridge University Pres Cambridge, 375-382.

Hamid, S. E., Marsden, B. G., and Whipple, F. L. (1968) 'Influence of a comet belt beyon Neptune on the motions of periodic comets'. *Astron. J.* **73**, 727-729.

Harrington, R. S. (1988) 'The location of Planet X'. *Astron. J.* **96**, 1476-1478.

Hartmann, W. K. (1969) 'Terrestrial, lunar, and interplanetary rock fragmentation'. *Icaru* **10**, 201-213.

Heggie, D. (1975) 'Binary evolution in stellar dynamics'. *Mon. Not. Roy. Astron. So* **173**, 729-787.

Heisler, J. (1989) 'Monte Carlo simulations of the Oort comet cloud', in preparation.

Heisler, J., and Tremaine, S. (1986) 'The influence of the Galactic tidal field on the Oo: comet cloud'. *Icarus* **65**, 13-26.

Heisler, J., and Tremaine, S. (1989) 'How dating uncertainties affect the detection of per odicity in extinctions and craters'. *Icarus* **77**, 213-219.

Heisler, J., Tremaine, S., and Alcock, C. (1987) 'The frequency and intensity of comε showers from the Oort cloud'. *Icarus* **70**, 269-288.

Hénon, M. (1970) 'Numerical exploration of the restricted problem. VI. Hill's case: noi periodic orbits'. *Astron. Astrophys.* **9**, 24-36.

Hills, J. G. (1981) 'Comet showers and the steady-state infall of comets from the Oo: cloud'. *Astron. J.* **86**, 1730-1740.

Hills, J. G. (1985) 'The passage of a Nemesis-like object through the planetary system *Astron. J.* **90**, 1876-1882.

Hoffman, A. (1985) 'Patterns of family extinction depend on definition and geologic timescale'. *Nature* **315**, 659-662.

Hoyt, W. G. (1980) Planets X and Pluto, University of Arizona Press, Tucson.

Hughes, D. W. (1982) 'Asteroidal size distribution'. *Mon. Not. Roy. Astron. Soc.* **19!** 1149-1157.

Hut, P., and Tremaine, S. (1985) 'Have interstellar clouds disrupted the Oort comet cloud *Astron. J.* **90**, 1548-1557.

Hut, P., Alvarez, W., Elder, W. P., Hansen, T., Kauffman, E. G., Keller, G., Shoemaker, E. M., and Weissman, P. R. (1987) 'Comet showers as a cause of mass extinctions'. *Nature* **329**, 118-126.

Ishida, K., Mikami, T., and Kosai, H. (1984) 'Size distribution of asteroids'. *Publ. Astr. Soc. Japan* **36**, 357-370.

Jackson, A. A., and Killen, R. M. (1988) 'Infrared brightness of a comet belt beyond Neptune'. *Earth, Moon, and Planets* **42**, 41-47.

Kerr, F. J., and Lynden-Bell, D. (1986) 'Review of galactic constants'. *Mon. Not. Roy. Astron. Soc.* **221**, 1023-1038.

Kirk, J. (1978) 'On companions and comets'. *Nature* **274**, 667-668.

Kitchell, J. A., and Pena, D. (1984) 'Periodicity of extinctions in the geologic past: deterministic versus stochastic explanations'. *Science* **226**, 689-692.

Kowal, C. T. (1989) 'A solar system survey'. *Icarus* **77**, 118-123.

Kowal, C. T., and Drake, S. (1980) 'Galileo's observations of Neptune'. *Nature* **287**, 277-278.

Kowal, C. T., Liller, W., and Marsden, B. G. (1979) 'The discovery and orbit of (2060) Chiron', in R. L. Duncombe (ed.), Dynamics of the Solar System, Reidel, Dordrecht, 245-250.

Kuijken, K., and Gilmore, G. (1989) 'The mass distribution in the galactic disc. III. The local volume density'. *Mon. Not. Roy. Astron. Soc.* **239**, 651-664.

Kuiper, G. P. (1951) 'On the origin of the solar system', in J. A. Hynek (ed.), Astrophysics: A Topical Symposium, McGraw-Hill, New York, 357-424.

Kuiper, G. P., Fujita, Y., Gehrels, T., Groeneveld, I., Kent, J., van Biesbroeck, G., and van Houten, C. J. (1958) 'Survey of asteroids'. *Astrophys. J. Suppl.* **3**, 289-427.

Lichtenberg, A. J., and Lieberman, M. A. (1983) 'Regular and stochastic motion', Springer-Verlag, New York.

Luu, J. X., and Jewitt, D. (1988) 'A two-part search for slow-moving objects'. *Astron. J.* **95**, 1256-1262.

Marsden, B. G. (1983), Catalog of Cometary Orbits, Enslow, Hillside.

Marsden, B. G., Sekanina, Z., and Everhart, E. (1978) 'New osculating orbits for 110 comets and analysis of original orbits for 200 comets'. *Astron. J.* **83**, 64-71.

McClintock, J. (1985). Private communication.

Morris, D. E., and Muller, R. A. (1986) 'Tidal gravitational forces: the infall of "new" comets and comet showers'. *Icarus* **65**, 1-12.

Oikawa, S., and Everhart, E. (1979) 'Past and future orbit of 1977 UB, object Chiron'. *Astron. J.* **84**, 134-139.

Oort, J. H. (1950) 'The structure of the cloud of comets surrounding the solar system, and a hypothesis concerning its origin'. *B.A.N.* **11**, 91-110.

Oort, J. H. (1986) 'The origin and dissolution of comets'. *The Observatory* **106**, 186-193.

Perlmutter, S., Burns, M. S., Crawford, F. S., Friedman, P. G., Kare, J. T., Muller, R. A.,

Pennypacker, C. R., and Williams, R. W. (1986) 'The Berkeley search for a faint stellar companion to the Sun', in M. C. Kafatos, R. S. Harrington, and S. P. Maran (eds.), Astrophysics of Brown Dwarfs, Cambridge University Press, Cambridge, 87-92.

Raup, D. M., and Sepkoski, J. J. (1984) 'Periodicity of extinctions in the geologic past'. Proc. Nat. Acad. Sci. **81**, 801-805.

Raup, D. M., and Sepkoski, J. J. (1986) 'Periodic extinction of families and genera'. Science **231**, 833-836.

Safronov, V. S. (1960) 'On the gravitational instability in flattened systems with axial symmetry and non-uniform rotation'. Ann. d'Astrophys. **23**, 979-982.

Seidelmann, P. K., and Harrington, R. S. (1988) 'Planet X—the current status'. Cel. Mech. **43**, 55-68.

Sepkoski, J. J., and Raup, D. M. (1986) 'Periodicity in marine extinction events', in D. K. Elliott (ed.), Dynamics of Extinction, Wiley, New York, 3-36.

Shapiro, S. L., and Teukolsky, S. A. (1983) Black Holes, White Dwarfs, and Neutron Stars, Wiley, New York.

Shoemaker, E. M., and Wolfe, R. F. (1982) 'Cratering time scales for the Galilean satellites', in D. Morrison (ed.), Satellites of Jupiter, University of Arizona Press, Tucson, 277-339.

Shoemaker, E. M., and Wolfe, R. F. (1986) 'Mass extinctions, crater ages and comet showers', in R. Smoluchowski, J. N. Bahcall, and M. S. Matthews (eds.), The Galaxy and the Solar System, University of Arizona Press, Tucson, **338-386**.

Smoluchowski, R., and Torbett, M. (1984) 'The boundary of the solar system'. Nature **311**, 38-39.

Standish, E. M. (1986) 'Numerical planetary and lunar ephemerides: present status, precision and accuracies', in J. Kovalevsky and A. Brumberg (eds.), Relativity in Celestial Mechanics and Astrometry, Reidel, Dordrecht, 71-83.

Standish, E. M., and Hellings, R. W. (1989) 'A determination of the masses of Ceres, Pallas, and Vesta from their perturbations upon the orbit of Mars'. Icarus **80**, 326-333.

Stevenson, D. J. (1982) 'Formation of the giant planets'. Planet. Sp. Sci. **30**, 755-764.

Stevenson, D. J. (1986) 'High mass planets and low mass stars', in M. C. Kafatos, R. S. Harrington, and S. P. Maran (eds.), Astrophysics of Brown Dwarfs, Cambridge University Press, Cambridge, 218-232.

Sussman, G. J., and Wisdom, J. (1988) 'Numerical evidence that the motion of Pluto is chaotic'. Science **241**, 433-437.

Talmadge, C., Berthias, J.-P., Hellings, R. W., and Standish, E. M. (1988) 'Model-independent constraints on possible modifications of Newtonian gravity'. Phys. Rev. Letters **61**, 1159-1162.

Taylor, J. H., and Weisberg, J. M. (1989) 'Further experimental tests of relativistic gravity using the binary pulsar PSR 1913+16'. Astrophys. J. **345**, 434-450.

Tholen, D. J., Buie, M. W., Binzel, R. P., and Frueh, M. L. (1987) 'Improved orbital and physical parameters for the Pluto-Charon system'. *Science* **237**, 512-514.

Tombaugh, C. W. (1961) 'The trans-Neptunian planet search', in G. P. Kuiper and B. Middlehurst (eds.), Planets and Satellites, University of Chicago Press, Chicago, 12-30.

Tremaine, S. (1986) 'Is there evidence for a solar companion star?', in R. Smoluchowski, J. N. Bahcall, and M. S. Matthews (eds.), The Galaxy and the Solar System, University of Arizona Press, Tucson, 409-416.

van den Bergh, S. (1989) 'Life and death in the inner solar system'. *Publ. Astron. Soc. Pac.* **101**, 500-509.

van Houten, C. J., van Houten-Groeneveld, I., Herget, P., and Gehrels, T. (1970) 'The Palomar-Leiden survey of faint minor planets'. *Astr. Astrophys. Suppl.* **2**, 339-448.

Weinberg, M. D., Shapiro, S. L., and Wasserman, I. (1987) 'The dynamical fate of wide binaries in the solar neighborhood'. *Astrophys. J.* **312**, 367-389.

Weissman, P. R. (1983) 'The mass of the Oort cloud'. *Astron. Astrophys.* **118**, 90-94.

Weissman, P. R. (1985) 'Dynamical evolution of the Oort cloud', in A. Carusi and G. B. Valsecchi (eds.), Dynamics of Comets: Their Origin and Evolution, Reidel, Dordrecht, 87-96.

Weissman, P. R. (1986) 'The mass of the Oort cloud: a post Halley reassessment'. *Bull. A. A. S.* **18**, 799.

Whipple, F. L. (1972) 'The origin of comets', in G. A. Chebotarev, E. I. Kazimirchak-Polonskaya, and B. G. Marsden (eds.), The Motion, Evolution of Orbits, and Origin of Comets, Reidel, Dordrecht, 401-408.

Whitmire, D. P., and Jackson, A. A. (1984) 'Are periodic mass extinctions driven by a distant solar companion?' *Nature* **308** 713-715.

Wisdom, J. (1982) 'The origin of the Kirkwood gaps: a mapping for asteroidal motion near the 3/1 commensurability'. *Astron. J.* **87**, 577-593.

Wisdom, J. (1983) 'Chaotic behavior and the origin of the 3/1 Kirkwood gap'. *Icarus* **56**, 51-74.

Yeomans, D. K. (1986) 'Physical interpretations from the motions of comets Halley and Giacobini-Zinner', in Proceedings of the 20th ESLAB Symposium on the Exploration of Halley's Comet (ESA SP-250), 419-425.

LOW-MASS STARS AND BROWN DWARFS

Lorne A. Nelson
Physics Department
Bishop's University
Lennoxville
Quebec J1M 1Z7
Canada

ABSTRACT. Very low luminosity stars are thought to constitute an important component of the baryonic dark matter. Only recently has significant observational and theoretical progress been made in understanding the nature of these objects. With the advent of more powerful instruments such as the Hubble Space Telescope, a wealth of new observational information will be forthcoming. Thus it is imperative that refinements to theoretical models keep pace with the observational advances. In this paper, the current status of the theoretical and observational investigations of very low luminosity stars is reviewed.

1. Introduction

A number of lines of evidence point to the fact that dark matter contributes significantly to the overall mass density of the Universe. If a large fraction of the dark matter is baryonic it becomes interesting to ask how much of this matter is in the form of very low luminosity "stars". Very low luminosity stars can be separated into two distinct classes, namely brown dwarfs and lower main-sequence stars. Unlike main-sequence stars, brown dwarfs do not achieve thermal equilibrium *via* hydrogen-burning and thus are destined to cool to their fully degenerate configuration.

The observable characteristics and spatial density of these low-mass objects may hold the key to a large number of important problems. These include: (i) the nature of massive dark haloes, (ii) star formation in cooling flows, (iii) the existence of missing mass in galactic disks, (iv) the physics of protostellar fragmentation and (v) the physics associated with brown dwarf cooling.

Although low luminosity stars are individually unspectacular, they are so numerous that they collectively account for a major portion of the mass of the Galaxy. The task of determining the actual contribution of very low-mass stars to the mass density is complicated by several factors. The major difficulties stem from (i) incomplete sampling, (ii) theoretical uncertainties in the luminosity-mass relation, and (iii) the colour calibrations. The uncertainty associated with (iii) can be minimized, however, by increasing the spectral coverage

D. Lynden-Bell and G. Gilmore (eds.), Baryonic Dark Matter, 67–85.

and with the creation of improved spectral syntheses for low temperature, high-gravity stars. The theoretical uncertainties will be discussed in greater detail later in the paper.

This paper is organized as follows: §2 contains a qualitative analysis of the various types of low-mass contributors to baryonic dark matter that can exist in the Universe; a description of the present status of the theory and observations of stars on the lower main sequence can be found in §3, while §4 deals with the evolution of brown dwarfs; the detectability of Population III brown dwarfs is discussed in §5; some of the observational strategies associated with the detection of very low luminosity objects and also some of the recent claims concerning the detection of brown dwarfs are reviewed in §6.

2. The Nature of Low-Mass Baryonic Dark Matter

By using simple scaling arguments, it is possible to obtain approximate expressions for the masses and lifetimes of "stars" in terms of fundamental constants. This approach has been employed in a wide range of astrophysical contexts by Weisskopf (1975), Carr and Rees (1979), Rees (1983), Press and Lightman (1983), and others. These results will be used extensively to delineate between various forms of baryonic dark matter (henceforth BDM) with masses $\lesssim 1 \mathcal{M}_\odot$. In particular, the nature of brown dwarfs and their relationship to other types of BDM will be examined.

Rocks and asteroids belong to the class of objects that comprise the least massive type of macroscopic BDM. The structure of these amorphous, non-metallic objects is dominated by their degeneracy energy. Since the matter is solid, the mean atomic separation $\langle r \rangle$ is approximately equal to the Bohr radius a_o. The gravitational energy associated with rocks is negligible. However, it is important in determining the maximum mass of a large rock (asteroid) since gravitational forces will break the molecular bonds and cause the asteroid to become spherical. At this mass, the object is more properly classified as a planet.

This limit can be determined by equating the gravitational energy with the shear modulus (tensile strength) of the object multiplied by its volume. A very approximate expression for the shear modulus μ_s has been derived by Press and Lightman (1983) and can be stated as

$$\mu_s \sim \varepsilon \, \frac{(e^2/2a_o)}{(2a_o)^3} \sqrt{\frac{m_e}{m_p}} = \varepsilon \, \frac{1 \text{ Rydberg}}{(2a_o)^3} \sqrt{\frac{m_e}{m_p}} \, , \tag{1}$$

where ε is a constant ($\lesssim 0.1$) that accounts for structural flaws. The symbols m_e and m_p denote the electron and proton mass, respectively.

For an object of mass M and radius R that contains N atoms with average separation d, we expect that $M \sim N m_p$ and $R \sim N^{1/3} \langle r \rangle$. By equating the gravitational binding energy ($\sim GM^2/R$) to $\mu_s V$, it can be shown that the maximum mass of an asteroid is

$$M \sim \frac{\varepsilon^{3/2}}{64} \left(\frac{e^2}{Gm_p^2} \right)^{3/2} \left(\frac{m_e}{m_p} \right)^{3/4} m_p \sim 3 \times 10^{24} \text{ g} \tag{2}$$

where we have taken $\varepsilon \sim 0.1$. The actual mass corresponding to the dividing line between planets and asteroids is somewhat more than an order of magnitude smaller than that given by equation (2), but it nonetheless serves as a useful guide.

The maximum mass of a planet can also be estimated using energy scaling arguments. As the mass of a planet is increased, the degeneracy energy becomes less dominant compared to the gravitational energy. In fact, when the mass increases sufficiently, the gravitational and degeneracy energies are nearly equal. At this point, the object should no longer be referred to as a planet but rather as a cold, degenerate dwarf. In terms of the equation of state, this argument is equivalent to saying that atomic interactions become relatively unimportant. The degeneracy energy of an object can be derived by first obtaining the Fermi momentum from the uncertainty principle ($P_F \sim \hbar/\langle r \rangle$) and recalling that the Fermi energy is given by $E_F \sim P_F^2/(2m_e)$; thus the degeneracy energy is $\sim N(\hbar/\langle r \rangle)^2/2m_e$. By equating the degeneracy and gravitational energies and taking $\langle r \rangle = a_o$, the maximum mass of a planet can be stated as

$$M \sim \left(\frac{\alpha}{\alpha_g} \right)^{3/2} \sim 2 \times 10^{30} \text{ g} , \qquad (3)$$

where α is the electronic fine structure constant, and α_g is the analogous gravitational fine structure constant ($\alpha_g \equiv Gm_p^2/\hbar c$). Note that the maximum mass of a planet is approximately one Jovian mass ($1\ M_J$). This estimate compares favourably with the work of Zapolsky and Salpeter (1969) who, using a sophisticated zero-temperature equation of state, showed that the maximum radius of cold objects is attained at a mass of $\sim 2M_J$.

The mass-radius relation of cold, degenerate dwarfs can be obtained by simply equating the degeneracy and gravitational energies. From this one finds that the radius is proportional to the mass to the $-1/3$ power. Specifically,

$$R \sim \left(\frac{\hbar^2}{2Gm_e m_p^{5/3}} \right) M^{-1/3} . \qquad (4)$$

However, for large masses ($\gtrsim 1\,\mathcal{M}_\odot$), the electrons become relativistic and the degeneracy energy approaches the electron's rest mass energy ($\sim Nm_e c^2$). For large enough masses, stable hydrostatic solutions cannot be found. This limiting mass is known as the Chandrasekhar mass M_{Ch} and can be found by equating the degeneracy, gravitational and relativistic electron energies. This yields

$$M_{Ch} \sim \left(\frac{\hbar c}{GM_P} \right)^{3/2} = \alpha_g^{-3/2} m_p^2 \sim 4 \times 10^{33} \text{g} . \qquad (5)$$

In the above analysis, we have assumed that the stars are essentially cold. However, if the internal temperatures are sufficiently large, the effect of the thermal energy on the structure of the stars cannot be ignored. Moreover, at high enough temperatures, nuclear fusion will be ignited allowing the star to achieve a state of thermal equilibrium. Stars which have achieved this state are referred to as main-sequence stars. By equating the gravitational energy of a star with its thermal energy, one can show that

$$\frac{M}{R} \sim \frac{k\langle T \rangle}{G} \sim 7 \times 10^{22} \text{ g/cm} , \qquad (6)$$

where we have taken $\langle T \rangle \sim 10^7 K$ (Clayton 1983). Of course $\langle T \rangle$ contains all the complications due to nuclear cross-sections and the radiative opacities. The value of $\langle T \rangle$ has been set equal to a constant in order to simplify the subsequent analysis. In fact, an approximately linear mass-radius relationship is appropriate on the lower main-sequence (Dorman, Nelson, and Chau 1989; hereafter DNC).

There exists a minimum hydrogen-burning main-sequence mass (M_{min}) for which stars cannot achieve thermal equilibrium. This corresponds to a situation where the nuclear energy generated by a star is less than the energy radiated from its surface (the luminosity deficit is made up by thermal energy stored within the star). Although a star can contract, thereby increasing its temperature and concomitantly its nuclear luminosity, electron degeneracy eventually halts the collapse and the star cools to its fully degenerate configuration. The minimum mass for which a star can achieve thermal equilibrium is estimated by equating the degeneracy and thermal energies and substituting the mass-radius relation given by equation (6). Therefore,

$$M_{min} \sim \left(\frac{m_p}{10 m_e} \right)^{3/4} \left(\frac{\alpha}{\alpha_G} \right)^{3/2} m_p \sim 1 \times 10^{32} \text{g} . \tag{7}$$

This mass corresponds to $\sim 0.05 \mathcal{M}_\odot$ or $\sim 50 M_J$. Objects with masses between about 2 and $50 M_J$ never become full-fledged stars (except for a brief period of deuterium burning) and are usually classified as brown dwarfs. The upper limit is actually closer to $80 M_J$ while the lower limit depends largely on the process by which the object was formed.

While the previous derivation for M_{min} was obtained from energy scalings, a more detailed analysis can be employed to obtain a more accurate evaluation of M_{min}. Several assumptions must be made, in order to make this problem tractable. The validity of these assumptions can be checked *a posteriori*. They include:

(i) the specific entropy between the center and surface is equal (*ie* efficient adiabatic convection);
(ii) the interior structure is characterized by a polytrope of index $n = 3/2$;
(iii) the surface species are predominantly molecular hydrogen and helium;
(iv) the nuclear luminosity and H_2 surface entropy can be well represented by a power law;
(v) the surface opacity is taken to be a constant parameter;
(vi) the condition for M_{min} is defined such that $(dM/dR) \cong 0$.

Numerical models have shown that most of the assumptions are quite good.

After some manipulation and approximation, it can be shown that

$$M_{min} \simeq 0.09 \mathcal{M}_\odot \left\{ \left(\frac{\kappa_s}{.03} \right)^{-1/10} \right\} \tag{8}$$

where $X = 0.7$ and $Y = 0.28$ are the hydrogen and helium mass fractions, respectively, and κ_s is the surface opacity is units of cm^2/g. According to equation (8), the minimum Population I main-sequence mass is about $0.09 \mathcal{M}_\odot$. Although the value of κ_s is very difficult to determine accurately, the dependence of M_{min} on κ_s is quite weak. However, it should be noted that κ_s is very sensitive to the metallicity Z. As Z decreases, κ_s must decrease and M_{min} goes up.

The luminosity of stars at the end of the main-sequence can also be estimated using the same method. It can be shown that the luminosity at M_{min} is

$$L_{min} \propto \left(\frac{\kappa_s}{.03} \right)^{-1} . \tag{9}$$

It is clear from equation (9) that L_{min} is quite sensitive to the choice of κ_s. This underscores one of the important points, namely that surface opacities (and other physical processes close to the surface) are important to the determination of the luminosity-mass relationship close to the end of the main sequence. These uncertainties, among others, can have a major impact on our determination of such things as the mass function. Thus it is important that the surface conditions be analyzed carefully.

3. Very Low-Mass Stars

Low-mass stars that lie on the lower main-sequence (LMS) are different from brown dwarfs in that the former achieve a state of almost complete thermal equilibrium on the order of the Hubble time. While brown dwarfs may constitute a major fraction of the missing mass in the galaxy, the study of stars on the LMS is nonetheless very important. Since the physical structure and properties of brown dwarfs and LMS stars are similar, more accurate LMS calibrations give us increased confidence in the reliability of brown dwarf model predictions. Moreover, ever-increasing numbers of observational searches are revealing more very low luminosity stars whose observationally inferred properties can be compared with the theoretical predictions.

Models of stars on the LMS have been constructed by numerous investigators. The early studies demonstrated that the size of the radiative stellar core decreases with decreasing mass and that at $\sim 0.3 \mathcal{M}_\odot$, the stellar interiors become completely convective. Moreover, it was shown that stars with masses of $\lesssim 0.1 \mathcal{M}_\odot$ could not achieve a state of thermal equilibrium by burning hydrogen. More recent investigations of the LMS have been carried out by Copeland, Jensen, and Jorgensen (1970); Grossman, Hays, and Graboske (1974); Sienkiewicz (1982); VandenBerg et al. (1983); Rappaport and Joss (1984); Neece (1984); D'Antona and Mazzitelli (1985) and Dorman, Nelson and Chau (1989: DNC). The improvements incorporated in these later studies included, among other things, more accurate evaluations of low-temperature radiative opacities and the equation of state.

In the recent paper by DNC, ZAMS models were calculated by constructing an evolutionary sequence of models starting from the Hayashi contraction phase. These ZAMS models were calculated for masses extending down to the end of the LMS (ie at M_{min}). While the main objective of this work was to calculate accurate stellar models, it also served to show which refinements in the input physics would have the greatest impact on the predicted model parameters. In terms of the equation of state, two recent determinations (Fontaine, Graboske, and VanHorn 1977 [FGVH]; Magni and Mazzitelli 1979 [MM]) were adopted. Thus a direct quantitative comparison of one of the more important sources of uncertainty in the input physics was carried out.

Dorman, Nelson, and Chau found that an accurate computation of the structure of the atmospheres and envelopes of LMS stars is very important to the determination of their overall properties and, moreover, to the evaluation of the minimum main sequence mass. The task of computing the structure of the outer layers of these stars is complicated by several factors. For example, the equation of state has to be evaluated in regions where partial pressure ionization is important. As well, the transport of energy in the atmosphere

and envelope is also difficult to calculate accurately. Finally, the opacities in the photospheric region are difficult to evaluate accurately. DNC used one of the latest versions of
the Alexander opacities that include grains and molecules.

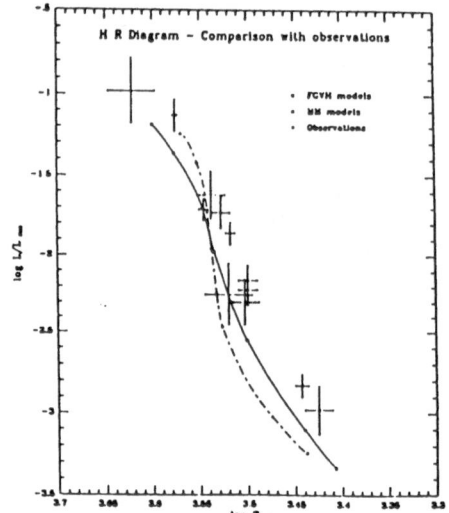

Figure 1. Position of the theoretical lower main
sequence in the H-R diagram. The sequence calculated
using the FGVH equation of state is represented by a
solid curve while that calculated using MM is denoted
by a dashed curve. The symbols correspond to stellar
models of mass equal to 0.55, 0.50, 0.40, 0.30, 0.20,
0.15, 0.10 and 0.09 $\mathcal{M}_\odot$. The filled circles indicate the
observationally inferred positions of low-mass binary
components taken from the data of Popper (1980).

Two series of standard models were constructed by DNC using the FGVH and MM
equations of state. In Figures 1, 2, and 3, the model results are plotted on the H-R diagram,
the mass luminosity and mass-radius planes, together with the observational results of
Popper (1980). A direct comparison of these two sets of results provides some insight into
the sensitivity of stellar models to the equation of state and gives us a quantitative estimate
of the magnitude of this source of theoretical uncertainty. The two sets of main-sequence
results are reasonably similar over most of the mass range that was studied. As expected,
the results show the greatest deviation at low masses.

The FGVH models become fully convective at a mass of slightly less than 0.31 $\mathcal{M}_\odot$, while
the corresponding value for the MM models is slightly less than 0.32 $\mathcal{M}_\odot$. Once the stars are
fully convective, they closely approximate polytropes of index 3/2 (the effective polytropic
index is altered slightly by the extent of the partial ionization zones and by non-ideal effects). In the fully convective regime, the MM models tend to be smaller in radius but more
luminous than the FGVH models. As a result, the MM models can attain thermal equilibrium at lower masses thereby extending the end of the main sequence. Specifically, DNC
found that the FGVH sequence extended down to a mass of $\sim 0.080\,\mathcal{M}_\odot$ and a luminosity of
$\log L/\mathcal{L}_\odot \sim -4.05$ while the MM sequence extended to $\sim 0.076\,\mathcal{M}_\odot$ and $\log L/\mathcal{L}_\odot \sim -4.14$.

After comparing the DNC results and other theoretical models, the following general
comments can be made:

1. For ZAMS models in the mass range of $0.1 - 0.5\,\mathcal{M}_\odot$, the effective temperatures seem
to be determined to within $\pm 10\%$ and the bolometric luminosity to within $\sim 60\%$.

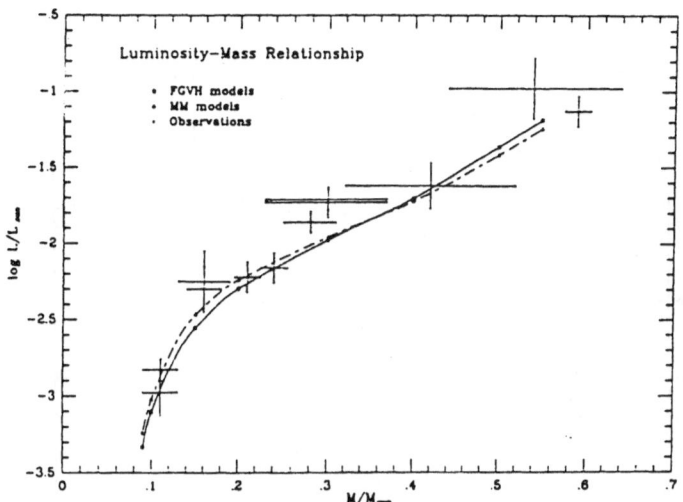

Figure 2. Dependence of the bolometric luminosity on mass for the FGVH and MM series of models. The observational points including error estimates are taken from the results of Popper (1980).

2. The largest deviations in effective temperature and luminosity occur for the smallest masses since models in this mass range have the largest uncertainties in their input physics.

3. The model parameters seem to be most sensitive to the treatment of the outer boundary conditions (in particular, the opacity and envelope equations of state).

4. The model radii are relatively well constrained, even for stars of higher mass that have radiative cores ($\Delta R/R \lesssim 10\%$).

5. A systematic displacement of the predicted positions of stars in the H-R diagram from older loci as a result of improvements to the implementation of the input physics is not observed. This seems to suggest that the theoretical results are quite robust and that it may be necessary to invoke large scale physical phenomena, such as strong magnetic fields or more realistic theories of convection, in order to change the theoretical results significantly.

The accuracy of the LMS model calculations can be estimated by comparing the theoretical predictions with the observational data. Substantial information concerning the properties of LMS stars has been obtained from the observations of visual, spectroscopically resolved, and eclipsing binaries. These binaries provide the data necessary to obtain the relationship between the mass and luminosity of LMS stars since the masses of the components of these systems can be directly determined. The accuracy of the inferred bolometric luminosity and the effective temperature of these stars can depend quite sensitively on the method used to calibrate the radiative flux, although in the case of eclipsing binaries, the radii of the components can be determined independently. Unfortunately, only two eclipsing binaries have so far been discovered with suitably small masses. These systems are YY Gem ($M_1 = M_2 = 0.59\ \mathcal{M}_\odot$, $\pm 0.015\ \mathcal{M}_\odot$) and CM Dra ($M_1 = 0.24\ \mathcal{M}_\odot$, $M_2 = 0.21\ \mathcal{M}_\odot$, $\pm 0.015\ \mathcal{M}_\odot$), see Popper 1980, and references therein). The radius-mass and luminosity-mass relations inferred from observations are contrasted with the theoretical

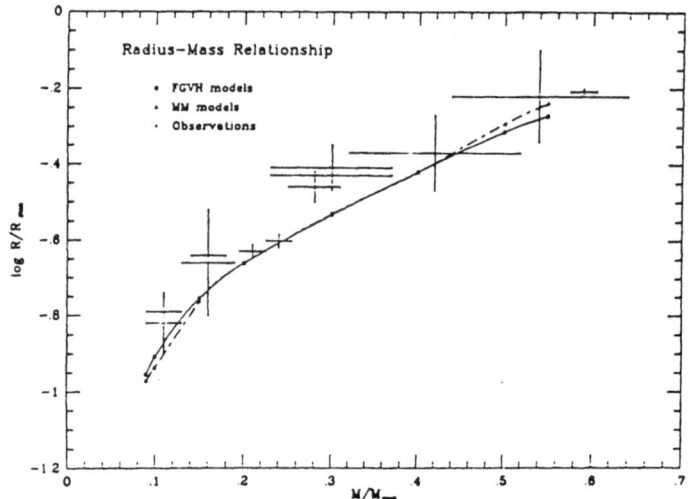

Figure 3. Dependence of the stellar radius on mass for the FGVH and MM series of models. The observational points including error estimates are taken from the results of Popper (1980).

determinations of DNC in Figures 2 and 3. The results of the eclipsing binary analyses are much more reliable than those of the visual binaries, as indicated by the relative sizes of the error boxes.

In general, there is relatively good agreement between models and observations for stellar masses of $\sim 0.5\ \mathcal{M}_\odot$, but poorer agreement near the end of the main sequence. It is precisely in this low-mass range ($\lesssim 0.2\ \mathcal{M}_\odot$) that one would like to obtain a reliable mass function. Also note that the theoretical models almost systematically under estimate the observationally inferred radii of the visual and spectroscopic binaries. These problems can be resolved only with further concerted observational and theoretical effort.

4. Evolution of Brown Dwarfs

It is assumed that brown dwarfs form in the same way as stars (by fragmentation without significant dissipation or chemical fractionation) but that they never achieve thermal equilibrium *via* hydrogen-burning. In fact, most brown dwarfs derive some of their luminosity from thermonuclear reactions (either hydrogen or deuterium burning). However, they are destined to cool to their fully degenerate configuration. Unlike the situation in planets, atomic and molecular interactions are relatively unimportant in brown dwarfs. For the reasons discussed in §2, we can place a lower mass limit of approximately $0.002\ \mathcal{M}_\odot (= 2M_J)$

on brown dwarfs. The maximum mass coincides with the end of the main-sequence and is $\lesssim 0.08 \mathcal{M}_\odot$.

The evolutionary history of brown dwarfs can generally be divided into three distinct phases: (i) an initial contraction stage (for ages $t \lesssim 10^6$ yr); (ii) a deuterium-burning phase (with a duration of $\lesssim 10^7$ yr); and, (iii) a degenerate cooling stage. The study of the initial contraction of brown dwarfs was first initiated by Hayashi and Nakano (1963). The first estimates of the properties of brown dwarfs during the late stages of degenerate cooling were made by Tarter (1975). Her estimates were based on numerical models that she then extrapolated to lower effective temperatures ($T_e < 1200$ K) and later ages than were covered by the models. Stevenson (1978) carried out the first semi-analytical calculations of the cooling of brown dwarfs to very low temperatures and old ages, but subject to the constraint of a constant stellar radius (in the limit of high degeneracy). It should be noted, however, that changes in the radius of brown dwarfs are quite important since the change in their gravitational potential is a significant source of energy.

More detailed calculations of the evolution have since been carried out by D'Antona and Mazzitelli (1985), Nelson, Rappaport, and Joss (1985: NRJ), Lunine, Hubbard, and Marley (1986), Stevenson (1986), and Lunine et al. (1989). A simplified (non-Henyey) code was used by NRJ to compute the evolution. One of the advantages of this approach is that it facilitates a systematic investigation of some of the major uncertainties in the input physics while yielding reasonably accurate evolutionary tracks. Some of the simplifying assumptions include: (i) representing the interior structure by a "modified" $n = 3/2$ polytrope; (ii) matching the specific entropy between the stellar surface and center (ie efficient convection throughout); and (iii) approximating the thin radiative layer above the surface by a simple gray atmosphere.

The most significant uncertainty arises in the evaluation of the radiative surface opacities for temperatures below 2500 K. At these temperatures, the contributions to the opacity due to various molecules, in particular H_2O, are difficult to estimate. Moreover, the formation of solid-phase matter (grains) begins to occur at these temperatures and densities. In order to parameterize the effects of grains and molecules on the radiative opacities, NRJ devised a scheme of constant extrapolation wherein the low-temperature opacity could be arbitrarily chosen.

NRJ carried out evolutionary calculations for masses in the range of $0.01 - 0.10 \mathcal{M}_\odot$. The evolution of the radius for models with $M = 0.01, 0.02, 0.04, 0.06$, and $0.08 \mathcal{M}_\odot$ is shown in Figure 4. The models all started with an initial radius of $\sim \mathcal{R}_\odot$, a value very large compared to the star's ultimate radius at late ages. For early ages, the stellar radius contracts with elapsed time, t, approximately as $t^{-1/3}$, as can be demonstrated analytically if one assumes a completely convective model for which the effective temperature remains approximately constant. For very early ages ($t < 10^6$ yr), the evolution is highly uncertain because of the possible effects of rapid stellar rotation and mass loss or accretion. For ages near 10^6 yr, there is an interval on each evolutionary curve (for $M \gtrsim 0.015 \mathcal{M}_\odot$), where R remains nearly constant; this corresponds to the deuterium-burning main-sequence phase. After deuterium exhaustion the star contracts, and for a period of time the internal temperatures increase (in accordance with the virial theorem). For stars with masses in excess of $\sim 0.08 \mathcal{M}_\odot$, the internal temperatures reach sufficiently high values such that nuclear fusion (p-p) generates enough energy to establish thermal equilibrium. The evolution of the effective temperature and bolometric luminosity are shown in Figures 5 and 6, respectively.

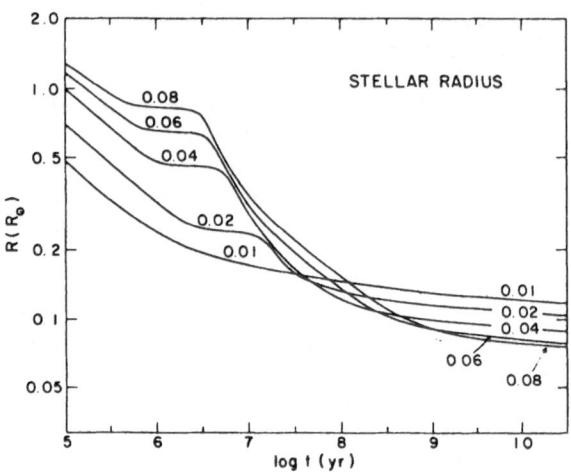

Figure 4. Evolution of the stellar radius, R, as a function of age, t, for models of very low mass stars with masses in the range $0.01 - 0.10 \; \mathcal{M}_{\odot}$. Each track is labeled with the corresponding stellar mass in units of solar masses.

From these calculations, NRJ were able to reach a number of secure conclusions. For late ages ($5 \times 10^8 \text{yr} < t < 2 \times 10^{10}$ yr) the effective temperatures of stars with masses in the range of $\sim 0.01 - 0.06 \, \mathcal{M}_{\odot}$ are given approximately by

$$T_e \simeq 1270 \left(\frac{M}{0.05 \, M_{\odot}} \right)^{0.68} \left(\frac{t}{10^9 \, \text{yr}} \right)^{-0.29} \text{K} \tag{10}$$

while the luminosity is given approximately by

$$L \approx 2.1 \times 10^{-5} \left(\frac{M}{0.05 \, \mathcal{M}_{\odot}} \right)^{2.34} \left(\frac{t}{10^9 \, \text{yr}} \right)^{-1.22} \mathcal{L}_{\odot} \; . \tag{11}$$

These fitting formulae should not be extrapolated to much earlier ages, or to much larger masses ($M \gtrsim 0.07 \, \mathcal{M}_{\odot}$). Upon testing the sensitivity of the evolutionary results to the input physics, NRJ found that the effective temperatures and bolometric luminosities were fairly well determined despite the residual uncertainties in the input physics. With reasonable range of choices for the opacity extrapolation and the entropy mismatch between the center and surface, they found that the uncertainties in the effective temperatures for stellar ages up to the age of the Galaxy were typically less than 10%. However, there still would be large theoretical uncertainties in the spectral features of such low-temperature stars even if the effective temperatures were known exactly. For ages up to the age of the Galaxy, crystallization and Debye cooling are unlikely to have any significant effect on the evolution.

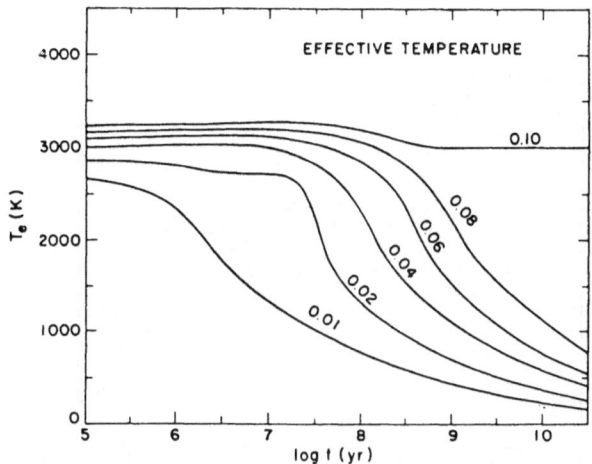

Figure 5. Evolution of the effective temperature, T_e, as a function of age t, for models of very low mass stars with masses in the range of $0.01 - 0.10 \, \mathcal{M}_\odot$. Each track is labeled with the corresponding stellar mass in units of solar masses.

NRJ also showed that the observable properties of brown dwarfs with masses in the range of $\sim 0.085 \, \mathcal{M}_\odot$ to $\sim 0.06 \, \mathcal{M}_\odot$ are quite sensitive to the treatment of the "surface" physics, especially to the surface opacity (see §3). It is in this "transition region" that the largest refinements to the theoretical models are likely to be made. Thus it is extremely important to have reliable observations of stars/brown dwarfs in this mass range. Finally, concerning the observability of brown dwarfs as a class, NRJ concluded that if brown dwarfs account for the local missing mass of the galactic disk the *theoretical* luminosity functions for such objects are reasonably well determined (except for the contribution of stars just below the end of the main sequence). These luminosity functions indicate that while IRAS is not a particularly useful tool to detect brown dwarfs, numerous brown dwarfs should be detected with the Hubble Space Telescope and SIRTF. The observed properties of individual "brown dwarfs" will be discussed in §6.

5. Detectability of Population III Brown Dwarfs

According to some models of galaxy formation, much of the primordial gas may have condensed at high redshifts to form Population III "stars". Population III objects may provide an explanation of the missing mass in clusters of galaxies and galactic haloes (see, eg., Rees 1980; Carr, Bond and Arnett 1984, and references therein). Indeed the work of Stahler, Palla, and Salpeter (1986) and Zinnecker (1986) seems to indicate that Popula-

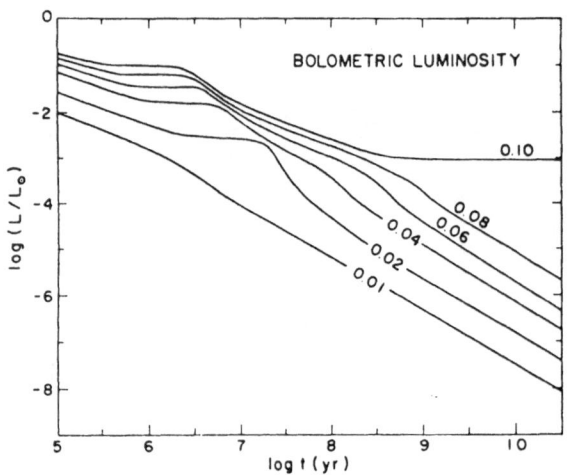

Figure 6. Evolution of the bolometric luminosity, L, as a function of age, t, for models of very low mass stars with masses in the range of $0.01\text{-}0.10\,\mathcal{M}_\odot$. Each track is labelled with the corresponding stellar mass in units of solar masses.

tion III objects with masses of $\lesssim 0.10\,\mathcal{M}_\odot$ can form. Thus it is possible that most of the primordial gas resides in Population III brown dwarfs.

If one accepts the existence of Population III brown dwarfs, then it is interesting to ask whether these objects could contribute to a detectable, isotropic electromagnetic background. An upper limit to this background can be obtained by assuming that $\Omega_B = 1$ and that the mass necessary for closure is composed solely of brown dwarfs. Recent models of inhomogeneous nucleosynthesis (see Fowler 1990, and references therein) imply that Ω_B could be as large as unity without causing a seriously disagreement between predicted and observed elemental abundances. Karimabadi and Blitz (1984) attempted to analyze the problem of detection by using the simple power-law cooling relations obtained by Tarter (1975) and Stevenson (1978) for Population I brown dwarfs. In fact, these relations are not very accurate during certain epochs. Moreover, the actual cooling of Population III brown dwarfs is quite different from that of Population I objects.

The evolution of Population III brown dwarfs has recently been calculated by Nelson (1989) and the results for the evolution of the effective temperature as a function of time is shown in Figure 7. These tracks were computed using the same code as that used by NRJ to calculate Population I evolution. The only major change to the input physics was the replacement of the radiative opacities. Specifically the zero-metallicity opacities of Stahler, Palla, and Salpeter (1986) ($X = 0.72, Y = 0.28$) were adopted. In many ways, these opacities are more reliable than those used for a Population I composition since grain formation is impossible and the only molecules that can form are those of molecular hydrogen. Stahler *et al.* found that the major contributor to the opacity at low temperatures was pressure-

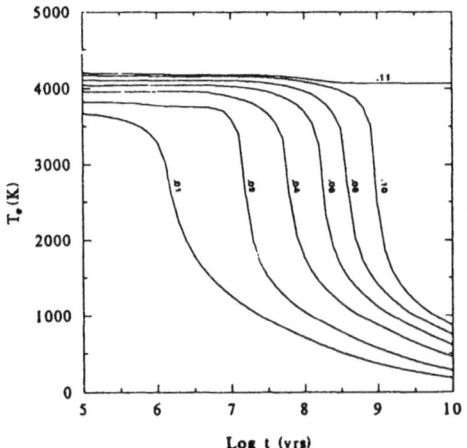

Figure 7. Evolution of the effective temperature, T_e, as a function of age t, for Population III stars in the mass range of $0.01 - 0.10\,\mathcal{M}_\odot$. Each track is labeled with the corresponding stellar mass in units of solar masses.

induced molecular hydrogen absorption. Since its contribution is approximately constant at low temperatures, the constant extrapolation technique mentioned in §4 was implemented for temperatures below 1000 K.

The evolutionary tracks of Population I and III brown dwarfs are qualitatively similar. Nevertheless, there are some important differences which should be noted. The end of the Population III main sequence has increased by $\sim 0.02\,\mathcal{M}_\odot$ to approximately $0.106\,\mathcal{M}_\odot$. Similarly, the end of the deuterium-burning main sequence has increased to $\sim 0.020\;\mathcal{M}_\odot$. More importantly, at early ages, Population III brown dwarfs are hotter (~ 4000 K) and more luminous than their Population I counterparts. This is to be expected since Population III opacities must necessarily be less than those for a Population I composition. Since Population III brown dwarfs have lower surface opacities, they can cool more quickly than a Population I brown dwarf of the same mass. This assertion is borne out on comparison of Figures 5 and 7.

Depending on the redshift at which they formed, the difference in cooling tracks can have a significant effect on the derived intensity function for an isotropic distribution of brown dwarfs. To examine the detectability of this background, Nelson (1989) chose the Einstein-de Sitter model with closure density being provided solely by brown dwarfs. Values of H_o between 50 and 100 km s^{-1} Mpc^{-1} were selected and various formation epochs (t_f) between 10^5 years and 10^9 years were assumed. In figure 8 the expected intensity (I_ν) is shown as a function of the observed frequency for three different masses (the mass function has been taken to be a delta function in each case). For this particular case, $H_o = 75$ km s^{-1} Mpc^{-1} and $t_f = 10^9$ years. The spectrum is seen to deviate significantly from a blackbody and thus has its own unique signature. The peak in the spectrum is, in general, rather broad and is found to lie in the frequency range of 10^{12} to 10^{14} Hz. It will also be noted that the maximum intensity depends rather weakly on the mass of the brown dwarfs.

In Figure 9, the spectral intensity is plotted against the observed frequency. The maxi-

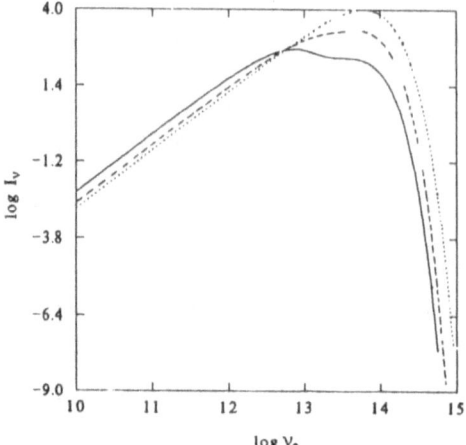

Figure 8. Integrated intensity, I_ν (Jy ster^{-1}, as a function of observed frequency, ν_0 (Hz), for Population III stars ($H_0 = 75$ km s^{-1} Mpc^{-1} and t_f $= 10^9$ yrs). The solid, dashed, and dotted lines denote masses of $0.005\,\mathcal{M}_\odot$, $0.02\,\mathcal{M}_\odot$, and $0.1\,\mathcal{M}_\odot$, respectively.

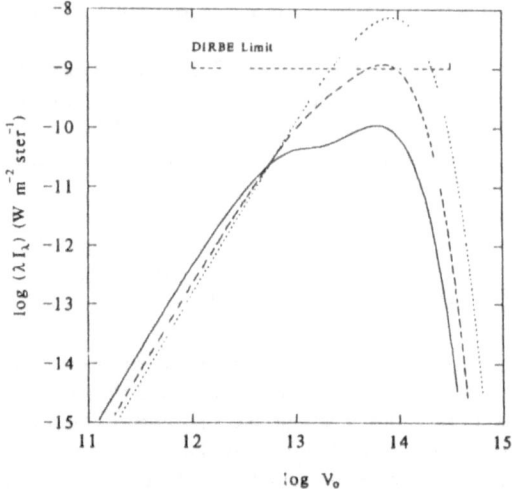

Figure 9. Spectral intensity, λI_λ (W m^{-2} ster^{-1}), as a function of observed frequency, ν_0 (Hz), for Population II stars ($H_0 = 75$ km s^{-1} Mpc^{-1} and $t_f = 10^9$ yrs). The solid, dashed, and dotted lines denote masses of $0.005\,\mathcal{M}_\odot$, $0.02\,\mathcal{M}_\odot$, and $0.1\,\mathcal{M}_\odot$ respectively. The DIRBE sensitivity limit is also shown.

mum spectral intensity is approximately 10^{-8} W m^{-2} ster^{-1}. Since the peak in the spectral intensity lies in the infrared (between $\sim 3\mu$m and 100μm) the most sensitive instruments that could be used to detect this radiation are COBE and SIRTF. On board COBE, the DIRBE (Diffuse InfraRed Background Experiment) could be used to try to detect this isotropic background. The DIRBE covers a wavelength band of 1 to 300 μm and will use ten different wavelength filters. The sensitivity of DIRBE is approximately 10^{-9} W m^{-2} ster^{-1} and thus, in principle, it may be possible to detect this background. Unfortunately, analysis of the DIRBE data is complicated by diffuse infrared radiation such as interplanetary and interstellar dust. Nonetheless, if one observes perpendicular to the ecliptic and Galactic plane it may be possible to model the "local" emission to within 1% precision. If a residual isotropic background can be detected then it may be possible to comment on the extent to which Population III brown dwarfs contribute to the cosmological BDM.

6. Observational Methods and Results

Although the intrinsic faintness of low-mass stars makes them difficult to study, enormous observational progress has been made during the past decade. Some of the more productive methods that are currently being used to detect low-luminosity objects include: (i) astrometry; (ii) high-precision radial velocity monitoring; (iii) speckle interferometry; (iv) infrared surveys; and, (v) deep CCD photometry. An excellent review of the observational methods including their strengths and limitations can be found in the review article by Liebert and Probst (1987).

Two basic strategies are being employed to search for very low luminosity objects. The first method involves searching for these objects as companions to observed stars. One of the main advantages of this method is that reasonably accurate masses of the binary components can sometimes be obtained. Not only does this method allow for a more accurate calibration of the lower main sequence, but it can also yield useful information concerning questions related to binary formation. The second strategy involves searching for very red field stars ("free floaters"). Although it is not possible to obtain a direct measure of the mass of these stars without appealing to the theoretical models, this is one of the few ways to obtain reliable information about the spatial densities of these objects.

Even if the mass of a star is known with some certainty, the age of the object must be estimated before theory and observation can be compared (this is especially true of brown dwarfs). One solution to this problem is to carry out a search in a cluster of stars of known age. Unless a cluster is quite close, it is important that young clusters be selected when searching for brown dwarfs. Brown dwarfs cool very quickly, and have reasonably large luminosities only when they are young ($< 10^8$ yrs). The other solution involves looking for a companion to a star whose age is reasonably well known. This approach is especially useful if the companion is a hot white dwarf since it stands to reason that the companion will necessarily be young and, moreover, the spectra of the individual components should be much easier to distinguish. Since, the cooling curves of white dwarfs are generally very reliable, accurate ages and spectra can be inferred. While the idea of searching for companions to hot white dwarfs is an attractive one, certain *caveats* should be noted. The most important of these is that mass transfer may have occurred when the white dwarf progenitor was in its giant phase. If this were the case, the observed companion star would not be pristine but rather would have experienced chemical contamination and started its evolution with a completely different set of initial conditions. Livio (1982) has examined this problem of "double-core" evolution and concludes that the low-mass companion can either spiral in or spiral out depending on the mass- accretion/mass-loss rates. Thus a Jovian-type planet may accrete sufficient to be subsequently classified as a brown dwarf. Another approach is to search for brown dwarfs in the vicinity of K or M stars that either exhibit flaring or have large rotational speeds or small spatial velocities. Thus, it would be reasonable to assume that the system is young and that a brown dwarf companion would be in its self-luminous phase.

6.1. SELECTED OBSERVATIONS

Van Biesbroeck 8B (VB 8B) was originally "discovered" in 1984 by McCarthy, Probst, and Low (1985). It was reported to be a very low-luminosity ($\sim 3 \times 10^{-5} \mathcal{M}_\odot$), companion to Van Biesbroeck 8 at an angular separation of ~ 1" and at a distance of about 7 parsecs. Since the effective temperature of the companion was thought to be ~ 1300 K, comparison to theoretical models indicated that the mass of the companion should be in the range of approximately $0.06 - 0.08$ $\mathcal{M}_\odot$. This corresponded to an age for VB 8 of between 1 and 5 Gyr. According to Harrington (1986), VB 8 is also an astrometric binary whose companion mass is thought to be between 0.01-0.02 $\mathcal{M}_\odot$. However, since 1985 the companion to VB 8 has not been observed (Perrier and Marriotti 1987; Skrutskie et $al.$ 1987) and it now seems highly unlikely that it was ever observed. Nonetheless, this reported detection has motivated a significant amount of observational and theoretical work.

Becklin and Zuckerman (1987) have been engaged in an ongoing program of trying to observe low-temperature companions to young white dwarfs. They looked for companions to young white dwarfs in the Pleiades and Hyades clusters, but were unsuccessful in detecting any. However, they continued with a photometric survey of 200 relatively young white dwarfs in the solar neighborhood and found an infrared excess peaked at $3.5\,\mu$m close to or coincident with the white dwarf Giclas 29-38. G29-38 is a ZZ Ceti star that has an effective temperature of 11,500 K, a bolometric luminosity of 2×10^{-3} $\mathcal{L}_\odot$ and is at distance of 14.1 pc. They have put forward two hypotheses, the first being that the infrared excess could be explained by orbiting particulate matter. In this case the radius of the dust shell would have to be $\sim 1\,\mathcal{R}_\odot$ and have a surface temperature of ~ 1200 K. However, the time scale for the dust to spiral into the white dwarf due to the Poynting-Robertson effect is on the order of 10 yrs. The second hypothesis is that the IR excess emanates from a low luminosity brown dwarf companion that has $T_e = 1200 \pm 200$ K and a luminosity of $\sim 5 \times 10^{-5} \mathcal{L}_\odot$. If we assume that the IR excess is due to a brown dwarf and further assume that the age of the brown dwarf is the same as that of the white dwarf companion ($\sim 6 \times 10^8$ yr) then the best fit to the results of NRJ corresponds to a brown dwarf mass of $\sim 0.045 \mathcal{M}_\odot$. However, it should be noted that for this mass and age the theoretical calculations yield a luminosity of $\sim 4 \times 10^{-5} \mathcal{L}_\odot$ and an effective temperature of ~ 1350 K. Thus, the agreement between theory and observation is not overly good. In fact, in a follow up paper, Tokunaga et $al.$ (1988) found that the IR excess could not be spatially resolved (< 5.6 AÜ.) and that no water vapour absorption bands could be detected. These findings indicate that an alternate explanation to the brown dwarf scenario must be sought.

In a subsequent paper, Becklin and Zuckerman (1988) spatially resolved a very low luminosity object at a distance of 120 AÚ. from the white dwarf GD 165 (this was one of the white dwarfs that was originally targeted in the 200 sample survey). As part of this same study they found seven other very low luminosity objects and suspect that the masses are typically $\lesssim 0.1\,\mathcal{M}_\odot$. The companion to GD 165 was detected in the JHKL bands using the 3m IRTF telescope. The absolute magnitude was measured to be about 11.8 in the K band for which an effective temperature of ≈ 2130 K and a luminosity of $\log(L/\mathcal{L}_\odot) \approx -4.1$ were estimated. By assuming an age of $\sim 6 \times 10^8$ yr we can conclude that the mass of the companion is $> 0.06\,\mathcal{M}_\odot$. The inferred mass is very dependent on the effective temperature and thus further observations are required before a definitive statement can be made. Nevertheless, these preliminary results strongly suggest that a "transition" brown dwarf has been detected.

Despite several unsuccessful searches around "normal" main sequence field stars (see, for example, Marcy and Benitz 1989, and references therein) a brown dwarf companion to HD 114762 was discovered by radial velocity monitoring (Latham et al. 1989). This system is at a distance of 28 pc has an 84 day orbital period. The mass function of this system has been measured yielding a mass of the companion $\approx 0.011\,\mathcal{M}_\odot/\sin i$. This implies that the mass of the companion is probably about $0.02\,\mathcal{M}_\odot$. If these observations are correct then this system is one of the most clear cut examples of a binary containing a brown dwarf component. Moreover, brown dwarf companions to main-sequence stars may not be as rare as previously thought.

Recently, Jameson and Skillen (1989) looked for low mass stars and brown dwarfs in the Pleiades. Seven fields, each 25 square arcmin in size, were imaged in the R and I bands using the CCD prime focus of the Isaac Newton Telescope. On the basis of theoretical isochrones fit to the Pleiades main sequence members, they were able to identify nine low mass candidates. Based on their proximity to theoretical isochrones of brown dwarfs, Jameson and Skillen claim that if the age of the Pleiades is taken to be 65 Myr then five good brown dwarf candidates have been identified and if the age is equal to 250 Myr then that number is reduced to three. If these results are correct then there may be as many as 5000 brown dwarfs in the entire cluster. In the future Jameson and Skillen plan to survey more of the cluster, perform JHK photometry and also confirm that the candidates are indeed members of the Pleiades by a proper motion analysis. From a theoretical standpoint, it should be noted that if the isochrones are too blue (which implies that the tracks should be shifted to the red) then only one or two cluster members would be potential candidates. Nonetheless, these results are very exciting and should be followed up as quickly as possible.

As a result of a recent survey in the K band, Forrest et al. (1989) claim that several potential brown dwarfs have been identified in the direction of Taurus. They originally looked for companions to approximately 60 emission-line field stars with low velocities and reported no detections. They also looked for companions to 57 Hyades stars and also obtained a null result. However, when they examined 26 naked T-Tauri stars they identified 6 or 7 extremely red "companions". These stars typically have values of H-K ≈ 0.5, which translates into effective temperatures between about 2000 and 3000 K. One possible explanation is that these objects are $0.01 - 0.02\,\mathcal{M}_\odot$ brown dwarfs with ages on the order of 10^6 to 10^7 yrs. Another possibility is that these stars are late M dwarfs with ages $> 10^8$ yrs. If this latter case were true, then there would have to be a background bias of M stars in the direction of Taurus. On the other hand, the proper motions of those candidates that have been measured are consistent with the proper motion of Taurus. It is interesting to note that the probability of detection at a given angular separation from the "companion" star indicates that the candidate stars are randomly distributed along the line of sight suggesting that they are not binary companions. Clearly, much more work needs to be done to confirm this analysis. The survey should be expanded and near-IR spectroscopy should be carried out.

The search for brown dwarfs in infrared databases has so far proved to be fruitless. Specifically, the IRAS database has been searched and no brown dwarf candidates were found (this includes the IRAS co-addition and serendipity survey). However, this null detection does not place any severe constraints on the local mass function. The SIRTF surveys, on the other hand, should yield statistically meaningful results and thus we will have a much better understanding as to the extent and nature of any local dark baryonic mass.

I would like to acknowledge many stimulating discussions on this subject with W.Y. Chau, B. Dorman, P. Joss, and S. Rappaport. I would also like to thank G. Gilmore and D. Lynden-Bell for their hospitality. This work was partially supported by a grant from NSERC (Canada).

REFERENCES

Becklin, E.E., Zuckerman, B., (1987) *Nature (Letter)*, **330**, 138.
Becklin, E.E., Zuckerman, B., (1988) *Nature (Letter)*, **336**, 656.
Carr, B.J, and Rees, M.J., (1979) *Nature*, **278**, 605.
Carr, B.J., Bond, J.R., and Arnett, W.D., (1984) *Ap. J.*, **277**, 445.
Clayton, D.D., (1983) *Principles of Stellar Evolution and Nucleosynthesis* (Chicago: University of Chicago Press).
Copeland, H., Jensen, J.O., and Jorgensen, H.E., (1970) *Astr. Ap.*, **5**, 12.
D'Antona, F., and Mazzitelli, I., (1985) *Ap. J.*, **296**, 502.
Dorman, B., Nelson, L.A., Chau, W.Y., (1989) *Ap. J.*, **342**, 1003 (DNC).
Fontaine, G. Graboske, H.C., Jr., and Van Horn, H.M., (1977) *Ap. J. Suppl.*, **35**, 293 (FGVH).
Forrest, W.J, Ninkov, Z., Garnett, J.D., Skrutzkie, M.F., and Shure, M., (1989) preprint.
Fowler, W., (1990) in *Baryonic Dark Matter*, ed. D. Lynden-Bell (Dordrecht: Kluwer Academic).
Grossman, A.S., Hays, D., and Graboske, H.C., Jr., (1974) *Ast. Ap.*, **30**, 95.
Harrington, R.S., (1986) in *Astrophysics of Brown Dwarfs*, ed. M. Kafatos (Cambridge: Cambridge University Press), p. 3.
Hayashi, C., and Nakano, T., (1963) *Prog. Theor. Phys.*, **30**, 460.
Jameson, R.F., and Skillen, I., (1989) *M.N.R.A.S., (in press)*.
Karimabadi, H, and Blitz, L., (1984) *Ap. J.*, **112**, 190.
Latham, D.W., Mazeh, T., Stefanik, R.P., Mayor, M., and Burki, G., (1989) *Nature (Letter)*, **339**, 38.

Liebert, J., and Probst, R.G., (1987) *Ann. Rev. Astr. Ap.*, **25**, 473.

Livio, M., (1982) *Astr. Ap.*, **112**, 190.

Lunine, J.I., Hubbard, W.B., Burrows, A., Wang, Y-P., Garlow, K., (1989) *Ap. J.*, **338**, 314.

Lunine, J.I., Hubbard, W.B., and Marley, S., (1986) *Ap. J.*, **310**, 238.

Magni, G., and Mazzitelli, I., (1979) *Astr. Ap.*, **72**, 134 (MM).

Neece, G.D., (1984) *Ap. J.*, **277**, 738.

Nelson, L.A., (1989) in preparation.

Nelson, L.A., Rappaport, S., and Joss, P.C., (1985) *Nature*, **316**, 42 (NRJ).

Perrier, C., and Mariotti, J.-M., (1987) *Ap. J. (Letters)*, **312**, L27.

Popper, D.M., (1980) *Ann. Rev. Astr. Ap.*, **18**, 115.

Press, W.H., and Lightman, A.P., (1983) *Phil. Trans. R. Soc. London A*, **310**, 323.

Rappaport, S., and Joss, P.C., (1984) *Ap. J.*, **283**, 232.

Rees, M.J., (1983) *Phil. Trans. R. Soc. London A*, **310**, 311.

Rees, M.J., (1980) in *Variability in Stars and Galaxies*, (Liège: Institut d'Astrophysique), p. G.2.

Sienkiewicz, R., (1982) *Acta. Astr.*, **32**, 275.

Skrutskie, M.F., Forrest, W.J., and Shure, M., (1987) *Ap. J.*, **312**, L55.

Stahler, S.W., Palla, F., and Salpeter, E.E., (1986) *Ap. J.*, **302**, 590.

Stevenson, D.J., (1986) in *Astrophysics of Brown Dwarfs*, ed. M. Kafatos, R.S. Harrington, and S.P. Maran (Cambridge: Cambridge University Press) , p. 218.

Stevenson, D.J., (1978) *Proc. Astr. Soc. Australia*, **3**, 227.

Tarter, J., (1975) Ph.D., thesis, University of California, Berkeley.

Tokunaga, A.T., Hodapp, K.-W., Becklin, E.E., Cruikshank, D.P., Rigler, M., and Toomey, D., (1988) *Ap. J.*, **332**, L71.

VandenBerg, D.A., Hartwick, F.D.A., Dawson, P., and Alexander, D.R., (1983) *Ap. J.*, **226**, 747.

Weisskopf, V.F., (1975) *Science*, **187**, 605.

Zapolsky, H.S., and Salpeter, E.E., (1969) *Ap. J.*, **158**, 809.

WHITE DWARFS AND THE LOCAL MASS DENSITY

Volker Weidemann
Institut fur Theoretische Physik und Sternwarte
Universität Kiel
D-2300 Kiel
Federal Republic of Germany

ABSTRACT. In order to estimate the contribution of cooled-down white dwarfs to dark baryonic matter we consider white dwarf luminosity functions, observed and theoretically calculated within models of Galactic evolution. Depending on the details of these models the invisible mass fraction in the solar neighbourhood is of the order of $2\,\mathcal{M}_\odot \mathrm{pc}^{-2}$ or 4% of the dynamical mass. A redetermination of the white dwarf birth rate yields a somewhat higher value than currently assumed.

1. Introduction

White dwarfs are known to be faint degenerate stars which after exhaustion of their nuclear fuel cool down towards invisibility, thus definitely contributing to baryonic dark matter. Together with neutron stars and black holes they constitute the final stages of stellar evolution.

At the present observational limit of about $M_v = 16.5$ they are hard to detect since the white dwarf cooling sequence merges with the faint end of the main sequence in the M_v, (B-V) plane (Jahreiss, 1987). Most cool degenerates have been found in proper motion surveys, such as those provided by Luyten, where degenerates have comparatively large reduced proper motions $H = m + 5 + 5\,\log\mu$ (Jones, 1972). However it is possible to separate them photometrically from cool M dwarfs only if red colour indices and magnitudes are used (Dahn *et al.* 1978) for which comparable numerous surveys are not yet available.

White dwarfs exist at higher effective temperatures in two main varieties: the majority with hydrogen-dominated atmospheres (spectral type DA) and a fraction of 10 – 20% with nearly pure helium atmospheres (DB or other non-DA), which form well separated cooling sequences in appropriate (Strömgren or multichannel) two-colour or colour-magnitude diagrams (*eg* Graham, 1969, Greenstein 1984). Extensive spectroscopic analysis (Koester *et al.* 1979) has demonstrated that DA white dwarfs are confined to a comparatively narrow range of radii (HR-diagram) or surface gravities which fulfil in the average the theoretically predicted mass-radius-(surface gravity) relation for completely degenerate (zero temperature) configurations with a hydrogen-free interior. The derived mass distribution is very narrow and centered at about $0.6\,\mathcal{M}_\odot$. Average masses for non-DA white dwarfs are equal

D. Lynden-Bell and G. Gilmore (eds.), Baryonic Dark Matter, 87–101.
© 1990 *Kluwer Academic Publishers.*

($0.55\,\mathcal{M}_\odot$) within the uncertainties (see Weidemann, 1987a). DA and non-DA white dwarfs are also similar in velocity distribution (Sion *et al.* 1988), therefore their differences - which are still unexplained - are not yet important as far as calculations and interpretation of luminosity functions, and derivation of space and mass densities are concerned.

2. Pre-White Dwarf Evolution

White dwarfs are the end product of evolution of stars with low and intermediate masses, from the Galactic turn-off ($\sim 1\,\mathcal{M}_\odot$ for population I or $\sim 0.8\,\mathcal{M}_\odot$ for population II) to an upper limit between 5 and 8 $\mathcal{M}_\odot$ depending on details of stellar evolution theory such as core overshooting and mass loss (see Weidemann, 1987b, for an extensive discussion of the initial-final mass relation). The white dwarf cooling history depends - at least through its first stages - strongly on the pre-WD evolution: the WD is actually prefabricated in the stellar core of a red giant during the double shell burning phase on the asymptotic giant branch (AGB). After sufficient mass loss the progenitor leaves the AGB and evolves at constant luminosity towards higher temperatures until, at T_{eff} about 30000 K, it becomes visible as a central star of a planetary nebula (see Iben, 1985, for a schematical summary of stellar evolution including nuclear burning stages up to this point).

Although the vast majority of white dwarfs is born *via* the planetary nebulae stage - as demonstrated by estimates of birth rates and the similarity of mass distributions (see Weidemann, 1989, for a discussion) there are other channels feeding the WD cooling sequence, *eg* sdB and sdO stars or horizontal branch stars which either do not reach the AGB or evolve from it so slowly that planetary nebulae are dissolved before ionization. The fraction of such progenitors with masses below about $0.55\,\mathcal{M}_\odot$ is not well determined but probably not larger than 20 - 30% (see Weidemann and Koester, 1984, Heber, 1986) as estimated from the WD mass distribution and sdB-evolution.

3. White Dwarf Cooling

Pre-white dwarf and white dwarf evolution is schematically displayed in Figure 1 (from Weidemann, 1975) which demonstrates the location of the chemical boundaries and the continuous contraction of individual mass shells in the temperature-density plane.

Conventional evolution theory predicts a highly degenerate carbon/oxygen core of about 99%, a remnant unburned helium layer with thickness of not more than 1%, overlayed by a hydrogen-layer of $10^{-4}\,\mathcal{M}_\odot$ at quiet burnout (but possibly much less in later stages due to continuing mass loss or ejection by late thermal pulses - see several recent papers on the thin (H) layer hypothesis published in Wegner (1989) and Philip *et al.* (1987). Earlier cooling theory is due to Mestel and Ruderman (1967) and considers the loss of energy from an isothermal interior through a non-degenerate envelope and predicts cooling times as a function of luminosity and mass proportional to $(M/L)^{5/7}$. Deviations from

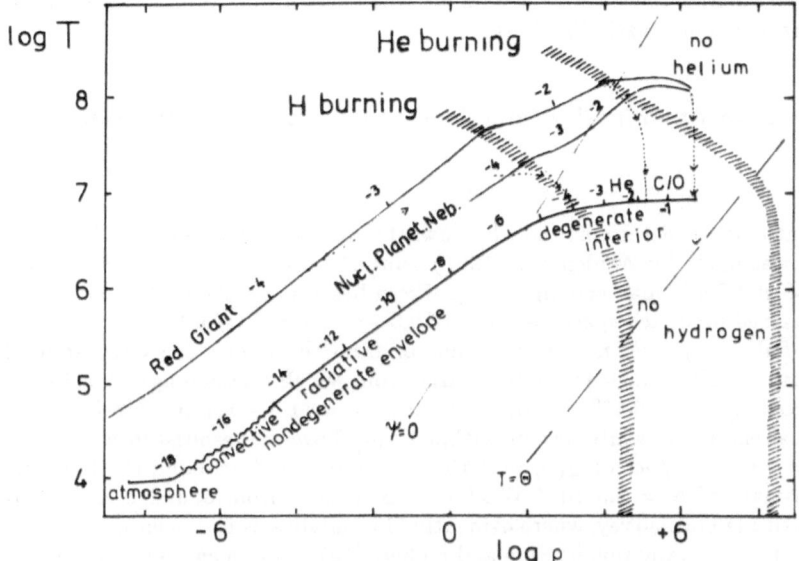

Figure 1. Structure of a typical white dwarf ($0.6\,\mathcal{M}_\odot$, $T_{\text{eff}} = 12000$ K) and of its immediate progenitors, with (log) mass fractions from the surface indicated, as well as the degeneracy limit ($\Psi = 0$) and Debye temperature θ.

Mestels law are considered in later investigations (*eg* Koester, 1972): delayed cooling due to crystallization, and accelerated cooling at the hot end due to neutrino leakage or at the cool end after the interior temperature sinks below the Debye temperature. The latter is reached comparatively early in white dwarf interiors (see Fig. 1), especially if the central density is high. Thus massive white dwarfs will cool down to invisibility more quickly. This can be seen *eg* in cooling curves calculated by Winget and coworkers (as displayed by Iben and Laughlin, 1989).

As mentioned above the starting models for white dwarf cooling calculations are important, thus cooling predictions will depend on pre-white dwarf evolution. Consistent calculations from the AGB down into the white dwarf region have been carried out by Iben and Tutukov (1984), Mazzitelli and D'Antona (1986) and Koester and Schönberner (1986), the latter demonstrating how the evolutionary track for white dwarfs only slowly approaches the zero temperature radius.

Winget and van Horn (1987) have demonstrated how different input physics and parameters (like thickness and chemistry of outer layers, interior composition and conductivities) cause quite different cooling ages down to log $L/\mathcal{L}_\odot = -4.5$ which vary between 5 and 13 Gyr. The question of delayed cooling due to chemical separation of carbon and oxygen in the interior - which would constitute an additional source of gravitational energy - is be discussed by Canal at this meeting (Canal 1990). It seems to be less important now, especially since the increased $C(\alpha,n)O$ cross sections used in stellar evolution theory predict mainly oxygen as the product of central helium burning phases.

Improved cooling calculations are under way (Wood and Winget 1989, Tassoul *et al.* 1989, D'Antona and Mazzitelli 1989).

4. White Dwarf Luminosity Functions and White Dwarf Birth Rates

A first attempt to determine the white dwarf luminosity function (WD-LF) and the birth rate χ_{WD} was made by Weidemann (1967) using the then available observational material and the Mestel-Ruderman cooling theory. From best fits on the well observed parts of the LF he obtained predicted space densities of $0.5 - 0.710^{-2}$ WD/pc^3 down to $M_{bol} = 15.5$ for a Luyten-Palomar proper motion ensemble or from the Eggen and Greenstein (1965) list of spectroscopically confirmed white dwarfs, from which – assuming a constant birth rate over the cooling age of $5\ 10^9$ yr – one obtains $\chi_{WD} = (1.0 - 1.4)10^{-12}$ WD/pc^3/yr.

The local ensemble, white dwarfs within 10 pc, however resulted in a somewhat higher value $N = 10^{-2}$ WD/pc^3 or $\chi_{WD} = 2\ 10^{-12}$ WD/pc^3/yr (including the binaries). A still higher estimate of $N = 2.5\ 10^{-2}$ WD/pc^3 was derived from Sandage and Luyten's high galactic latitude blue survey, where evidently blue subdwarfs had been mistaken for hot DB white dwarfs. A reevaluation by Sion and Liebert (1977) for an enlarged ensemble extending down to fainter apparent magnitudes essentially confirmed these results. These first studies showed the difficulties in extending the observed range of the LF down to fainter luminosities or bolometric magnitudes M_{bol} due to the run of the observability function. This function is shown in Figure 2, which demonstrates how the expected number of observable faint cooled-down degenerates falls-off extremely rapidly towards increasing M_{bol} and even more for increasing M_v – due to the effects of bolometric corrections. Taking into account that even today the absolutely faintest degenerates are at $M_v \sim 16$ mag one sees that progress at the faint end must be (and is) very cumbersome.

Concerning the observed LF two steps must be marked during the last decade: first, a redetermination of the LF using the supposedly complete Palomar-Green survey ensemble resulted in a reduced space density of $5\ 10^{-4}$ DA-WD/pc^3 down to $M_v = 12.75$ and a corresponding birth rate of only $\chi_{DA} = (4 - 6)10^{-13}$ WD/pc^3/yr (depending on cooling age differences) or total WD birth rate, by adding 25% non-DA stars of $\chi_{WD} = (5 - 7.5)10^{-13}$ WD/pc^3/yr (Fleming *et al.* 1986).

Second, the observed steep decline at the faint end of the LF was interpreted as caused by the finite age of the Galactic disk, which Winget *et al.* (1987) determined to be 9.3 ± 2 Gyr, or an "age of the universe" of 10.3 ± 2 Gyr. Aside from the questionable assignment of only a 1 Gyr pre-disk age of the whole Galaxy (including halo), and crude modeling of the theoretical LF (Weidemann and Yuan, 1989) the result is based on the assumption that the steep decline is real and not caused by observational incompleteness implying that practically *no white dwarfs have cooled down to invisibility.*

That other authors (D'Antona and Mazzitelli, 1986, 1989) have reached different conclusions – *eg* 50% of all white dwarfs cooled down to invisibility – is partly due to the use of different cooling curves, and partly due to extremely different assumptions on Galactic modeling. An example of this is Larson's famous attempt to explain the then believed 50% local missing mass totally by cooled-down remnants, most of which were assumed to be

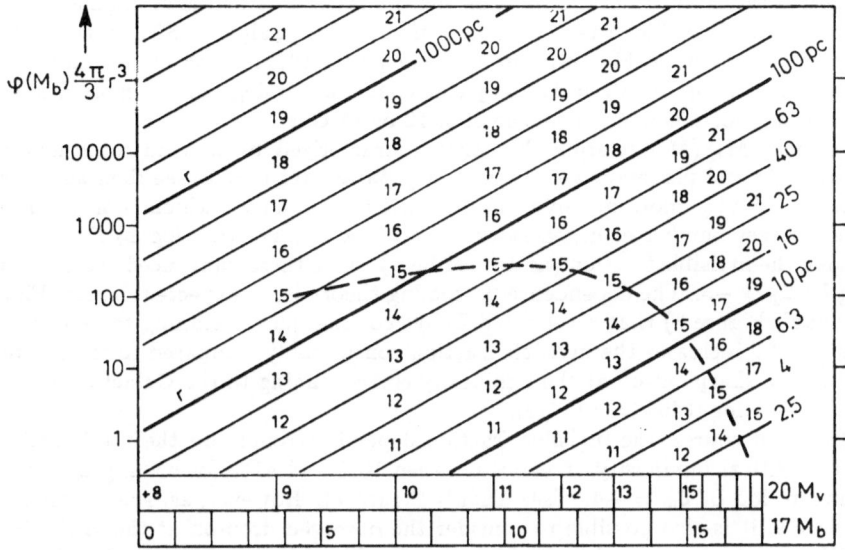

Figure 2. Observability of degenerates, giving numbers of stars expected down to an apparent magnitude limit for a Mestel cooling law.

massive and copiously produced during early phases of Galactic evolution. It seemed therefore advisable to check the dependence of white dwarf luminosity functions theoretically, *ie* to predict their shape and the fraction of invisible degenerates, for different assumptions about the parameters determining stellar and Galactic evolution, and compare these with the observations.

The most recent observed LF's have been obtained by the efforts of Liebert, Dahn and Monet (1988, 1989) and Dahn *et al.* (1989) for a Seven–Tenth (proper motion) ensemble. The steep fall-off at the faint end has thereby already been reduced. It must be kept in mind, however, that the detailed shape depends partly on binning and partly on the assumed bolometric corrections.

5. Theoretical White Dwarf Luminosity Functions and Models of Galactic Evolution

For the theoretical calculations of Galactic evolution models we have to specify and consider different initial mass functions (IMF), star formation rates as a function of time, SFR(t), time scales for nuclear evolution, main sequence to the AGB, t(evol), initial-final mass relations, M_f (M_i), and ages of the Galactic disk, t(disk). We furthermore need cooling curves in order to connect luminosity, M_f and t(cool) (see Weidemann, 1979, for a schematic diagram). For details we refer to the publication by Yuan (1989). Here we concentrate on the situation at the faint end of the LF. Considering the mass fraction of degenerates below log L = −4.2 to bè invisible, Yuan's standard model (IMF: Salpeter, SFR = const, t(evol) for moderate overshooting, M_f (M_i) from Weidemann, 1987b, with upper limit of white dwarf production, M_{WD}, at $M_i = 8\,\mathcal{M}_\odot$) yields cooled-down degenerates mass fractions of 12%, 27% or 39% for a Galactic disk age of 9, 12 or 15 Gyr.

For a steeper M_f (M_i) relation, like that characterized by a wind mass loss factor η(Reimers)= 1/3 and planetary nebula efficiency parameter b = 1 (see Iben and Renzini, 1983), $M_{WD} \sim 5\,\mathcal{M}_\odot$, more massive white dwarfs are produced which cool down faster, predicting a larger number of degenerates at very low luminosities, log L/ $\mathcal{L}_\odot$ < −5, but not changing the invisible fraction (as defined here), since these "advanced" stars are missing at log L/ $\mathcal{L}_\odot$ > −5. The dependence on cooling theory is, as expected, larger. First one notices (Yuan's Figure 3) that accelerated Debye cooling, just beginning to be effective at log L/ $\mathcal{L}_\odot$ = −4.2 increases the invisible fraction considerably compared to the Mestel law which predicts a sharp cut-off at the luminosity corresponding to the Galactic disk age (at log L/ $\mathcal{L}_\odot$ = −4.5 for t(disk)= 12 Gyr).

In addition, of course, the invisible fraction depends strongly on the total cooling age down to the critical luminosity, it varies between 10% and 47% for cooling ages of 13 or 5 Gyr down to log L/ $\mathcal{L}_\odot$ = −4.5 (see Yuan's Figure 7). However, as Yuan demonstrates in her Figure 9, it is also possible to consider the observed drop-off at the visibility limit as real (Winget et al. approach) and find best fits for disk ages between 6 and 14 Gyrs, more generally at disk ages which are about 1 Gyr larger than the cooling age down to log L/ $\mathcal{L}_\odot$ = −4.5. The invisible fraction is then, nearly by definition, small, of the order of 10%, and insignificant for the local mass budget.

As expected, extreme fractions of invisible mass in degenerates can be obtained by a large overproduction of massive white dwarfs in an early epoch of galactic evolution: the bimodal model of Larson (1986). Yuan's model predicts 62% dark matter in degenerates for this case, with IMF and SFR from Larson but for t(disk)= 12 Gyr. Larson's original model assumed an age of 15 Gyr in which case the invisible fraction goes up to 78%. Together with a large amount of dark matter coming from supernova remnants – Larson extrapolated a linear M_f (M_i)–relation from the white dwarf range to the largest M_i – he thus was able to explain the local missing mass, considered to be 50% of the total at that time. However the recent results by Kuijken and Gilmore (1989; cf. Gilmore 1990) as well as the investigations by Bienaymé et al. (1987) remove the need for so much local dark matter. Besides the Larson model predicts a much stronger increase of the WD-LF already before log L/ $\mathcal{L}_\odot$ = 4.2 (see Yuan's Figure 5) which is not compatible with the observations. Without the bimodal scheme, however, a decreasing SFR or even an initial 1 Gyr "burst" of star formation produces only a slightly larger number of cool visible white dwarfs, which cannot be ruled out by present observations. The invisible fraction, of course, is larger and increases from 27 to 46% compared to the standard model for a SFR ten times stronger during the first billion years (see Yuan, Figure 4).

On the other hand, Yuan's models show that shorter period ($2\ 10^8$ yr) bursts during Galactic history, as proposed by Barry (1988), and evaluated by Noh and Scalo (1989) are not distinguishable in the WD-LF, due to the fact that white dwarfs which enter the cooling tracks are coming from progenitors which were born in quite different epochs in the past: the majority comes from (old) turn-off stars and only a few from massive progenitors, the average age being larger than 3 Gyr as shown in Figure 3. Thus SFR variations are

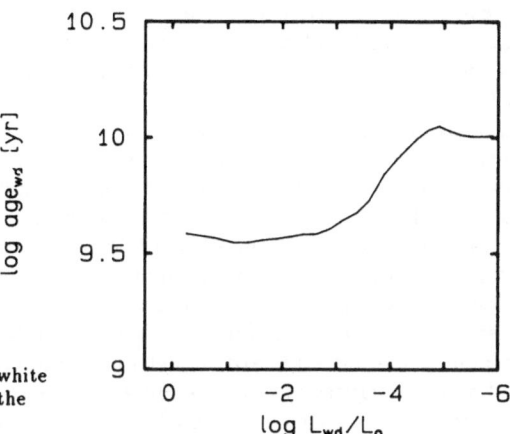

Figure 3. Average total age, including pre–white dwarf evolution, of cooling degenerates, for the standard model described in the text.

extremely smoothed on the WD-LF, and it is practically impossible to derive information about the past history of star formation from the WD-LF, the only exception being a very recent SFR burst (see Yuan, Figure 13). Young white dwarfs are not younger stars: this has been mistakenly assumed *eg* for a search strategy to discover young and therefore still relatively bright brown dwarfs in the neighborhood of hotter white dwarfs. It is also reflected in the fact that the average scale height of white dwarfs on the LF begins already at about 150 pc, then increases slowly to about 300 pc from $\log L/\mathcal{L}_\odot < -3$ to the present observation limit at $\log L/\mathcal{L}_\odot \approx -4.5$. This increase of scale height (calculated with different relations for H(t), *eg* by Scalo (1986)) causes a depression of the LF at the cool end, since the LF refers to the number of stars per pc^3, whereas the model evolution considers primarily column densities. This so-called "inflation effect" (see Yuan, Figure 6) has to be kept in mind if one considers mass or number fractions of cooled-down degenerates (and even more for neutron stars which diffuse to much larger heights, cf. Lyne at this Meeting: Lyne 1990).

6. Surface and Volume Densities of Dark Stellar Remnants

For the purpose of this conference J.W. Yuan has kindly adapted her evolution program to provide tables which contain number and mass surface and space densities, for all degenerates as well as above and below certain luminosities, separately also for supernova remnants, main sequence stars, giants and all stars, finally the ratio of white dwarf to supernova birth rates, and the returned mass to the interstellar medium, from degenerates, supernovae and total – all data tabulated for the present epoch, but also evaluated as a function of time during Galactic evolution.

In order to have a trustworthy standard model, the IMF was taken to be either Salpeter down to $M = 0.25\,\mathcal{M}_\odot$ and constant below, or Miller-Scalo (1979) or Scalo (1986) which lock up even less mass than the adapted Salpeter function. The different white dwarf M_f (M_i)-relations were supplemented by the assumptions that from the upper limit for WD-production up to $30\,\mathcal{M}_\odot$ neutron stars of $1.4\,\mathcal{M}_\odot$ are produced and that stars between 30 and $50\,\mathcal{M}_\odot$ leave $10\,\mathcal{M}_\odot$ black holes.

All models are normalized to provide a present white dwarf birth rate of $\chi_{WD} = 1\ 10^{-12}$ WD/pc^2. Extracts of the results are given in the following tables. Table 1 is intended to show our model data in comparison with the much discussed results of Larson (1986) and Kuijken and Gilmore (1989) and Gilmore (1990) for surface (or column) mass densities.

Our standard model has t(disk) = 12 Gyr, IMF modified Salpeter, M_f (M_i) from Weidemann (1987b) for white dwarfs and supernova remnants as above, and SFR(t) = const. It provides a total mass in visible (main sequence + giant) stars of $17\,\mathcal{M}_\odot$/pc^2, less than what seems to be a better established observational value which we take to be $28\,\mathcal{M}_\odot$/pc^2. We therefore scale our standard model up by increasing $\chi_{WD}/10^{-12}$ pc^{-3} yr^{-1} from 1 to 1.65 and give the figures in boldface. In a similar way we deal with the results from other models: they are all scaled by a constant factor so as to provide 28 $\mathcal{M}_\odot$/pc^2 in "living" stars and make the data more comparable. Column 1 designates author, model or model changes compared to standard. Column 2: χ_{WD}, column 3: surface mass densities of stars, column 4: in remnants, column 5: in white dwarfs (degenerates), column 6: in neutron stars and black holes. Column 7 gives the white dwarf/supernova birth rate (pc^{-2}) which can be used to estimate the galactic supernova frequency if an effective disk volume is assumed (see below).

From Table 1 we see that nearly independent of the model details, predicted surface mass densities of remnants are between 3 and $6\,\mathcal{M}_\odot$, in agreement with earlier estimates. The differences caused by IMF variations eg are due to smaller number of massive progenitors for Scalo (1986) vs Salpeter, thereby lowering the SN remnant production and raising the WD/SN ratio. Differences caused by M_f (M_i) relations are similarly understood: the steeper these are (the smaller the limiting mass M_{WD}) the larger the SN production and the relative mass fraction of SN remnants.

Concerning the history of the star formation rate, one notices that the mass fraction of degenerates increases more as a higher production of stars occurred in the early galactic history, however the effects are minor compared to the standard model. Finally the increase of the remnant surface density with galactic age is expected, but it is relatively smaller than the age increase, due to our normalization of χ_{WD} to a fixed value at the present age: if the latter is large the scaling factor for the whole population has to be lowered (otherwise χ_{WD} would increase with galactic age, since numerous low mass progenitors come into play).

With Table 2 we turn to the more relevant question for this Meeting: the amount of baryonic dark matter in the form of cooled-down degenerates, and also in form of supernova remnants. Table 2 gives for the same models as in Table 1: the mass fraction of cooled-down degenerates (column 3), the corresponding mass (column 4) and the total amount of dark matter, including neutron stars and black holes (column 5), all in solar masses per pc^2, column 6 gives the fraction and mass of cooled-down degenerates per 1000 pc^3. The local mass of supernova remnants is comparatively smaller, since the scale height for these is much larger . Thus column 8, "total", is more uncertain and was calculated assuming six times stronger inflation than that for degenerates. Columns 9 and 10 give the total predicted number of degenerates and supernova remnants within 10 pc.

As can be seen from Table 2 the dark matter fraction of degenerates goes up from 16%

TABLE 1

SURFACE MASS DENSITIES

Author	Stars $\mathcal{M}_\odot \text{pc}^{-2}$	Remnants $\mathcal{M}_\odot \text{pc}^{-2}$	ISM $\mathcal{M}_\odot \text{pc}^{-2}$	Total $\mathcal{M}_\odot \text{pc}^{-2}$
Larson (1986) Model	22	37	8	67
Kuijken/Gilmore (1989) Data	28	4	15	46

No.	Model	χ_{WD} $10^{-12}\text{pc}^{-3}\text{yr}^{-1}$	Stars $\mathcal{M}_\odot \text{pc}^{-2}$	Total Remnants	WD $\mathcal{M}_\odot \text{pc}^{-2}$	NS+BH $\mathcal{M}_\odot \text{pc}^{-2}$	WD/SN no.pc^{-2}
1	Yuan Standard	1.0	17	2.1	1.6	0.5	18.0
	12 Gyr, SFR const	1.65	28	3.4	2.6	0.8	
2	Standard, but	1.0	9	1.8	1.6	0.2	31.0
	IMF Miller-Scalo	3.11	28	5.6	5.0	0.6	
3	Standard, but	1.0	13	1.7	1.6	0.14	74.0
	IMF Scalo (1986)	2.13	28	3.6	3.4	0.2	
4	but M_f (M_i)	1.0	18	2.6	1.8	0.8	9.5
	Iben/Renzini	1.53	28	4.0	2.7	1.3	
	(1983), $\eta = 1$						
5	but M_f (M_i)	1.0	21	3.3	2.0	1.3	5.4
	Iben/Renzini	1.35	28	4.5	2.7	1.8	
	(1983), $\eta = 3$						
6	but SFR	1.0	25	3.5	2.7	0.8	23.7
	initial burst	1.13	28	4.0	3.1	0.9	
7	but SFR	1.9	24	3.2	2.5	0.7	13.7
	exp (-t/10 Gyr)	1.17	28	3.7	2.9	0.8	
8	Standard,	1.0	14	1.5	1.1	0.4	16.0
	but 9 Gyr	2.00	28	3.0	2.2	0.8	
9	Standard	1.0	19	2.6	2.0	0.6	20.4
	but 15 Gyr	1.43	28	3.7	2.9	0.8	

TABLE 2

Space densities of cooled, dark remnant matter

Model	χ_{WD} 10^{-12}pc^{-3}yr^{-1}	Surface mass density			Volume mass density			no. < 10pc		
		%	$\mathcal{M}_\odot$pc^{-2}	total	%	$\frac{\mathcal{M}_\odot}{(1000pc^3)}$	Total	Total	WD's	SNR
1	1.65	34%	0.9	1.7	27%	1.5	1.8	34	0.6	
2	3.11	34%	1.7	2.3	27%	2.7	3.0	65	0.6	
3	2.13	27%	0.9	1.2	23%	1.5	1.6	45	0.2	
4	1.53	34%	0.9	2.2	27%	1.5	2.0	32	1.0	
5	1.35	33%	0.9	2.7	27%	1.5	2.2	58	1.6	
6	1.13	55%	1.7	2.6	48%	2.7	3.0	37	0.5	
7	1.17	43%	1.3	2.1	36%	2.1	2.4	36	0.5	
8	2.00	16%	0.4	1.2	12%	0.6	1.0	32	0.5	
9	1.43	46%	1.3	2.1	39%	2.2	2.5	37	0.5	

to 34% to 46% for disk ages from 9 to 15 Gyr, is little dependent on IMF, but strongly on the SFR (55% for initial burst model). The local mass fraction of invisible degenerates is relatively smaller due to the inflation effect, and the total mass density amounts to about 5 $10^{-3}\,\mathcal{M}_\odot$/pc^3 for the standard model with $\chi = 1.65$. This is of the order of 5% of the total local mass, rather than the 1% of the dynamical mass estimated by Liebert $et\ al.$ (1988). The fraction of dark remnant matter in the local column amounts to typically 1.7 $\mathcal{M}_\odot$/pc^2, ie about half of the total remnant matter (including white dwarfs) or 3.5% of the revised dynamical value of 50 $\mathcal{M}_\odot$/pc^2. Within 10 pc we typically expect 34 degenerates, of which 9 are below log L/ $\mathcal{L}_\odot = -4.2$, and at most one neutron star: this prediction will be compared with observations in the next section.

We may shortly, and with larger uncertainty, estimate figures for the whole Galaxy. First, assuming an effective disk radius of 15 kpc we find from the range of our models a total degenerate mass of typically 2 10^9 $\mathcal{M}_\odot$ and an additional 5 $10^8\,\mathcal{M}_\odot$ in supernova remnants, or a total of 2.5 $10^9\,\mathcal{M}_\odot$ in stellar remnants. The corresponding numbers are 3 10^9 degenerates and 2.5 10^8 SN remnants, the latter figure is the more model dependent. This is reflected in the variation of the WD/SN birth rate ratio (see Table 1). The galactic WD birth rate is $\chi_{WD}/10^{-12} = 0.35$ yr^{-1}, for the standard model with $\chi = 1.65$. The

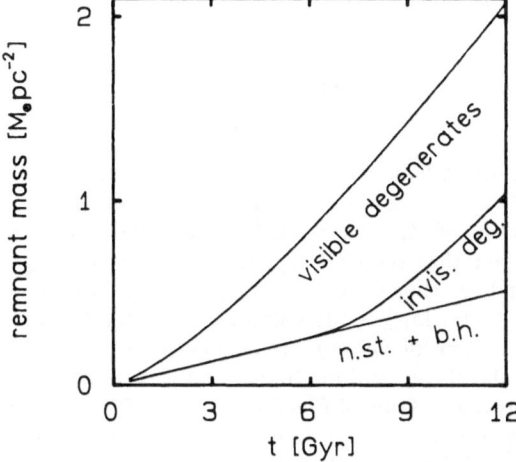

Figure 4. Remnant masses and dark matter surface densities as a function of time for Yuan's standard model.

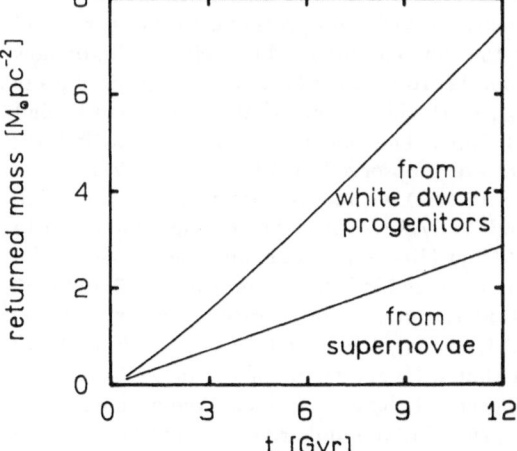

Figure 5. Mass returned to the interstellar medium as a function of time for the standard model adopted in figure 4.

corresponding rate for SN is $0.35/18 = 1/51$ yr^{-1}. Anticipating, that the best value for an effective total WD birth rate (including binaries) - see next section - is $\chi_{WD} = 2 \times 10^{-12}$ WD pc^{-3} yr^{-1} we obtain for the disk 0.5 WD/yr and SN intervals between 10 and 50 years, which covers well the range of earlier estimates. In looking at these figures one must keep in mind that we only considered an effective disk - a bulge, thick disk and halo model is not yet available. As far as the evolution during galactic history is concerned we show in Figure 4 how the invisible remnant mass increases (for column densities in the standard model): first there are only supernova remnants, white dwarfs do not contribute before the minimum cooling time to the visibility limit, *ie* after about 5 billion years. For the same model we also show in Figure 5 the mass returned to the interstellar medium, data which are important for models of galactic evolution which include the gas budget (*eg* infall, chemical yields) which we do not consider at this time.

7. A New Determination of the White Dwarf Birth Rate

We only shortly mention the method and results (which shall be published in detail elsewhere). The method goes back to Liebert (1978) who considered comparisons of observations and predictions of WD numbers as a function of M_{bol} –intervals for single white dwarfs in a restricted 10 pc volume. In view of the fact that the cool degenerates below 13 mag have been detected by large proper motions and are thus selected for high tangential velocities he applied a velocity correction factor of 2.6 for $M_{bol} > 13$, which he estimated from a comparison with the tangential velocity distribution for Gliese/McCormick (late K and M dwarfs) or Van de Kamp (< 5 pc) stars. For the theoretical predictions he used a 3 times larger birth rate (from Green, 1977) than the revised of one of Fleming *et al.* (1986). In the meantime many more cool degenerates have been found in the LHS ensembles, the results of which are discussed by Liebert *et al.* (1989) and Dahn *et al.* (1989). Although Liebert now feels that a velocity correction factor is not important, I repeated his 1978 study with new data and determined a velocity correction factor as a function of M_{bol} from a comparison with a predicted increase of the average tangential velocity proportional to the increase of scale height and the observed increase with the new ensemble. The velocity factor now increases from 1.3 for $M_{bol} = 13 - 14$, to 1.5 for 14–16 to 2 for $M_{bol} > 16$. Restricting my calculations to the more certain range of $M_{bol} < 15$ this implies that one expects to find another three low velocity degenerates within 10pc. The calculation then goes as follows: in the restricted (sky coverage) volume there are 9 observed or 12 expected WD within 10 pc, or, for the full sky (factor 1.79) N(10) = 16 or 21 without or with velocity correction. This yields an expected space density of 0.005 WD/pc². Since the cooling time down to $M_{bol} = 15$ ($\log L / \mathcal{L}_\odot = -4.1$) is between 4.2 10^9 yr (Koester and Schönberner, 1986) and 6 10^9 yr (Wood and Winget, 1989) the resulting average birth rate is between 1.25 and 0.83 10^{-12} WD/pc³. However our models show that the present birth rate is higher than the average over the past 5 Gyr by a factor of 1.14, so we get $\chi = (0.94 - 1.43)10^{-12}$ WD/pc³ yr for single white dwarfs only. It is agreed that a binary factor of the order 5/3 has to be applied in order to take into account the closer WD binaries (which have been excluded in the above 10 pc statistics. Jahreiss's (1987) cumulative number versus parallax-diagram clearly shows the deviations from completeness already beyond 5 pc (beyond 13 pc for single WD). Jahreiss estimates the binary factor to be 1.6 similar to the one for all stars within 5 pc. Applying this factor we obtain finally an (mass) effective WD birth rate of 1.5 to 2.3 10^{-12} WD/pc³/yr. These relatively higher values agree much better with the planetary nebula birth rate, considered to be (2.4 ± 0.3) PN/pc³/yr according to Phillips (1987), and see also Weidemann (1989).

8. Summary of Main Results and Problems

Recent observational determinations of the white dwarf luminosity function as well as theoretical modeling show that the turn-down of the LF at low luminosities does not yet allow a reliable determination of the age of the galactic disk. Even if the observed turn-off is considered to be real it is possible to fit the LF with models whose age is about 1 Gyr larger than the cooling age to $\log L/\mathcal{L}_\odot \approx -4.5$. The cooling theory thus definitely needs improvement.

As shown by galactic evolution models, the surface mass density of degenerates is between 2 and $5\,\mathcal{M}_\odot/\text{pc}^2$ depending on details of IMF, initial-final mass relation, SFR and disk age.

The total amount of local baryonic dark matter, cooled-down white dwarfs plus supernova remnants, is about $2\,\mathcal{M}_\odot/\text{pc}^2$ and thus contributes 4% to the local dynamical mass of $50\,\mathcal{M}_\odot/\text{pc}^2$.

The mass fraction of degenerates is 4-10% rather than the 1% estimate by Liebert et al. (1988).

The white dwarf birth rate is almost certainly higher than the established value of Fleming et al. (1986), about $1.2\ 10^{-12}$ WD/pc^3 for single and about $2\ 10^{-12}$ WD/pc^3 for all white dwarfs, including binaries.

Remaining issues which could influence and change these conclusions are:

the galactic disk age and the history of its formation, especially in connection with the halo, and the discrepancies between large globular cluster ages and disk ages;

the shape and the universality of the IMF (see Gilmore and Roberts, 1988) especially its low mass and high mass parts;

the time dependence of the SFR —increasing, decreasing, burst-like?

the stellar evolution parameters, especially the nuclear evolution time scale as a function of initial mass (and other parameters) and its dependence on overshooting;

the initial-final mass relation and limiting mass for white dwarf production, again dependent on main sequence core overshoot (see Weidemann, 1987b);

binary evolution towards its final stages (see eg Iben and Tutukov, 1987);

the evolution of scale heights for progenitors, white dwarfs, neutron stars and black holes;

a better observational determination of the cool end of the white dwarf luminosity function, especially in view of the fact that new degenerates have been recently detected (see Wegner (1989) and Ruiz et al. 1988) and that selection effects operate against low velocity white dwarfs.

References

Barry, D.C. (1988). *Astrophys. J.* **334**, 446.

Bienaymé, O., Robin, A.C., Crézé, M. (1987). *Astron. Astrophys.* **180**, 94.

Canal, R. 1990, in *Baryonic Dark Matter* eds D. Lynden-Bell and G.Gilmore (Kluwer: Dordrecht) p 103

Dahn, C.C., Hintzen, P.M., Liebert, J.W., Stockman, H.S., Spinrad, H. (1978). *Astrophys. J.* **219**, 979.

Dahn, C.C. *et al.* 1989, in *Wegner 1989, op cit.*

D'Antona, F. and Mazzitelli, I. (1986). *Astron. Astrophys.* **162**, 80.

D'Antona, F. and Mazzitelli, I. 1989 preprint

Eggen, O., Greenstein, J.L. (1965). *Astrophys. J.* **141**, 83.

Fleming, T.A., Liebert, J., Green, R.F. (1986). *Astrophys. J.* **308**, 176.

Gilmore, G. 1990, in *Baryonic Dark Matter* eds D. Lynden-Bell and G.Gilmore (Kluwer: Dordrecht) p137

Gilmore, G., Roberts, M.S. (1988). *Comments on Astrophys.* **12**, 123.

Graham, J.A. (1969) in *Low Luminosity Stars*, Ed. S.S. Kumar, Gordon and Breach, New York, p. 139.

Green, R.F. (1977), Ph. D. Thesis Calif. Inst. Technology, Pasadena

Greenstein, J.L. (1984). *Astrophys. J.* **276**, 602.

Heber, U. (1986). *Astron. Astrophys.* **155**, 33.

Iben, J., Jr. (1985). *Quart. J. Roy. Astron. Soc.* **26**, 1.

Iben, I., Jr., Renzini, A. (1983). *Ann. Rev. Astron. Astrophys.* **21**, 271.

Iben, I., Jr., Tutukov, A.V. (1984). *Astrophys. J.* **282**, 615.

Iben, I., Jr., Tutukov, A.V. (1987). *Astrophys. J.* **313**, 727.

Iben, I., Jr., McLaughlin, G. (1989). *Astrophys. J.* **341**, 312.

Jahreiss H. (1987). *Mem. Soc. Astr. Italia* **58**, 53.

Jones, E.M. (1972). *Astrophys. J.* **173**, 671.

Koester, D. (1972). *Astron. Astrophys.* **16**, 459.

Koester, D., Schulz, H., Weidemann, V. (1979). *Astron. Astrophys.* **76**, 262.

Koester, D., Schönberner, D. (1986). *Astron. Astrophys.* **154**, 125.

Kuijken, K., Gilmore, G. (1989). *Mon. Not. Roy. Astron. Soc.* **239**, 605.

Larson, R.B. (1986). *Mon. Not. Roy. Astron. Soc.* **218**, 409.

Liebert, J. (1978). *Astron. Astrophys.* **70**, 125.

Liebert, J., Dahn, C.C., Monet, D.G. (1988). *Astrophys. J.* **332**, 891.

Liebert, J., Dahn, C.C., Monet, D.G. 1989 in Lecture Notes in Physics **328**, Springer, p. 15

Lyne, A. 1990, in *Baryonic Dark Matter* eds D. Lynden-Bell and G.Gilmore (Kluwer: Dordrecht) p111

Mazzitelli, I., D'Antona, F. (1986). *Astrophys. J.* **308**, 706.

Mestel, L., Ruderman, M.A. (1967). *Mon. Not. Roy. Astron. Soc.* **136**, 513.

Miller, G.E., Scalo, J.M. (1979). *Astrophys. J. Suppl.* **41**, 513.

Noh, H.R., Scalo, J. (1989) preprint

Philip,A.G.D. D.S. Hayes and J. Liebert 1987 (editors) *The Second Conference on Faint Blue Stars. IAU Coll. 95* (L. Davis Press, Schenectady, 1987).

Phillips, J.P. (1989), in *Planetary Nebulae*, ed. S. Torres-Peimbert, Kluwer, p. 425

Ruiz, M.T., Maza, J., Mendez, R., Wischnjewsky, M. (1988), *ESO Messenger* No. 53, p36

Scalo, J. (1986). *Fundamentals of Cosmic Physics* **11**, 1.

Sion, E.M., Liebert, J.L. (1977). *Astrophys. J.* **213**, 468.

Sion, E.M., Fritz, M.L., McMullin, J.P., Lallo, M.D. (1988). *Astron. J.* ,, 96.251

Tassoul, M., Fontaine, G., Winget, D.E. (1989) preprint

Weidemann, V. (1967). *Zeitschrift Für Astrophysik* **67**, 286.

Wegner G. 1989 (editor) *White Dwarfs. IAU Coll. 114* Lecture Notes in Physics **328**, Springer, 1989.

Weidemann, V. (1975), in *Problems in Stellar Atmospheres and Envelopes*, eds. B. Baschek, W.H. Kegel, G. Traving, Springer, Berlin, p. 173.

Weidemann, V. (1979), in *White Dwarfs and Variable Degenerate Stars*, ed. H.M. Van Horn and V. Weidemann, Univ. Rochester, p. 206.

Weidemann, V. (1987a) *in The Second Conference on Faint Blue Stars op cit.* p. 19

Weidemann, V. (1987b). *Astron. Astrophys.* **188**, 74.

Weidemann, V., Yuan, J.W. in *Wegner 1989* p. 1

Weidemann, V. (1989). *Astron. Astrophys.* **213**, 155.

Weidemann, V., Koester, D. (1984). *Astron. Astrophys.* **132**, 195.

Winget, D.E., Van Horn, H.M. (1987) *in The Second Conference on Faint Blue Stars op cit.* p. 363

Winget, D.E., Hansen, C.J., Liebert, J., Van Horn, H.M., Fontaine, G., Nather, R.E., Kepler, S.O., Lamb, D.Q. (1987). *Astrophys. J.* **315**, L77.

Wood, M.A., Winget, D.E. (1989), in *Wegner 1989, op cit.* p. 282

Yuan, J.W. (1989). *Astron. Astrophys.* **224**, 108.

COOLING RATES AND NUMBERS OF FAINT WHITE DWARFS

R. CANAL
Departament d'Astronomia i Meteorologia
Universitat de Barcelona
08028 Barcelona
Spain

ABSTRACT. White dwarf cooling theory is reviewed with particular emphasis on the role of different phase diagrams for the C+O plasma at high densities. Theoretical luminosity functions for both disk and halo white dwarfs, based on those diagrams, are then described. We conclude that faint, still undetected white dwarfs are unlikely to be major contributors to baryonic dark matter, neither in the disk nor in the halo of the Galaxy.

1. White Dwarf Cooling

1.1. INTRODUCTION

The theory of white dwarf cooling is an essential ingredient in constructing theoretical white dwarf luminosity functions. Comparison of the last with observations can give us estimates of the ages of both the galactic disk and the galactic halo, as well as of the mass in the form of low-luminosity degenerate remnants that could still be hidden in those two components of our Galaxy.

The basis of white dwarf cooling theory was laid out by Mestel (1952). Later contributions include those of Schatzman (1953), Van Horn (1968), Koester (1972), Lamb and Van Horn (1975), Shaviv and Kovetz (1976), Kovetz and Shaviv (1976), Sweeney (1976), D'Antona and Mazzitelli (1978), and Iben and Tutukov (1984). Recently, other authors have revived the subject by discussing the influence of different phase diagrams for the Coulomb plasma on cooling times: Mochkovitch (1983), García-Berro et al. (1988a,b), Isern et al. (1989), and Mochkovitch et al. (1990). Following, for instance, García-Berro et al. (1988a), we have that the total (photon plus neutrino) luminosity of a cooling white dwarf is the sum of different contributions coming respectively from the release of thermal, gravitational, and nuclear energies:

$$L + L_\nu = L_{th} + L_{grav} + L_{nuc} \tag{1}$$

(where L stands for the photon luminosity). For $L \leq 10^{-1} \mathcal{L}_\odot$ both L_ν and $L_{nuc} \lesssim 0$ (the last if the white dwarf has a helium-rich envelope: see Iben and Tutukov 1984). Assuming uniform temperature throughout the whole interior:

$$L/\mathcal{L}_\odot = \mathcal{L}(T)m_{WD}/\mathcal{M}_\odot \tag{2}$$

D. Lynden-Bell and G. Gilmore (eds.), Baryonic Dark Matter, 103–110.
© 1990 *Kluwer Academic Publishers.*

(Van Horn 1968), and we can use, for $\mathcal{L}(T)$, a numerical fit to known internal temperature-luminosity relationships (where the physics of the envelope plays the main role).

As for the thermal contribution to luminosity, we have:

$$L_{th} = -\frac{dE_{th}}{dt} = -\left[\int_0^{m_{WD}} c_v dm\right]\frac{dT}{dt} + \frac{dm_{sol}}{dt}\ell \tag{3}$$

where E_{th} is the total thermal energy contents, c_v the specific heat at constant volume, and the last term corresponds to the release of latent heat at crystallization (ℓ is the specific latent heat and m_{sol} the solid mass).

1.2. PHASE TRANSITION

Crystallization of a Coulomb plasma is a first-order phase transition. It was predicted by Kirzhnits (1960), Abrikosov (1960), and Salpeter (1961). Its influence on the cooling rate was first considered by Mestel and Ruderman (1967) and later by the authors cited in the previous paragraph. Crystallization has been studied by Monte Carlo simulations starting with the work of Brush, Sahlin, and Teller (1966). Further simulations have been performed by Pollock and Hansen (1973), Slattery, Doolen, and DeWitt (1981), and Ogata and Ichimaru (1987). They concern the one-component plasma (OCP). The basic parameter is the plasma coupling constant Γ, defined as:

$$\Gamma = Z^2 e^2 / kTa \tag{4}$$

where $a = [3/(4\pi n)]^{1/3}$ is the ion-sphere radius: the radius of the sphere containing, on average, one ion. Phase transition takes place for $\Gamma \lesssim 155$, and the crystal structure should be a body-centered cubic lattice (bcc). The latent heat released is $\lesssim kT$ per ion (Pollock and Hansen 1973). We thus have:

$$\ell \lesssim \mathcal{R}T/\mu \tag{5}$$

with $\mathcal{R}$ the gas constant and μ the molecular weight of the ions. In the solid phase Debye's formula (Landau and Lifshitz 1958) should be used for c_v in the first term of (3). In the "Coulomb liquid" phase, the expression given by Hansen, Torrie, and Vieillefosse (1977) can be adopted. The mass in solid phase in the second term of (3) varies according to:

$$\frac{dm_{sol}}{dt} = \frac{dm_{sol}}{dT}\frac{dT}{dt} \tag{6}$$

If white dwarf interiors could be adequately modelled as OCPs, this would pose no further problem. Most white dwarfs, however, are C+O white dwarfs (Iben and Tutukov 1985). Their interiors are not OCPs, but binary ion mixtures (BIM) (if one considers only the two most abundant ion species). The problem of what phase diagram applies to a C+O BIM thus arises. And more in particular: are carbon and oxygen miscible in any proportion, both in fluid and in solid phase? The first part of the problem was studied by Stevenson (1976), who found that carbon and oxygen were indeed completely miscible in fluid phase.

The second part was first addressed by Loumos and Hubbard (1973) by means of a direct Monte Carlo simulation. Their result suggested that carbon and oxygen were also entirely miscible in solid phase and that the phase diagram could be approximated by a simple generalization of the OCP case. Crystallization would occur for $\overline{Z^{5/3}}\Gamma_0 = 155$, with $\overline{Z^{5/3}} = \sum_i x_i Z_i^{5/3}$ (x_i and Z_i respectively being the fraction by number and charge of carbon and oxygen) and Γ_0 defined as in (4) but with n_0 (number density of electrons) now replacing n (number density of ions) (Pollock and Hansen 1973). For $X_c = X_0 = 0.50$ (X_i being mass fraction), this gives $T/T_c = 1.26$ at crystallization, where $T_c = 2.3 \times 10^4 \rho^{1/3}$ is the freezing temperature of pure carbon. $m_{sol}(T)$, for a 0.6 $\mathcal{M}_\odot$ white dwarf, can be seen in Figure 1 of García-Berro et al. (1988a). In this case, the crystallization process makes no contribution to L_{grav} in equation (1).

A completely opposite phase diagram was later predicted by Stevenson (1980). In his model he assumed ideal entropy of mixing in both fluid and solid phases plus the electrostatic energy of a random alloy for the crystal. He obtained an eutectic phase diagram (Figure 2 in Stevenson 1980), with a temperature minimum for $X_c^0 \approx 0.6$. The diagram can be approximated by:

$$T/T_c = 0.925 X_c^l + 0.075 \ [X_c^l \geq X_c^0]$$
$$T/T_c = 1.615 - 1.642 X_c^l \ [X_c^l \leq X_c^0] \tag{7}$$

where X_c^l is carbon mass fraction in the fluid phase (Mochkovitch 1983). Carbon and oxygen would thus be inmiscible in solid phase (but maybe for the eutectic proportion given by X_c^0). Oxygen "snowflakes" would first form for $X_c = X_0 = 0.50$. Being denser than the remaining fluid, they should fall to the center and a solid, pure oxygen core would start to grow. When $X_c \geq X_c^l$, carbon crystals would also form. Being less dense than the surrounding fluid, they should move towards the surface, where they would melt again. "Salt finger"-like instabilities should rehomogenize the fluid (Stevenson 1980; Mochkovitch 1983). Depending on the details of the process, two outcomes might be envisaged:

i) Formation of a central solid O core ($m = 1/6 m_{WD}$) surrounded by also solid C+O layers with the eutectic composition, plus external fluid C+O layers;

ii) Central solid O core ($m = 1/2 m_{WD}$), plus intermediate solid C layers, plus external solid layers. Expressions for $m_{sol}(T)$ in both cases are given in García-Berro et al. (1988a).

1.3. GRAVITATIONAL CONTRIBUTION TO THE LUMINOSITY

When complete miscibility in the solid phase is assumed, $L_{grav}/L_{th} \sim kT/E_F \ll 1$, E_F being the Fermi energy (Lamb and Van Horn 1975). This is no longer the case if carbon and oxygen separate when crystallizing. The binding energy of the white dwarf then changes by an amount that is proportional to the mass of the oxygen core and we have:

$$L_{grav} = e_{grav} \frac{dm_{ox}}{dt} \tag{8}$$

e_{grav} can be evaluated in two different ways: i) By comparing the binding energies of white dwarf models with different oxygen core masses (Mochkovitch 1983); ii) By treating chemical separation as a local change in composition (García-Berro et al. 1988a). Both methods give $e_{grav} \sim 10^{14}$ erg g^{-1}. The significance of this contribution can be easily seen by comparing it to the latent heat for oxygen: $\ell \lesssim 5 \times 10^{12}(T/10^6 K)$ erg g^{-1}. The total energy to be radiated is: $E_{grav} = m_{WD} X_0 e_{grav}$. It thus depends on the oxygen mass fraction X_0 and it is maximum for $X_0 = X_c = 0.50$. In this case $E_{grav} \lesssim 6 \times 10^{46}$ erg. Previous stratification of the chemical composition would change those amounts. Depending on progenitor star mass and also on the still uncertain rate of the $^{12}C(\alpha, \gamma)^{16}O$ reaction plus the amount of overshooting in convective cores, Mazzitelli and D'Antona (1986, 1987) do indeed predict nonhomogeneous chemical profiles for C+O white dwarfs: the central layers would be more oxygen-rich than the outer ones. E_{grav} would then be lower. As a first approximation, however, one can still assume chemically homogeneous (fluid) white dwarfs with $X_c = X_0 = 0.50$ (Isern et al. 1990).

The evolution of the luminosity of a cooling white dwarf of $0.6\,\mathcal{M}_\odot$ has been calculated, with the last simplifying assumptions, by García-Berro et al. (1988a,b), for three different hypotheses as to the phase diagram of the C+O plasma: complete miscibility in solid phase, partial miscibility (with the eutectic composition), and complete inmiscibility. In the first case, their results agree with those of Iben and Tutukov (1984): crystallization (without any chemical differentiation) introduces some time delay in the cooling process (at $L \lesssim 10^{-3.5}\,\mathcal{L}_\odot$), that afterwards accelerates when reaching the stage of Debye cooling. Taking this case as a reference, deposition of oxygen at the center delays the cooling process by $\Delta t \lesssim 1.6$ Gyr per $0.1\,\mathcal{M}_\odot$ of deposited oxygen. Thus, in the second case (limited miscibility) chemical separation would be completed in 5.9 Gyr, while in the last case (total inmiscibility) it would take $\lesssim 10$ Gyr.

The preceding could have important implications as to the age of the galactic disk. If one interprets the paucity of white dwarfs with luminosities $L < 10^{-4}\,\mathcal{L}_\odot$ as due to the finite age of the galactic disk (Winget et al. 1987), the first case corresponds to their estimate of $t_D = 9.3$ Gyr. The second case, however, would imply an age $t_D \lesssim 10$ Gyr, while the third one would mean that $t_D \geq 15$ Gyr (García-Berro et al. 1988a,b). We now see the importance of the phase diagram for C+O mixtures in constructing theoretical luminosity functions. There have been more recent developments in this field but the problem, as we will see, is not yet completely settled.

Barrat, Hansen, and Mochkovitch (1988) have used a density functional approach to the free energy to calculate the diagram. They find a diagram of the *spindle* form: the solid phase is only slightly more oxygen-rich than the fluid one. Thus, after complete crystallization the total gravitational energy release would only be: $E_{grav} \lesssim 1.86 \times 10^{46}$ erg (that is $\lesssim 1/3$ of the amount obtained from total separation of oxygen and carbon). But since crystallization would start at higher temperatures and thus at higher luminosities ($\langle L \rangle \lesssim 2 - 3 \times 10^{-4}\,\mathcal{L}_\odot$ during crystallization) and the time delay introduced by the separation process is $\Delta t \lesssim E_{grav}/\langle L \rangle$, now $\Delta t \lesssim 0.5 - 0.75$ Gyr only. Transition to an ordered alloy with the structure of ClCs (a bcc lattice with carbon and oxygen ions forming simple cubic sublattices: see Dyson 1971) is predicted at lower temperatures and that would mean a further energy release. But the whole calculation involves a somewhat drastic simplification and the last prediction is only qualitative.

Independently, Ichimaru, Iyetomi, and Ogata (1988) have shown by Monte Carlo simulations that *linear mixing* formulae are more accurate than the *random mixing* hypothesis in calculating the free energies of solid alloys. They deduce from it an *azeotropic* phase diagram. Again, carbon and oxygen are partially miscible in solid phase and the solid is more oxygen rich than the fluid. Crystallization happens at slightly lower temperatures than in the phase diagram of Barrat, Hansen, and Mochkovitch (1988) and thus Dt is about 50% larger.

All the preceding concerns BIM only. In the (mainly) C+O plasma of white dwarf interiors, less abundant species can also play a noticeable role. ^{22}Ne, for instance, amounts to 1-2% of the total. Its sedimentation would induce an extra delay in the cooling process, Δt being $\lesssim 2$ Gyr (Isern et al. 1990).

The whole problem of chemical composition changes with crystallization can hardly be considered as solved. Godon et al. (1989) have also recently approached it with solid state techniques (linear Muffin-Tin orbitals) and they conclude that, within ~ 1% accuracy, chemical separation requires no energy and thus that the question as to its occurrence cannot be reliably answered.

2. Luminosity Functions

2.1 THE GALACTIC DISK

From the cooling curves corresponding to a given phase diagram for the C+O plasma, theoretical luminosity functions can be derived. Comparison between the luminosity functions obtained from complete miscibility in solid phase and for either partial miscibility or total immiscibility, the last two with Stevenson's (1980) phase diagram, has been done by García-Berro et al. (1988a,b). They construct the luminosity functions in an approximative way, assuming that all white dwarfs have the same mass ($0.6\,\mathcal{M}_\odot$) plus either a constant white dwarf birth rate ($\nu_{\rm WD} = 0.5$ yr^{-1}) or a simplified galactic evolution model. They obtain reasonable fits to the observed luminosity function (Liebert, Dahn, and Monet 1988) even in the case of complete separation of carbon and oxygen at crystallization. In this case, the long delay introduced in the cooling process produces a "piling up" at luminosities close to the apparent low-luminosity cutoff if all white dwarfs are assumed to be distributed with the same scale height above the galactic plane. The amplitude of such a "bump" can nonetheless be significantly reduced by taking into account that old disk population objects have larger scale heights over the galactic plane than young disk population ones. Even in the case of a bimodal initial mass function (Larson 1986; Wyse and Silk 1987), the local mass density of white dwarfs dimmer than $L \lesssim 10^{-4}\,\mathcal{L}_\odot$ deduced is only: $\rho_{\rm WD}(L < 10^{-4}\,\mathcal{L}_\odot) \lesssim 0.006\,\mathcal{M}_\odot$ pc^{-3}, much less than the local density of dark matter inferred by Bahcall (1984) from the vertical distribution of F stars: $\rho_{\rm DM} \lesssim 0.1\,\mathcal{M}_\odot$ pc^{-3}.

A similar study of the dependence of the white dwarf luminosity function on the age of the galactic disk has recently been performed by Iben and Laughlin (1989), taking into account the main sequence lifetimes of the white dwarf progenitors. They adopt a Salpeter-like initial mass function and the cooling curves of Iben and Tutukov (1984), that do not assume any compositional change at crystallization. Iben and Laughlin (1989) do not predict a simple abrupt falloff after the space density maximum but a "shoulder" in the tail of the luminosity function. Such a "shoulder", however, seems to result from an incorrect numerical fit to the cooling of the 1 $\mathcal{M}_\odot$ white dwarf (Mochkovitch et al. 1989). The space density of white dwarfs inferred by Iben and Laughlin (1989) would thus be an overestimate.

The dependence of the theoretical space density of disk white dwarfs beyond the present observational cutoff on galactic evolutionary model characteristics is further discussed by Yuan (1989, see also Weidemann 1990).

2.2. THE GALACTIC HALO

The recent discovery of six white dwarfs with tangential velocities larger than 250 km s^{-1} (Liebert, Dahn, and Monet 1989) has allowed the construction of a very preliminary luminosity function for halo white dwarfs. The local space density of white dwarfs with luminosities $L > 10^{-4.35}\mathcal{L}_\odot(M_v < 16)$ would be $1.3(\pm0.6) \times 10^{-5}$ pc^{-3}. Again for an average white dwarf mass of $0.6\,\mathcal{M}_\odot$, we would have: $\rho_{WD}(L > 10^{-4.35}\mathcal{L}_\odot) \lesssim 8 \times 10^{-6}\,\mathcal{M}_\odot$ pc^{-3}. But since the galactic halo might be significantly older than the disk, the luminosity function of halo white dwarfs might continue to grow well beyond the present detection limit and the contribution of those very faint stars to dark matter in the halo might thus be much larger than the above value. This problem has been studied by Mochkovitch et al. (1989). They use several initial mass functions (appropriate for a burst of star formation) compatible with the observed segment of the halo white dwarf luminosity function, and they consider halo ages in the range 10 Gyr $\leq t_H \leq$ 16 Gyr. For the cooling curves they adopt the results of Barrat, Hansen, and Mochkovitch (1988). Mochkovitch et al. (1989) find, indeed, that the corresponding luminosity functions peak at lower luminosities than that of the disk (assumed to have an age of 9 Gyr). But the inferred mass density of low-luminosity halo white dwarfs is: $\rho_{WD}(L < 10^{-4.5}\mathcal{L}_\odot) \sim$ a few times $10^{-5}\,\mathcal{M}_\odot$ pc^{-3} only, in all cases. This is more than a factor of 1000 below the estimated local density of dark matter: $0.07\,\mathcal{M}_\odot$pc^{-3}. Only very peculiar initial mass functions, with a sharp low-mass cutoff at $M \leq 2\,\mathcal{M}_\odot$ and forming essentially intermediate-mass stars (Tamanaha *et al.* 1989) might account for such a density without overproducing white dwarfs with $L \geq 10^{-4.5}\mathcal{L}_\odot$ at the same time (see also Silk 1990).

Mochkovitch *et al.* (1989) have equally computed discovery functions for halo white dwarfs. They predict that deep surveys in the near infrared (up to $m_K = 20$) should allow the construction of the halo luminosity function up to the peak, even for an age t = 14 to 16 Gyr.

3. Summary

A fit of theoretical white dwarf luminosity functions to observed ones, both for the disk and the halo, should allow the deduction of the respective ages of these components of the Galaxy and also put limits on the contribution of faint, still undetected white dwarfs to dark matter.

Theoretical luminosity functions strongly depend on the rate of white dwarf cooling. There are two main physical problems involved in this process: mass and composition of the envelopes (which presently give uncertainties of at least 1 Gyr in the cooling times), and the phase diagram of the high density interiors (much larger uncertainties: up to 10 Gyr if chemical separation of carbon and oxygen are discarded).

Even within the preceding uncertainties, white dwarfs do not seem to make a significant contribution to dark matter neither in the disk nor in the halo of our galaxy. Deep surveys in the red and in the infrared should, in principle, allow determination of the halo white dwarf luminosity function up to its theoretical peak and thus settle the question of whether an initial mass function very different from that of the disk may have given rise to a significant population of very faint white dwarfs there.

It is a pleasure to thank Prof. V. Weidemann for enlightening talks during this meeting. This work has been supported in part by DGICYT Grant No. PB87-0147-C02-01.

4. References

Abrikosov, A.A. 1960 *Soviet Phys. JETP* **12** 1254.

Bahcall, J.N. 1984. *Astrophys. J.* **276**, 169.

Barrat, J.L., Hansen, J.P., and Mochkovitch, R. 1988. *Astron. Astrophys.* **199**, L15.

Brush, S.G., Sahlin, H.L., and Teller, E. 1966, *J. Chem. Phys.* **45**, 2102.

D'Antona, F., and Mazzitelli, I. 1978. *Astron. Astrophys.* **66**, 453.

Dyson, F. 1971. *Ann. Phys.* **63**, 1.

García-Berro, E., Hernanz, M., Mochkovitch, R., and Isern, J. 1988a. *Astron. Astrophys.* **193**, 141.

García-Berro, E., Hernanz, M., Isern, J., and Mochkovitch, R. 1988b. *Nature* **333**, 644.

Godon, P., Shaviv, G., Ashkenazi, J., and Kovetz, A. 1989, in *White Dwarfs*, ed. G. Wegner (Springer Verlag, Berlin), p. 85

Hansen, J.P., Torrie, G.M., and Vieillefosse, G.P. 1977. *Phys. Rev. A* **16**, 2153.

Iben, I., and Laughlin, G. 1989. *Astrophys. J.* **341**, 312.

Iben, I., and Tutukov, A.V. 1984. *Astrophys. J.* **282**, 615.

Iben, I., and Tutukov, A.V. 1985. *Astrophys. J. Suppl.* **58**, 661.

Ichimaru, S., Iyetomi, H., and Ogata, S. 1988. *Astrophys. J. Lett.* **334**, L17.

Isern, J., García-Berro, E., Hernanz, M., and Mochkovitch, R. 1989, in *White Dwarfs*, ed. G. Wegner (Springer Verlag, Berlin), p. 278

Isern, J., Garc 1a-Berro, E., Hernanz, M., and Mochkovitch, R. 1990, in *New Windows to the Universe*, ed. M. Vázquez (Cambridge Univ. Press, Cambridge), in press

Kirzhnits, D.A. 1960, *Soviet Phys. JETP* **11**, 365.

Koester, D. 1972. *Astron. Astrophys.* **16**, 459.

Kovetz, A., and Shaviv, G. 1976. *Astron. Astrophys.* **52**, 403.

Lamb, D.Q., and Van Horn, H.M. 1975. *Astrophys. J.* **200**, 306.

Landau, L.D., and Lifshitz, E.M. 1958, *Statistical Physics* (Pergamon, Oxford).

Larson, R.B. 1986. *Mon. Not. Roy. Astron. Soc.* **218**, 409.

Liebert, J., Dahn, C.C., and Monet, D.G. 1988. *Astrophys. J.* **332**, 891.

Liebert, J., Dahn, C.C., and Monet, D.G. 1989, in *White Dwarfs*, ed. G. Wegner (Springer Verlag, Berlin), p. 891

Loumos, G.L., and Hubbard, W.B. 1973. *Astrophys. J.* **180**, 199.

Mazzitelli, I., and D'Antona, F. 1986. *Astrophys. J.* **308**, 706.

Mazzitelli, I., and D'Antona, F. 1987, in *The Second Conference on Faint Blue Stars*, ed. A.G.D. Philip, D.S. Hayes, and J.W. Liebert (Davis Press, Schenectady), p. 351

Mestel, L. 1952. *Mon. Not. Roy. Astron. Soc.* **112**, 583.

Mestel, L., and Ruderman, M.A. 1967. *Mon. Not. Roy. Astron. Soc.* **136**, 27.

Mochkovitch, R. 1983. *Astron. Astrophys.* **122**, 212.

Mochkovitch, R., García-Berro, E., Hernanz, M., Isern, J., and Panis, J.F. 1990, preprint.

Ogata, S., and Ichimaru, S. 1987. *Phys. Rev. A* **36**, 5451.

Pollock, E.L., and Hansen, J.P. 1973. *Phys. Rev. A* **8**, 3110.

Salpeter, E.E. 1961. *Astrophys. J.* **134**, 669.

Schatzman, E. 1953, *Ann. D'Astrophs.* **16**, 162.

Shaviv, G., and Kovetz, A. 1976. *Astron. Astrophys.* **51**, 383.

Silk, J. 1990, in *Baryonic Dark Matter* eds D. Lynden-Bell and G. Gilmore (Kluwer: Dordrecht) p279

Slattery, W.L., Doolen, G.D., and DeWitt, H.E. 1982. *Phys. Rev. A* **21**, 2087.

Stevenson, D.J. 1976, *Phys. Lett* **58A**, 282.

Stevenson, D.J. 1980, *J. Phys. Suppl.* No 3, 41, C2-61.

Sweeney, M.A. 1976. *Astron. Astrophys.* **49**, 375.

Tamanaha, F., Silk, J., Wood, M.A., and Winget, D.E. 1989, in preparation.

Van Horn, H.M. 1968. *Astrophys. J.* **151**, 227.

Weidemann, V. 1990, in *Baryonic Dark Matter* eds D. Lynden-Bell and G. Gilmore (Kluwer: Dordrecht) p87

Winget, D.E., Hansen, C.J., Liebert, J., Van Horn, H.M., Fontaine, G., Nather, R.E., Kepler, S.O., and Lamb, D.Q. 1987. *Astrophys. J. Lett.* **315**, L77.

Wyse, R.F.G., and Silk, J. 1987. *Astrophys. J. Lett.* **313**, L11.

Yuan, J.W. 1989. *Astron. Astrophys.* **224**, 108.

THE GALACTIC DISTRIBUTION OF NEUTRON STARS

A. G. Lyne
University of Manchester
Nuffield Radio Astronomy Laboratories
Jodrell Bank
Macclesfield, Cheshire SK11 9DL
United Kingdom

ABSTRACT. Pulsars provide us with a brief glimpse of neutron stars immediately following their formation. It is believed that they are mostly formed in the collapse of massive stars and reveal themselves as pulsars for perhaps a few million years before they fade into obscurity, unless they are given a further burst of life by accretion of external material. In this case we may see them as low-mass X-ray binary systems and then millisecond pulsars or else we may see them as γ-ray bursters. During their short pulsar lifetime they stand up to be counted and we can establish the rate at which they are formed and their kinematics, allowing a prediction of their future galactic distribution and space density. In this paper, we summarise our knowledge of the pulsar population and extrapolate this to give a picture of the population of dead pulsars.

1. The Galactic Distribution of Radio Pulsars

The galactic distribution of pulsars has been the subject of a number of papers during the past 20 years. Here I mostly rely on the results of the work by Lyne, Manchester and Taylor (1985). This is based upon a statistical study of nearly 400 pulsars detected in the major surveys at Jodrell Bank, Arecibo, Greenbank and Molonglo. These were all conducted at frequencies around 400 MHz and had well defined selection effects. The observed population of about 400 represents only about 10^{-3} of the active pulsars in the Galaxy, primarily because of the inverse square law and other selection effects which permit much better sampling of the Solar neighbourhood than the rest of the Galaxy. However as we shall see, the extended form of the pulsar luminosity function allows us to study the population throughout most of the galactic plane.

It is valuable to examine the pulsar population as a function of three variables: luminosity, galactocentric radius and galactic z-distance. These can be calculated for each pulsar because estimates of distance are possible through the dispersion measures and a model of the galactic electron distribution. I do not here describe the analysis technique whereby the true or corrected distributions of pulsars are derived from the observed distributions. The results of this analysis are shown in figures 1-3.

The luminosity function is closely power-law with a slope of -1 spanning about 4 decades of luminosity (Fig.1). There seems to be a cut-off in luminosity below about 1 mJy kpc^2. The distribution in galactocentric radius (Fig.2) shows a falling density of pulsars outside the solar circle, with an increase towards the galactic centre. From the data used in this

111

D. Lynden-Bell and G. Gilmore (eds.), Baryonic Dark Matter, 111–115.
© 1990 *Kluwer Academic Publishers.*

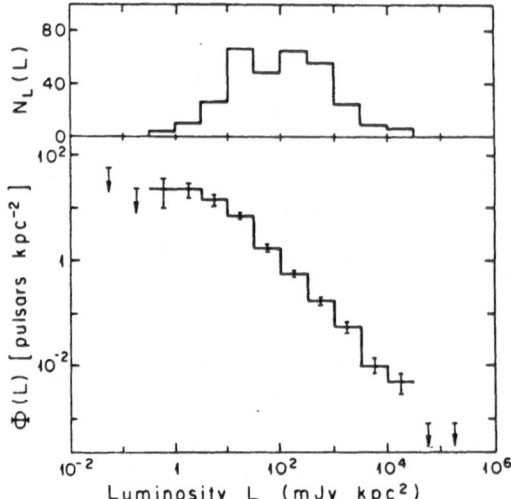

Figure 1: The observed distribution of pulsar luminosity is shown above the distribution corrected for selection effects.

analysis, there is much uncertainty about the density within about 5 kpc of the galactic centre. This is because of the large distances and because the high galactic noise background at 400 MHz in this direction greatly impairs the sensitivity of the receivers. A recent survey at the much higher frequency of 1400 MHz (Clifton and Lyne 1985) has penetrated this region of the Galaxy and shows that the density falls inside about 4 kpc of the galactic centre. The radial distribution is very similar to that of most population I stellar species, supernova remnants and HII regions. This provides good circumstantial evidence that pulsars are formed in the supernova events of massive stars which are formed in HII regions. This hypothesis is of course also supported by the association of young pulsars with the Crab and Vela supernova remnants.

On the other hand, the distribution in galactic z-distance (Fig.3) shows that pulsars have a scale height of 400 pc, about 5 times that of most population I species. At first sight, this is most surprising. However, Gunn and Ostriker (1969) suggested that this may arise from high velocities imparted to the pulsars at birth in the supernova explosions. This has subsequently been confirmed by proper motion measurements using a sub-set of the MERLIN array (Lyne, Anderson and Salter, 1982) which indicate that pulsars have high space velocities, typically 200 km sec^{-1}. Moreover they are mostly moving away from the galactic plane (Fig.4) as one would expect if they were given kicks in random directions by birth events near the galactic plane.

2. The Evolution of Pulsars

Since pulsars do not suffer significant deceleration by the galactic gravitational field during their lifetimes, extrapolation of their paths back onto the galactic plane allows a direct

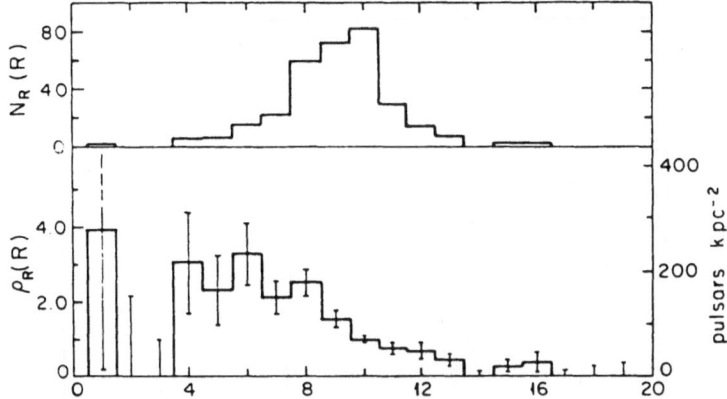

Figure 2: The observed galactic radial distribution of pulsars is shown above the corrected distribution. The Sun is assumed to be at a radius of 10kpc. Note the large uncertainty in density in the inner regions of the Galaxy.

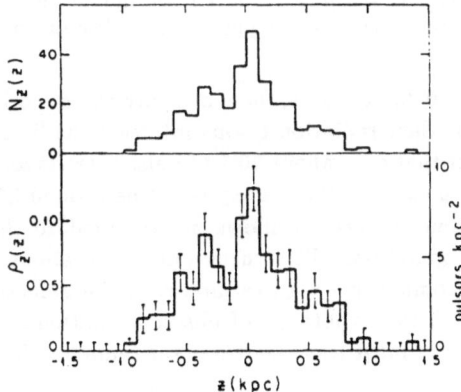

Figure 3: The observed galactic z-distance distribution of pulsars is shown above the corrected distribution. The curve represents the predicted form of this, based upon the observed luminosity evolution and space velocity distribution.

estimate of their ages. The comparison of these ages with the characteristic, or slow-down, ages ($\tau_c = P/2\dot{P}$) indicates that for young pulsars, there is reasonable agreement, but for older objects, the characteristic ages are too large. This can be accounted for by a decay of the effective magnetic dipole moments responsible for the slow-down. This seems to occur on a timescale of perhaps 5-10 million years. The luminosity of pulsars also decays, on about half this timescale. The combination of the high velocities and luminosity decay are responsible for the observed scale-height seen in Fig. 3. The smooth curve in this figure is the result of a modelling of this evolution, providing a satisfactory independent check of the model.

The total number of pulsars in the Galaxy with luminosity of greater than 1 mJy kpc^{-2}

114

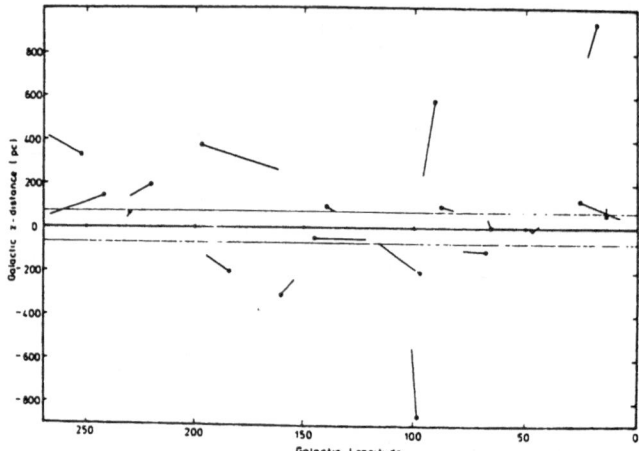

Figure 4: The observed velocities of pulsars relative to the galactic plane. The pulsar positions are shown as filled circles and the lines represent the approximate distance travelled by the pulsars in the last one million years.

is calculated to be about 50,000. However, these are only the minority of pulsars which are oriented in such a way that their radiation beams intersect the lines-of-sight to the Earth. Pulse-width studies suggest that only about 20% of pulsars are favourably oriented, so that the total number of active pulsars in the Galaxy could be around 250,000.

Knowing how the luminosity decays allows us to calculate the birthrate of pulsars required to sustain this population. This turns out to be about one new pulsar every 20-50 years. This rate is comparable with estimates of the galactic supernova rate, and is reasonably consistent with the hypothesis of pulsar formation in supernovae. However, most type II supernova events must leave neutron star remnants for the figures to match.

3. The Galactic Distribution of Dead Pulsars

What will become of these pulsars? A minority of them (perhaps 10%) are likely to have large enough velocities to escape from the Galaxy altogether, the escape velocity being about 300 km sec^{-1} relative to the local standard of rest in the Solar neighbourhood. This is somewhat dependent upon the direction of the velocity because of the Sun's galactic orbital motion. The majority will continue their present motion away from the galactic plane and rise to heights of a few kpc above or below the plane in about 30-60 million years, before galactic gravitation pulls them back. By this time, of course, they will have ceased emitting any significant amount of radio emission and are essentially dead.

If the current rate of galactic pulsar production is typical of that throughout the lifetime of the Galaxy, then a galactic halo containing about 3×10^8 dead neutron stars will have accumulated during this time. Since the formation rate of massive stars was likely to have been substantially greater in earlier times, the total number of dead neutron stars is likely to lie somewhere between 3×10^8 and 3×10^9.

We can also estimate the local space density of such objects. The local birthrate of active pulsars is about 30 kpc^{-2} My^{-1}, this being the rate per square kpc, integrated over z-distance. This leads to a total local population of dead pulsars of around 3x10^5 kpc^{-2}, spread over a scale height of $\pm$ 4 kpc. Thus the mean space density is around 40,000 kpc^{-3}. However, low-velocity objects will accumulate at low z-distance. Since about 30% of pulsars are seen to have z-components of velocity of less than about 30 km sec^{-1}, the mean space density in the galactic plane near to the Sun is probably about 100,000 kpc^{-3}. Clearly, this density of matter is several orders of magnitude less than that of visible matter and the contribution of dead pulsars to the missing mass must be small.

Once ordinary radio pulsars have died, there are two possible ways in which they may be seen again. Firstly, as millisecond pulsars, if they remained in binary systems which were not disrupted by the supernova events of formation. Spin-up to millisecond periods can then occur if the companion star overflows its Roche lobe and the magnetic field of the pulsar has decayed sufficiently (Alpar et al. 1982). It is believed that this spin-up is observed in the low-mass X-ray binary systems in which the accreting material emits copious X-rays. The rapid rotation rate will give rise to enhanced electrodynamic forces and produce radio emission again. Such objects are likely to live for a substantial fraction of the galactic lifetime. The second manifestation of the existence of dead neutron stars may be as γ-ray bursters, in which we possibly see the surface conflagration of accreted interstellar material. These bursters are likely to have the same halo-like distribution as that of the dead neutron stars.

References

Alpar, A., Cheng, A.F., Ruderman, M.A. and Shaham, J., 1982. *Nature* 300, 728.
Clifton, T.R. and Lyne, A.G., 1986. *Nature* 320, 43.
Gunn, J.E. and Ostriker, J.P., 1970. *Astrophys. J.* 160, 9.
Lyne, A.G., Anderson, B. and Salter, M.J., 1982. *Mon. Not. Roy. Astr. Soc.* 201, 503.
Lyne, A.G., Manchester, R.N. and Taylor, J.H., 1985. *Mon. Not. Roy. Astr. Soc.* 213, 613.

WIDE BINARIES AND MASS LIMITS ON DARK MATTER

M. D. Weinberg
Institute for Advanced Study
Princeton, NJ 08540
USA

ABSTRACT. I review the data and theoretical arguments relevant for mass and density limits on unseen objects in the Galactic disk. The physical process of wide binary evolution is described and illustrated for simple binary star birthrate models. Owing to selection effects, the physics of the evolving distribution, and the paucity of well-selected data, I unfortunately conclude that no limits on the mass of dark objects are yet possible. Finally, I present some results from a recent analysis of the Woolley catalog. We find 9 pairs with $4.5 < M_V < 9$ and separations $0.01\,\mathrm{pc} < s < 1\,\mathrm{pc}$. Although the detection of these 9 pairs provides an interesting constraint on the binary birthrate function, still no useful limits on the dark matter are obtained. In order to investigate the important widely separated pairs $\sim 0.1\,\mathrm{pc}$, an order of magnitude more data may be necessary.

1. Introduction

With a precise description of the wide binary population in the solar neighborhood, it may be possible to limit the density and masses of non-luminous objects that can not be otherwise detected.

The underlying physical principle is easy to understand. For example, take a wide binary system with a separation of $a = 0.1\,\mathrm{pc}$. For two stars in circular orbit with combined mass of $1\,\mathcal{M}_\odot$, the relative velocity will be $\approx 0.2\,\mathrm{km\,s^{-1}}$. Let us now imagine a disk population of perturbers (stars or putative black holes, if you will). This situation is shown schematically in the top half of Figure 1. Since the perturbers encounter the binary at a speed ($\approx 20\,\mathrm{km\,s^{-1}}$) roughly two orders of magnitude higher than the binary's relative velocity, the binary will tend to gain energy and expand. Eventually, the pair disrupts, its separation too large to remain bound in the presence of the Galactic tide. Since the evolution is rate dependent, the larger the perturber density and mass (for sufficiently low mass, see §4), the shorter the disruption lifetime. Thus, the distribution of wide binary separations holds a clue to the population of perturbers; the more perturbers, the fewer wide binary stars.

How does one infer the mass of such unseen objects from the distribution of wide binary semimajor axes? First, one needs to know the rate of binary formation per unit volume and semimajor axis or, in other words, the binary birth rate function (BBF). Secondly, one needs to quantify the known perturbers, their mass, density, velocity dispersion, size, etc. If

117

D. Lynden-Bell and G. Gilmore (eds.), Baryonic Dark Matter, 117–136.
© 1990 *Kluwer Academic Publishers.*

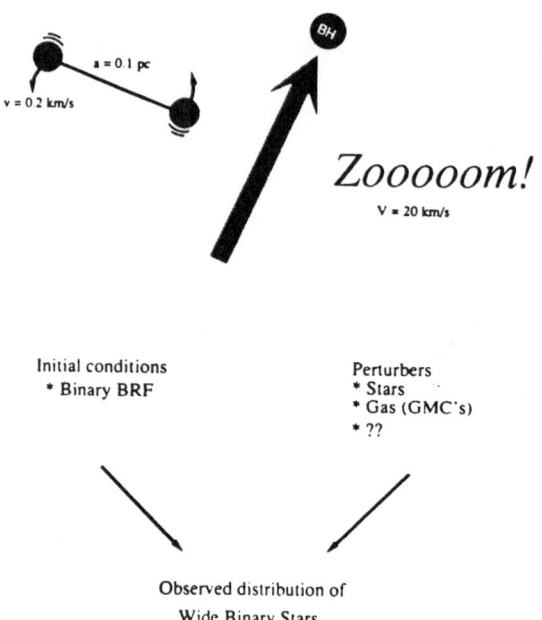

Initial conditions
* Binary BRF

Perturbers
* Stars
* Gas (GMC's)
* ??

Observed distribution of
Wide Binary Stars

Figure 1. The top part of the figure depicts an encounter between a binary system and a perturber with fiducial parameters marked. The bottom indicates the philosophy behind the approach: if the initial conditions and species of perturbers are known, the present-day distribution of wide binaries may be inferred. Conversely, given the observed distribution, the initial conditions and perturber populations may be constrained.

we knew the properties of *all* perturbers, their effects on the population of binaries could be computed directly and the distribution of binary pairs today could be predicted (see lower half of Figure 1). In reality, the situation is more complicated. Our knowledge of possible perturbers is incomplete, although we have a good description of the stellar distribution and some understanding of the gas content in the solar neighborhood. We know very little about the BBF except some overall constraints based on the average stellar birthrate. However, there is some information on the binary distribution. Our goal is to constrain the unknown initial conditions and possible perturbers using the observed distribution of binaries and theoretical models of their possible evolution.

This field has a short history. Similar arguments were first applied to the dark matter dilemma by Bahcall, Hut, and Tremaine (BHT, 1985). Based on the findings of Bahcall & Soneira (B&S, 1982) and Latham et al. (1984), they concluded that there was a cutoff in binary separations at $\approx 0.1\,\mathrm{pc}$. It turns out that the disruption lifetime for a $0.01\,\mathrm{pc}$ binary due to known perturbers is $\lesssim 10^{10}$ years. Therefore the fact that one sees binaries as wide as $0.1\,\mathrm{pc}$ suggests that the limits on possible mass given a significant mass density are severe. Given the potential importance of this result, it is worth studying the available data and methods in detail.

In order to appreciate the difficulties in obtaining rigorous limits, I will first review some of the major contributions to our knowledge of wide binaries (§2) and their interpretation (§3). After a brief description of the physics in §4, we will readdress the problem at hand, limits on the mass of unseen disk objects with an example in §5. Although tangential to our topic, I will also briefly discuss what we can learn about the binary population itself with this approach.

2. What We Know About Wide Binaries

Although the study of binary star systems is one of the oldest disciplines in astronomy, most researchers have been concerned with relatively close binaries—systems for which orbital parameters can be determined. Wide binaries, such as in the introductory example, must be identified by their common proper motion, physical separation and line–of–sight velocity since the orbital period may be many orders of magnitude longer than the observer's lifetime. Consequently, identifying wide binary systems is difficult and spurious identifications are always a worry. Wide binaries would have remained curiosities had it not been realized that they are easily affected by their environment and their distribution and frequency may be a clue to star formation. With this motivation, there has been a number of recent systematic searches and studies which I will briefly discuss below. This section is not intended to be a comprehensive review but to provide observational background for the interpretation to follow, and to indicate the possible pitfalls in detecting wide binaries.

2.1. ABT'S FREQUENCY STUDY

Abt and Levy (1976) studied the distribution of binary periods among F3–G2 stars selected from the Yale Bright Star catalog (shown in Figure 2). This work was summarized by Abt (1983). Abt and Levy identified two populations: a population with $a < 100\,\mathrm{AU}$ and one with $a > 100\,\mathrm{AU}$. The latter includes wide binaries. The distribution appears to be a smoothly decreasing function of period and therefore separation. After making selection corrections, Abt's findings indicate that nearly all stars in the subpopulation have a wide binary companion with $a > 100\,\mathrm{AU}$. Furthermore, the distribution of binary members by type is consistent with that of single stars. This result was corroborated by Halbwachs (1983) who also analyzed a subset of the Yale Bright Star catalog and found that 60% of all stars are in systems wider than $3\,\mathrm{AU}$ and also that there is no significant bias in spectral class of the components.

One of the inherent difficulties in the studies above is the proper inclusion of selection effects. A different tack has been taken in the the recent work of Eggleton, Fitchett, and Tout (1989). They incorporate all selection effects in the generation of a 'synthetic' catalog and compare the statistics of the synthetic and actual catalogs. In disagreement with the above studies, they find that component masses are correlated. Regardless of the resolution of this point, it appears safe to conclude that: 1) a large fraction of stars in the solar neighborhood are members of wide systems; and 2) the distribution is a decreasing

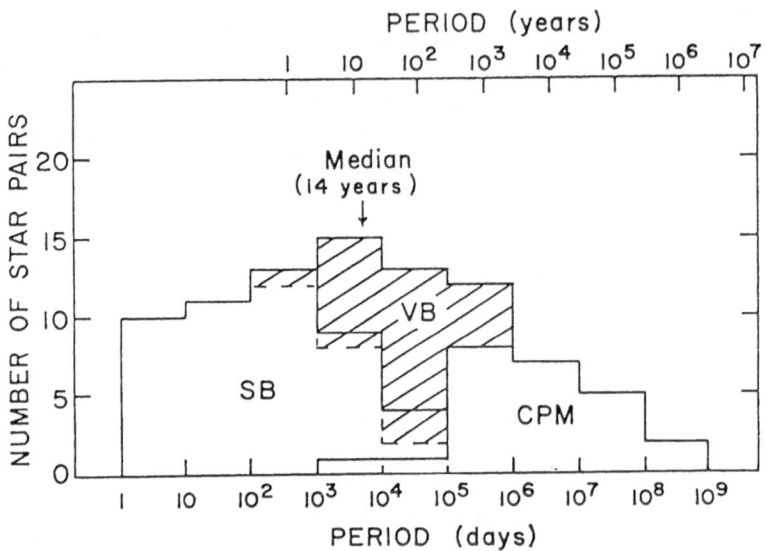

Figure 2. Histogram of binary periods for solar-type stars from Abt and Levy (1976).

function of separation for $s \gtrsim 100\,\text{AU}$.

2.2. BAHCALL & SONEIRA CORRELATION ANALYSIS

Taking a different tack, B&S computed angular correlations for sources in the Weistrop (1972) catalog. The Weistrop catalog is a 13.5 square degree survey in the direction of the NGP. B&S used a subsample between apparent magnitudes 8 and 12 consisting of 244 stars. They found significant correlation on scales less than 2ʹ of arc, identifying 19 candidate pairs. These 19 pairs were subsequently observed by Latham et al. (1984) who found that 6 of the 19 were likely to be physically bound.

Given the narrow spectral range of main sequence stars to which the survey was sensitive, this is a surprisingly large number of wide binary detections, significant even with the small sample. The results of B&S, of course, generated considerable interest in wide binaries, including today's talk. To obtain a larger sample Saarinen and Gilmore (1989) observed 800 square degrees in the direction of both the NGP and SGP, including the region containing the Weistrop survey. Also using an angular correlation analysis, they found that the binary density in the 13.5 square degrees about the NGP is a large fluctuation and not representative of the NGP as a whole. As an aside, they also found large variations from field to field, perhaps indicating substructure such as clustering, which at the very least should challenge our conception of the phase-space distribution of the local solar neighborhood. Nevertheless, since the analysis of the Saarinen and Gilmore sample is not yet complete and the Weistrop catalog played an historical role in the development of this field, I will come back to the interpretation of the B&S pairs in the next section.

2.3. OTHER CATALOGS

2.3.1. *FIDS*. Even though systematic studies of wide binaries are in their infancy, astronomers have observed and confirmed binary pairs over many decades. There are several compendia of binary systems, one of the most extensive being the Index Catalog of Visual Double Stars (IDS, Jeffers et al. 1963). The IDS is really a registry and is therefore not suitable for statistical studies. Poveda, Allen and Parrao (1982) attempted to remedy the situation by filtering the IDS, allowing only components below a certain flux limit. Nonetheless, the filtered IDS (or FIDS) has 58201 entries! Although this is potentially a gold mine of information, in my estimation too many systematic biases probably remain to allow rigorous analysis. Since the data come from diverse sources, the selection effects are practically impossible to quantify.

2.3.2. *AGK 3*. Ideally, one would like to have complete phase-space information for a well selected sample, which suggests a search in astrometric catalogs should be profitable. Full sky astrometric source catalogs do exist, a principal one being AGK3. Halbwachs (1986) searched for common proper motions stars and identified $\approx$ 100 such systems and estimated that 60% of those would be physical pairs.

2.3.3. *Yale Bright Star Catalog*. The Yale Bright Star catalog (YBS, Hoffleit 1982) is full sky and flux-limited so it is another natural place to begin. Weinberg and Wasserman (1988) searched the YBS for wide pairs and found 46 with separations below 0.1 pc. Unfortunately, since YBS is a *bright star* catalog the modal spectral type is early (A in fact) and therefore one might expect that the these stars have not phase mixed but retained a signature of their birthplace. This seems to be the case as we found clustering on scales too wide to be attributed to binaries using correlation analysis. Again, this is interesting in and of itself but does not help in the detection of the widest pairs.

2.3.4. *Woolley Catalog*. Two other possible astrometric source catalogs are the Woolley (1970) and Gliese (1969) which contain only objects within the limiting parallax 0.04, that is within 25 pc. Since these catalogs are moderately deep and consist of nearby old stars, they should not have the problem found in YBS. Wasserman and I have been studying the Woolley catalog, and I will give a preliminary report on our findings in §4.

2.4. MOVING GROUPS

Finally, I would be remiss if I did not mention the work of Eggen. Eggen performed some of the earliest work on correlations in the solar neighborhood in his identification of moving groups (see Blaauw and Schmidt 1965 for a review). More recently, he pointed out the ambiguity in identifying wide binaries as such (Eggen 1986), namely the problem of multiplicity. An isolated wide binary is likely to be rather rare, the more common situation being hierarchical pairs, or in the worst case a Trapezium system. In current analyses, all multiplicities must be dealt with on a case by case basis and there are frequent difficulties. In addition, selection effects such as unseen close companions which may result in large apparent relative motions must be identified and modeled if necessary.

3. Explanation of Known Results

Now that I have given an overview of the observational material with which we must work, I will discuss the theoretical progress in understanding the B&S results and the line of interpretive reasoning that led to limits on the mass of unseen dark objects.

Soon after B&S published their candidate list, Retterer and King (1983) had an explanation. They pointed out that a wide binary is bombarded by passing field stars whose net effect on the binary is an asymmetric random walk in energy; the binary gains energy and increases its separation on average. According to their calculations, a binary with semi-major axis of $a = 0.1$ pc will disrupt in roughly a Galactic age ($t_{gal} \approx 10^{10}$ yr). Thus, Retterer and King claimed that a cutoff in number for separations $s \gtrsim 0.1$ pc should be expected. This is essentially the situation described in §1. BHT extended these arguments to include a population of perturbers with unknown mass. By fixing the density of the unknown perturbers (at the value found by looking tracers of the potential eg Bahcall 1984a,b), one finds that the effect of these perturbers grows with their mass up to a critical value, beyond which it is insensitive to the mass (more about this in §4). Since the time scale for binary disruption was found to be accounted for with stars alone by Retterer and King, the maximum allowed mass by this argument is fairly small; BHT found $M_{unseen} \leq 2 \mathcal{M}_{\odot}$.

Weinberg, Shapiro and Wasserman (WSW, 1987) developed a theory for computing the evolution of binary systems under the perturbations by stars and GMC's. Having done this, the next logical step was to analyze and compare with the observed binary distribution. We planned to begin by reexamining the B&S data and the BHT limit. However, we found that we could not place limits on the mass of unseen disk objects with these data for the following two reasons. First, in addition to an upper and lower flux limit, the B&S search had a lower angular detection limit of $10''$ owing to resolution and positional accuracy and an upper limit of $2'$ owing to confusion. Combined with the stellar luminosity function which increases with M_V, the flux limit turns out to imply a characteristic distance or 'depth' for the sample and the angular limits then imply characteristic separations. The B&S survey was dominated by detections in the 0.01 pc $\lesssim s \lesssim 0.1$ pc range and the apparent cutoff is consistent with selection effects (see WW). Secondly, since the evolution of the binary systems is largely a process of diffusion, on physical grounds one does not expect a sharp cutoff or break in the separations but a slower power law distribution. With only 6 pairs, the observed cutoff may be partially a sampling effect. Although these additional problems mitigate the original BHT limit on the mass of unseen disk objects, the basic idea is a good one. It is worth understanding what is necessary to reinstate a mass limit. I will do this by way of example in §5. But first, to get a physical understanding of the binary disruption process, I will review the essential physical arguments in the next section.

4. Review of Physical Processes

4.1. TIME SCALES AND PHYSICAL REGIMES

As in Figure 1, let us assume the existence of a binary system with semimajor axis a and total mass M and a perturber of mass M_p approaching the system at velocity V. The encounter may occur in one of two limiting regimes. First, if the impact parameter of the perturber b is much less that a, one component of the binary pair receives a dominant velocity kick. We call this the *single-kick* regime. On the other hand, we may have $b \gg a$ in which case the effect is due to the differential or tidal force of the perturber. We call this the *tidal* regime. After averaging over an ensemble of binaries with fixed binding energy E and over all orientations of the encounter, the mean energy change using the impulse approximation is

$$\langle \Delta E \rangle = \begin{cases} 2 \left(\dfrac{GM_p}{bV} \right)^2 & \text{single kick } (b \ll a) \\[3ex] \dfrac{7}{3} \left(\dfrac{GM_p}{bV} \right)^2 \left(\dfrac{a}{b} \right)^2 & \text{tidal } (b \gg a) \end{cases} \tag{1}$$

We further divide each regime into 2 cases depending on the magnitude of the perturbation that the binary receives. If $|\Delta E/E| > 1$ we say that the encounter is *catastrophic* since a single event is sufficient to disrupt the pair. Similarly, if $|\Delta E/E| \ll 1$ we say that the encounter is *diffusive* since in general it will take many encounters to make a significant change in the pair's binding energy. As you might expect, the latter is a random walk process, may be described using a Fokker-Planck formalism, and was the regime explored by Retterer and King (1982) for encounters with stars.

It is useful to identify these various regimes by their critical length scales. For given M, M_p, V, and a, there is a critical impact parameter dividing the diffusive and catastrophic regimes,

$$b_{FP} \approx \left(\frac{16GM}{3aV^2} \right)^{1/2} \left(\frac{M_p}{M} \right) a \sim 2.4 \times 10^{-3} \, \text{pc} \qquad \text{if} \left(\frac{GM}{aV^2} \right) \ll \left(\frac{M}{M_p} \right) \tag{2}$$

and

$$\approx \left(\frac{8GM}{3aV^2} \right)^{1/4} \left(\frac{M_p}{M} \right)^{1/2} a \sim 10 \, \text{pc} \left(\frac{M_p}{6 \times 10^5 \, \mathcal{M}_\odot} \right) \qquad \text{if} \left(\frac{GM}{aV^2} \right) \gg \left(\frac{M}{M_p} \right). \tag{3}$$

Unless explicitly stated, formulae are evaluated for the fiducial parameters used in the introduction, namely $a = 0.1 \, \text{pc}$, $M = 1 \, \mathcal{M}_\odot$, and $V = 20 \, \text{km s}^{-1}$. (The subscript FP stands for Fokker-Planck). Alternatively, for fixed a, V and M, there is a critical perturber mass dividing two regimes:

$$M_0 \equiv M \sqrt{\frac{3aV^2}{16GM}} \sim 40 \, \mathcal{M}_\odot. \tag{4}$$

Thus, for fiducial parameters, encounters with stars ($\sim 1 \, \mathcal{M}_\odot$) are in the single-kick regime while encounters with GMC's or massive black holes are in the tidal regime. An overview of the hierarchy of scales is shown in Figure 3 in the form of a decision tree.

Clearly, binary stars are disrupted if an encounter delivers a kick $\Delta E > -E$. However, a binary also feels a tidal force due to the entire galaxy. This tidal field sets an upper limit to the possible separations, independent of perturbers. Calculations show that this maximal separation is roughly $a_T \simeq 0.93 (M/\mathcal{M}_\odot)^{1/3} \text{pc}$ (Antonov and Latyshev 1972, Torbett and

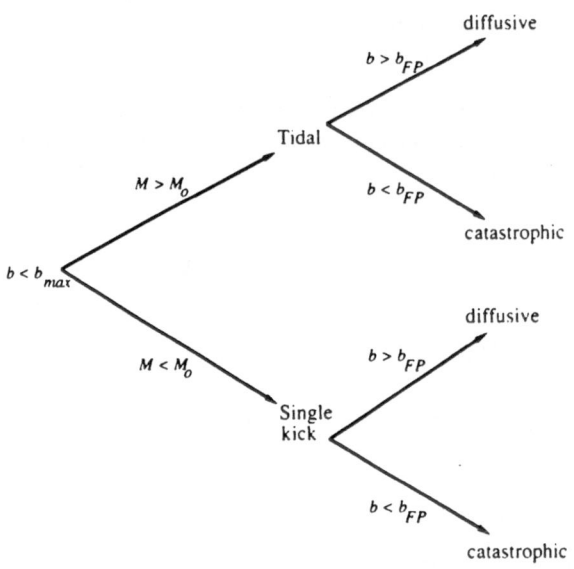

Figure 3. A binary system responds to passing objects according to their impact parameters and masses. The various regimes are shown as a decision tree.

Smoluchowski 1984, WSW). This limit may be also be written as a tidal energy: $E_T = -GM/2a_T$; a binary is then disrupted if $\Delta E > E_T - E$.

So far, I have assumed that the impulse approximation is reasonable since the orbital velocity v for wide binaries tends to be much smaller than the encounter velocity V. However, for very large impact parameters, the impulse approximation may be invalid since the duration of the encounter may be quite long. A precise condition for validity requires that

$$b \lesssim b_{max} \equiv \frac{VP(a)}{2\pi} = \frac{V}{v}a \sim 10\,\mathrm{pc}. \tag{5}$$

The binary orbit is adiabatically invariant if $b > b_{max}$. If $b_{FP} \geq b_{max}$, all encounters are in the catastrophic regime. This leads to another scale, the mass at which $b_{FP} \geq b_{max}$:

$$M_{crit} = \left(\frac{3}{8}\right)^{1/2} \left(\frac{V_{rel}}{v}\right)^3 M \sim 6 \times 10^5 \, \mathcal{M}_\odot, \tag{6}$$

where equation (3) has been used for b_{FP}. We see that for fiducial parameters that M_{crit} is quite large.

Now that we have defined the relevant regimes and length scales, let us see what is the most important pathway in the tree for evolution of binary systems. Let us begin by defining $t_*(a)$ to be the diffusive lifetime for binaries with semimajor axes a due to encounters with stars (for fiducial parameters, $t_*(a) \sim 5 \times 10^9\,\mathrm{yr}$) and q_p to be the density of perturbers in units of the stellar density, ρ_*. The case for perturbers with $M_p < M_0$ was discussed in WSW who argued that encounters with stars were largely diffusive. Since we know that stars alone will drive the evolution of binaries, is interesting to estimate the importance of unknown perturbers over $t_*(a)$. The case of $M_p < M_0$ is similar to that of stars so let us look at $M_p > M_0$. In this regime, it is easily shown that the number of catastrophic encounters during $t_*(a)$ is given by

$$N_{cat}\,(t_*(a)) = \frac{0.01}{\ln \Lambda} q_p \frac{V}{v} \sim \frac{10}{\ln \Lambda} q_p. \tag{7}$$

Note that N_{cat} is independent of M_p! The number of diffusive encounters over the same time period is

$$N_{diff}\,(t_*(a)) = \frac{1}{16 \ln \Lambda} \left(\frac{V}{v}\right)^4 q_p \frac{M_*}{M_p} \sim 6 \times 10^6 q_p M_*^{-1} M_p^{-1} \frac{1}{\ln \Lambda}. \tag{8}$$

For the typical value $\ln \Lambda \approx 3$, there are only few catastrophic events on average during $t_*(a)$ but many diffusive events. Let us now compare the diffusive timescale for perturbers with $M_p > M_0$ to those of stars. One finds

$$\begin{aligned}
T_p(a) &\equiv t_p(a)/t_*(a) \\
&= \frac{48}{7} \left(\frac{2}{3}\right)^{1/2} M_*/Mv/V \ln \Lambda (1 - M_p/M_{crit})^{-1} \\
&\sim 6 \times 10^{-2} \left[q_p(1 - M_p/M_{crit})\right]^{-1}.
\end{aligned} \tag{9}$$

Thus for abundant $(q_p \sim 1)$, low-mass $(M_p \ll M_{crit})$ perturbers, the diffusive effects of tidal encounters are more important than those of stars. Since $T(p) << 1$ we are also justified in ignoring the effects of catastrophic encounters, at least to lowest order.

This fortuitous circumstance allows us to treat both the effect of stars and the effect of point-like dark disk objects with a diffusive theory. This ignores the other major constituent of the local solar neighborhood, namely gas especially that in the form of GMC's. Since GMC's typically have masses of $5 \times 10^5\,\mathcal{M}_\odot$ and characteristic sizes of $25\,\mathrm{pc}$, we see from equation (6) that $b_{FP} \sim b_{max}$ so that GMC's will be catastrophic perturbers only. Furthermore, GMC's have significant substructure which enhances their effect at destroying wide binaries. Thus encounters with GMC's have the effect of destroying binaries that have been 'primed' by diffusive encounters with stars. These interactions have been discussed at length in WSW to which I refer anyone interested in the details. In short, we found that the GMC's decreased the binary lifetime by roughly a factor of 2. The surviving distribution is largely determined by diffusive encounters and is qualitatively unchanged by the GMC perturbations. For ease of discussion, I will ignore the effects of GMC's it what follows, keeping in mind that a full treatment must really include *all* perturbers.

4.2. FOKKER-PLANCK MODELS

In this section, I will briefly describe the computation of Fokker-Planck models for binary evolution. Over a timescale long compared to the encounter rate but small enough so that the total change in binding energy is still small, a binary feels many small energy-changing encounters with a well-defined distribution. The distribution of the small changes ΔE will approach a Gaussian (owing to the central limit theorem) whose mean $\langle \Delta E \rangle$ and the dispersion $\langle (\Delta E)^2 \rangle$ may be calculated directly from the scattering crossection. The differential equation describing this process is is known as the Fokker-Planck equation. For the evolution of an ensemble of binaries of particular binding energy ($E = -GM/2a$), this equation is

$$\frac{\partial N(E,t)}{\partial t} = -\left\{ \frac{\partial}{\partial E}\left(\epsilon_1 N(E,t)/P(E)\right) - \frac{1}{2}\frac{\partial^2}{\partial E^2}\left(\epsilon_2 N(E,t)/P(E)\right)\right\}, \tag{10}$$

where ϵ_1 and ϵ_2 are the mean and mean-squared change in energy per period, $P(E)$, and the tidal boundary condition is $N(E_T, 0) \equiv 0$. The quantity ϵ_1 denotes the motion or advection of the ensemble to higher or lower energies and ϵ_2 denotes spread of the distribution in energy about the mean. The functional forms of ϵ change depending on whether the perturbations are in the single-kick or tidal regimes. However, in both regimes, the Green functions solutions may be computed analytically. I will write the solution in terms of a dimensionless energy $x \equiv E/E_T = a_T/a$ which simplifies scaling for different conditions and in terms of a for comparison with data. In addition, there is a characteristic time scale which depends on the diffusion coefficients and E_T. For $M_p < M_0$, the characteristic time is proportional to $(M_p \rho_p)^{-1}$ and for $M_p > M_0$, it is proportional to ρ_p^{-1}. Let us call τ the time in units of the characteristic time. The relationships between physical time and τ in the two regimes are:

$$\frac{\tau}{t} = \begin{cases} 8\pi \ln \Lambda \dfrac{a_T}{V} \dfrac{M_p}{M} G\rho_p & \text{single kick,} \\[4mm] \sqrt{\dfrac{2a_T^3}{3GM}G\rho_p} & \text{tidal.} \end{cases} \tag{11}$$

The above time scales apply to perturbations by a single species of mass M_p and density ρ_p. For several species, the expressions in equation (11) may be explicitly summed over all perturber types, including unknown species. We adopt $t_{gal} \simeq 9$ Gyr following Iben and Laughlin (1989) which leads to $\tau = 7.4$ for stars alone in the the the single kick regime and $\tau = 25$ for $q_p = 1$ in the tidal regime.

We may find the distribution at τ for an initial distribution of binaries by energy by first finding the Green function solutions of equation (10); that is, the function G satisfying equation (10) such that $G(x, x_0, \tau = 0) = \delta(x - x_0)$. Then $N(x, \tau) = \int dx_0\, N(x_0, 0)G(x, x_0, \tau)$. The function $G(x, x_0, \tau)$ for successive values of τ in the single kick regime is given in Figure 4 for $x_0 = 10$. Integrating over x yields the probability that a binary survives a time t, $P(\tau) = \int dx\, G(x, x_0, \tau)$ and is shown in Figure 5 for the single kick regime. For the BBF given by $B(x, \tau)$, the theoretical distribution $F(x)$ is then given by

$$\tilde{F}(x) = \int_0^\tau d\tau' \int_{x_{min}}^{x_{max}} dx'\, B(x', \tau')G(x, x', \tau - \tau'). \tag{12}$$

The theoretical distribution in terms of semimajor axes is related to $\tilde{F}(x)$ as follows: $F(a)da = \tilde{F}(x)x^2/a_T da$.

In the section that follows, I will give an example of the overall program outlined in the introduction making simple assumptions about the BBF. First, I assume that the energy and time dependence in the BBF are separable, $B(x, \tau) = g(x)h(\tau)$. Secondly, I assume that binaries are formed with a distribution in semimajor axes of the form $f(a)da \sim a^{-\lambda}da$

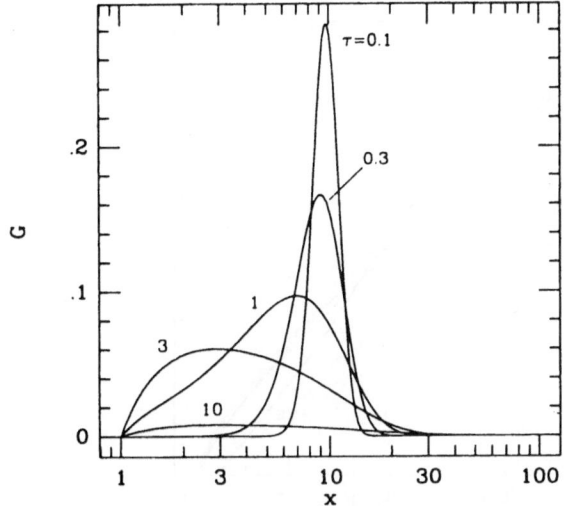

Figure 4 . Evolution of a single energy ensemble of binaries with $x_0 = 10$ at $\tau = 0$ is shown for the single kick regime (Green function). The new distribution of semimajor axes after perturbations by stars is shown for several successive values of τ as labeled.

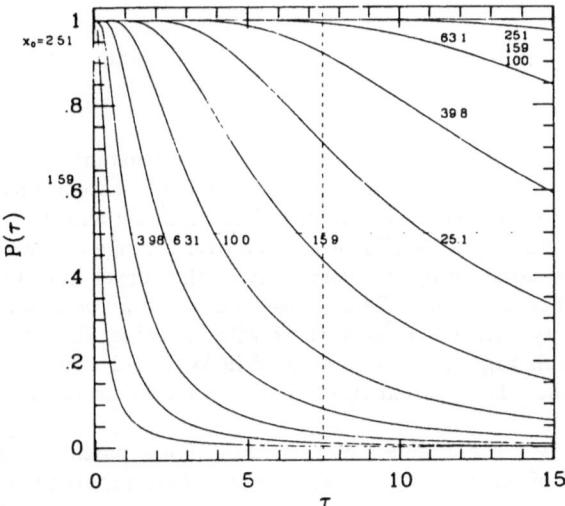

Figure 5 . The probability that a binary with a particular energy at $\tau = 0$ will survive until τ is given for several values of x_0 for the single kick regime. The point at which this probability reaches 0.5, the *halflife*, is the time at which the curve crosses the dotted line. The dashed line indicates the value of τ at t_{gal} for perturbations by stars alone.

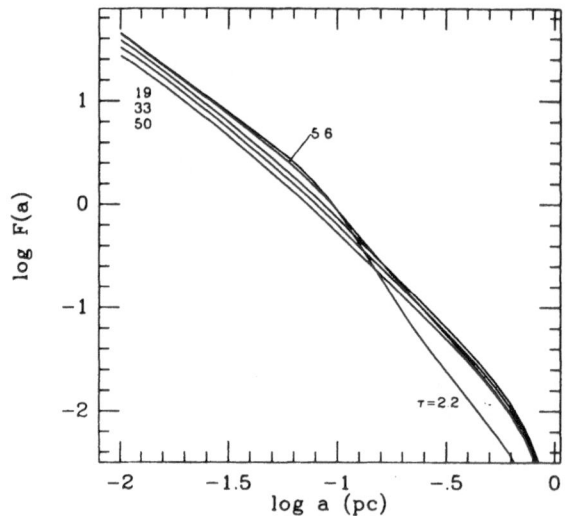

Figure 6. The run of $F(a)$ at various times for $\lambda = 1.5$, $x_{min} = a_T/a_{max} = 10$ and $x_{max} = a_T/a_{min} = 250$ with the continuous creation model in the single kick regime. The results are shown for $a_T = 1\,\mathrm{pc}$ with a in pc. The labels indicate the value of τ for each curve.

so $g(x)dx \sim x^{\lambda-2}dx$. Finally, I will assume that either $h(\tau) = $ constant, a "continuous-creation model", or $h(\tau) = \delta(\tau)$, a "burst" model. Given the model, there are two types of limits possible: 1) limits based on the shape of the distribution, $F(a)$; and 2) limits based on the normalization of $F(a)$. Examples of $F(a)$ evaluated over a range of τ are shown in Figure 6 under continuous creation. Except for very early τ, the curves have very similar shapes that show little evidence of breaks. Thus, based on *shape alone* it may be very difficult to constrain the density and type of perturbers without being able to resolve the structure of $F(a)$ in detail. This argument was presented in WW and combined with the selection problems indicated that the apparent cutoff found by B&S does not imply a mass limit.

However, we may also use overall normalization as a limit. For a particular model, the fraction f of the total number of binary pairs ever created may be computed from $F(a)$. If we observe n_{WB} pairs in a given volume of space containing $n_{catalog}$ stars altogether, then we have the obvious limit $n_{catalog} \geq 2n_{WB}/f$. (In general, we do not directly observe a well-defined volume, in which case this relation will have an additional dependence on the selection parameters.) Examples of possible limits on the mass and density of unseen dark objects will be presented in the next section.

5. Example: Woolley Catalog

In this section, I will apply the analysis discussed above to the Woolley catalog (Woolley et al. 1970). The Woolley catalog contains 1744 parallax selected (with $\pi < 0.040$ arc sec) systems. Most have both proper motions and radial velocity which allows the whole phase-space to be determined, at least in principle. Furthermore, in a volume-limited catalog, detections may be directly interpreted in terms of space density. In our previous work on the Weistrop sample (WW), the analysis was complicated by the necessity of modelling a flux limited catalog. Since the number of stars per volume is a decreasing function of intrinsic luminosity and the number of systems detected is an increasing function of luminosity for a flux-limited catalog, there is a bias toward stars from a particular distance and spectral type. This selection function is further complicated by the criteria for selecting candidate wide binaries. These problems plagued our earlier attempts and the hope of eliminating them motivated this study. These results discussed below are a preliminary report on work in progress.

5.1. SELECTING THE PAIRS

The objects in the Woolley catalog are not flux-limited and thus are incomplete at some limiting magnitude. To quantify the completeness, we reject all but main sequence stars which may be more simply typed and establish a flux limit by comparison with the local luminosity function (Bahcall and Soneira 1980). The catalog appears complete to $M_V = 9.0$ which leaves 1006 stars. To simplify our work, we further elect to eliminate all stars with ages less than that of a Galactic age (which leads to $M_V = 4.9$ for $t_{gal} = 9$ Gyr). The remaining 689 stars are flux and volume limited and should not be biased by age.

The candidate binary pairs were selected by the criteria

$$\Delta s \leq \sqrt{s_0^2 + 4\sigma_s^2},$$
$$\Delta v_\perp \leq \sqrt{v_0^2 + 4\sigma_v^2}, \tag{13}$$

where $s_0 = 1\,\mathrm{pc}$ and $v_0 = 1\,\mathrm{km\,s^{-1}}$. The quantities σ_s and σ_v are computed from the measurement errors in the quantities determining s and $v_\perp$, respectively. Although most stars in the catalog have their six phase-space values determined, the determinations are not of comparable quality. Since the radial velocities appear to be the least accurate, they are not used except in computing σ_v. Performing the search, we find that there are 10 pairs with $0.01\,\mathrm{pc} < s < 1\,\mathrm{pc}$. One of the pairs has very large relative radial velocity which we tentatively reject for the purposes of this example, leaving 9 pairs. We estimated the number of spurious background detections in two ways. We first used the luminosity function to compute the background assuming uniform space density and isotropic velocity ellipsoids. Secondly, we randomly swapped all velocities in the catalog and then reran our search algorithm. The second method gave 1.4 pairs at random on average, approximately a factor of two larger than the first estimate. We consider this good correspondence given the level of idealization in the first determination and therefore adopt the second determination. Anyway, the 9 pairs are a significant signal and we expect only a few at most to be spurious. In the examples that follow, we will assume all 9 pairs are physical although this will not be known without additional observations.

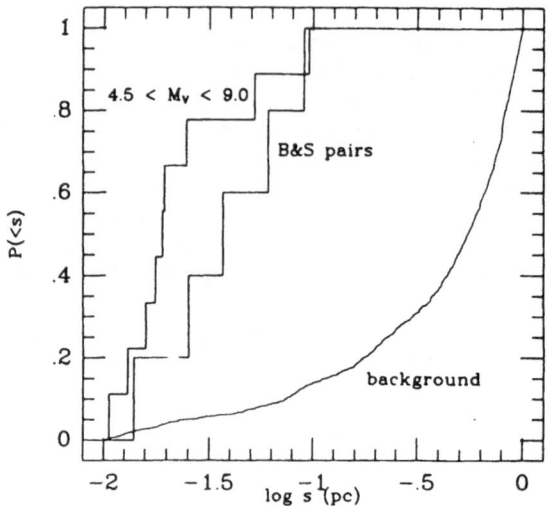

Figure 7. The cumulative distribution in separation of the 9 pairs is shown. The lower curve shows the expected background computed by randomizing the velocities of the catalog entries. The distribution of the 6 Bahcall and Soneira pairs is shown for comparison.

5.2. ANALYSIS

The cumulative distribution of the 9 pairs together with the expected background are shown in Figure 7. Examining Figure 7, it is striking that there are no pairs with $s \gtrsim 0.1\,\mathrm{pc}$ although the search criteria should have been sensitive to these very wide pairs. The distribution bears a striking resemblance to that for the 6 B&S pairs which is shown for reference. Before applying any of the theoretical models discussed in the last section, let us begin with a phenomenological question: does the distribution have a break? To answer this, I fit a continuous and broken power law to the data using the following two estimators for the differential distribution in s, $q(s)$:

$$q(s) \propto s^a \exp\left[-(s/s_0)^\mu\right] + s^b \left\{1 - \exp\left[-(s/s_0)^\mu\right]\right\}, \qquad (14)$$

$$q(s) \propto s^l, \qquad (15)$$

where a, b and l are power law indices, s_0 is the position of the break and μ controls the sharpness of the break (typically taken to be 20). Maximum likelihood gives $a = -1.64 \pm 0.51$, $b = -17.2 \pm 6.4$ and $s_0 = 0.112 \pm 0.014$ in the first case and $l = -1.99 \pm 0.51$ in the second. The first case confirms the visual impression that the data imply a break. However, a Kolmogorov-Smirnov (KS) test indicates that the single power law model may not with confidence be rejected. This may be restated as follows. If the cumulative distribution found for the 9 pairs is taken to be exact, the rejection of the single power law model at the 5% level would require 300 pairs! Although a break is suggested, no break is also consistent with the data.

Let us now explore the limits on the perturbers based on the evolutionary models discussed in §4. First we will look at limits based on the *shape* of the evolving distribution and then limits based on the overall *normalization* implied by the disrupting binary pairs. We will use the same BBF introduced in §4. Although the power law distribution $\sim a^{-\lambda} da$ is consistent with the observed distributions as we will see below, other distributions are certainly possible. However, it is also worth remarking that the size distribution of clumps in GMC's also appear to be well described by a power law (Blitz 1989). To assess the significance of any particular model given by equation (12), we compute both its maximum likelihood (ML) and KS significance. The value of $x_{max} = a_T/a_{min}$ is chosen so that there is little evolution for a binary with that semimajor axis for all τ of interest. For values of $\lambda \gtrsim 1$, the lower cutoff $x_{min} = a_T/a_{max}$ is not terribly important to the shape of the distribution although it will affect the normalization. However, very large values of x_{min} lead to inconsistencies which will be discussed below.

We will discuss the two cases $M < M_0$ and $M > M_0$ separately beginning with the former. Figure 8 shows both the ML and KS values for a continuous creation model with $x_{min} = 1$, $a_{max} \approx 1 \, \text{pc}$.

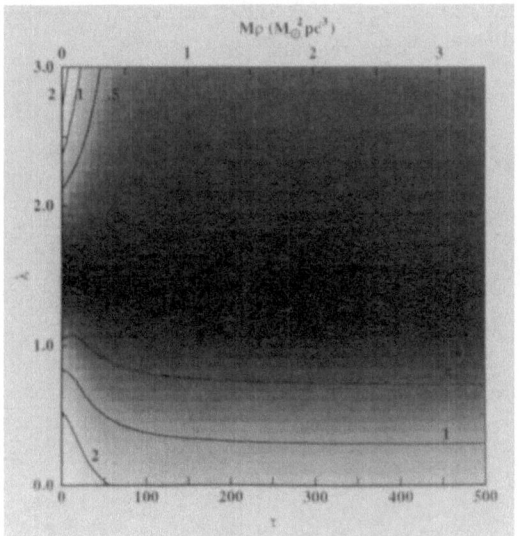

Figure 8. Fits to evolutionary models in the single kick regime. Contours of constant log likelihood are shown for the 9 data points and the continuous creation evolutionary models of §4. The contours are .5, 1 and 2 offsets from the log ML value, the lower offsets toward the center of the figure. At $\tau = 0$, ML occurs at $\lambda \approx 1.5$. The KS significance is shown as greyscale from 0 (white) to 1 (black). The upper abscissa shows $M_p \rho_p$ in units of $\mathcal{M}_\odot^2 \, \text{pc}^{-3}$ for fiducial values.

The KS significance is shown as greyscale from 0 (white) to 1 (black). The curves describe lines of constant log likelihood with values 0.5, 1 and 2 down from the maximum. The lower abscissa describes the value of τ, and the upper abscissa gives the value of $\rho_p M_p$ in units of $\mathcal{M}_\odot^2 \, \text{pc}^{-3}$ for fiducial values of V and M and $a_T = 1 \, \text{pc}$. Ones sees that at all τ there are evolutionary models that are acceptable fits to the binary distribution. In addition, there

132

is very little variation with τ in the best fit λ. In other words, *there are no limits on any unseen perturbers based on the shape of the distribution.* The situation is similar for burst models which are not shown.

However, following the discussion in §4, some of the models in the $\tau-\lambda$ plot may be ruled out based on normalization. For example, consider a model with small λ at large τ. For small λ, many pairs will be created at large separation where their lifetimes are very short. Thus, the fact that we observe pairs at large separation today implies that the binary birthrate must be high. Since disrupted pairs become single stars, a model may imply the existence of more single stars than actually observed in the catalog, $n_{catalog}$, hence ruling out the model. For every $\tau-\lambda$ one may easily compute the fraction f of all binaries disrupted. Thus, if there are n_{WB} pairs observed, the limit will be $2n_{WB}/f < n_{catalog}$. To get an idea of what models are ruled out, I plot lines of constant $1/f$ in Figure 9. For the 9 pairs in the Woolley catalog, models with $1/f > 38.3$, those in the lower right hand part of the diagram, are ruled out. By comparing with Figure 8 , we see that the best fit models with $\lambda \sim 1.5$ are acceptable and the normalization constraint does not yield a mass limit on unseen dark objects. However, if we assert, for example, that $\lambda = 0.5$, we would infer that $M_{unseen} \lesssim 20 \, \mathcal{M}_\odot$ for $\rho_{unseen} = 2\rho_*$. I might also add that models with $x_{min} = 10$, do not have $1/f > 38.3$ over the entire range of λ and τ surveyed. We will see below that models with $x_{min} \lesssim 50$ are consistent with the data.

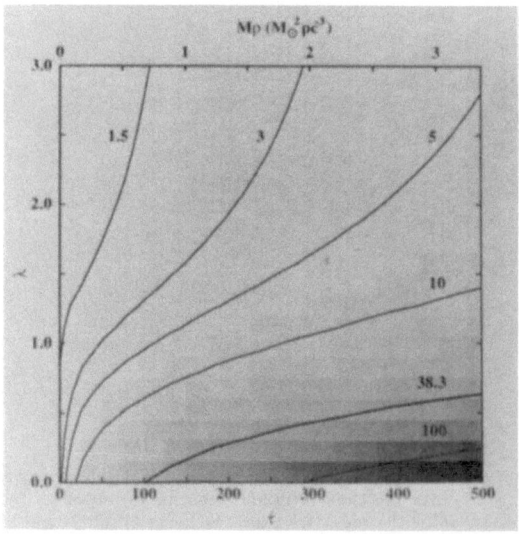

Figure 9. Contours of the disruption factor $1/f$ for the single kick regime for the same parameters as Fig. 8 . The heavy line corresponds to $1/f = 38.3$.

The situation for perturbers $M > M_{crit}$ is similar to that for the single kick regime. Figure 10 shows the ML and KS contours for black holes and Figure 11 shows the $1/f$ contours for the tidal case. As expected the shape of the evolving distribution provides

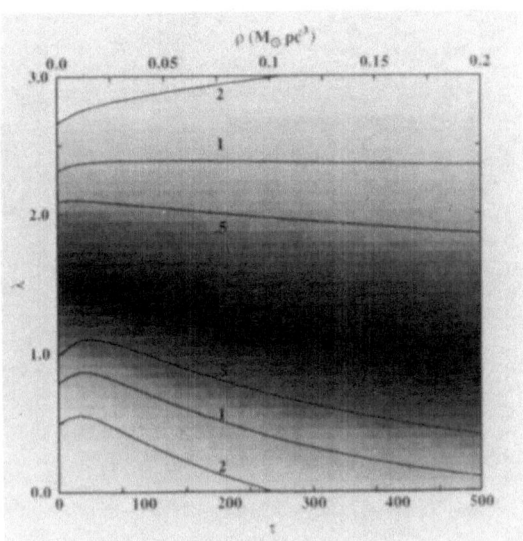

Figure 10 . Fits to evolutionary models in the tidal regime. The upper abscissa shows the density of perturbers in $\mathcal{M}_\odot$ pc^{-3}. Otherwise as in Fig. 8

little constraint. Normalization does not provide a constrain for any value of λ and τ shown.

So far we have been using the distribution of wide binaries to investigate the existence of possible species of perturbers. If we assume that there are no unseen perturbers, then we may use Figures 7 and 8 to infer the fraction of stars in wide binary systems, the multiplicity Q. Due to encounters with stars alone, $\tau = 7.4$ at t_{gal}. If we then choose the ML solution at this fixed τ, we find $\lambda = 1.5 \pm 0.5$ using Figure 8 . Then, $Q = 2n_{WB}/(fn_{catalog}) = 0.029$ in this case. Thus 3% of the main sequence stars with $4.5 < M_V < 9$ are formed in wide binaries for a continuous creation model with $\lambda = 1.5$.

Although we have had to assume a form for the BBF, if we again abandon the goal of limiting q_p and M_p, we may ask if we can limit the parameters of the model. In particular, we may limit the value of $x_{min} = a_T/a_{max}$ from above. It is fairly obvious that this is possible. In all likelihood, a binary with $s \sim 0.1$ pc today was born with a smaller separation since the lifetime of the pair is smaller than the age of the Galaxy. If we make x_{min} larger, then a larger fraction of the observed pairs must have been made at small separation and evolved to larger separations by the action of perturbers. If x_{min} were so large that a pair with semimajor axis a_T/x_{min} would evolve only negligibly in a Galactic age, we would see no wide binaries at all. We may use the evolutionary analysis to place limits on the maximum value of x_{min} if we assume encounters with stars alone. Again, τ is fixed and we may fit the models given by equation (12) on the x_{min}—λ plane. The ML and KS contours are shown in Figure 12 for a continuous creation model with perturbations by stars. For large x_{min}, it is clear that λ must drop; in order to get a significant population advected to large

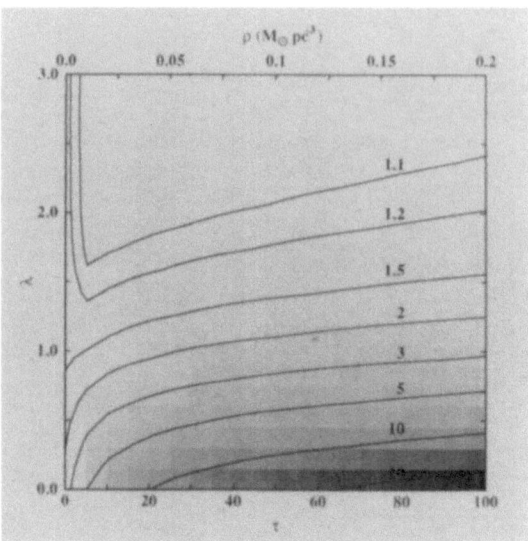

Figure 11 . Contours of the disruption factor $1/f$ for the tidal regime.

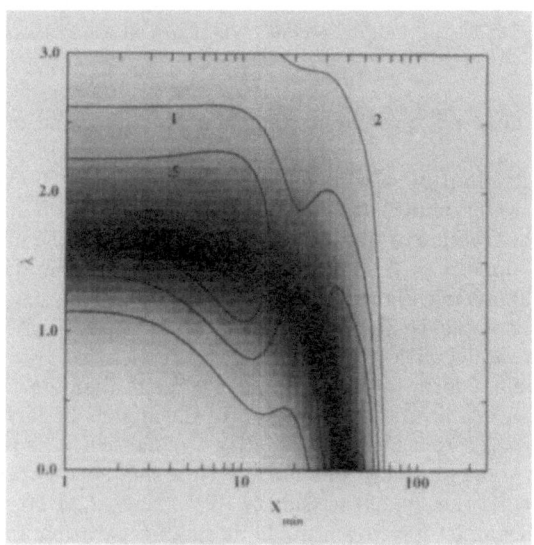

Figure 12. Fits to continuous creation evolutionary models for stars alone at τ corresponding to 9×10^9 years. Contours of log likelihood are shown as a function of x_{min} and λ. The KS significance is shown as greyscale from 0 (white) to 1 (black).

separation, one must populate the semimajor axes near the cutoff. Assuming $a_T = 1\,\mathrm{pc}$, we see that no models fit for $a \lesssim 0.02\,\mathrm{pc} \approx 4 \times 10^3\,\mathrm{AU}$ and in most cases $1 \lesssim \lambda \lesssim 2$.

Summary

We have seen that placing limits on any population of unseen perturbers requires several steps. First, we must make an assumption about the form of the binary birth rate function. The parameters of the model may be constrained by the shapes of the actual and theoretical distributions. Secondly, once a best fit model is determined, there is a maximum density and mass for the unseen perturbers that implies an inconsistency in the total stellar number density, hence a limit. Although it may be possible to place limits based on the shape of the distribution alone, this requires large amounts of data so that the large separation tail is statistically well resolved. Variations in the shape of the distribution with varying perturber mass and density are small (e.g. Fig. 6).

For a particular observed distribution of wide binary separations, cutoffs may be deceptive. Cutoffs may be caused by selection effects in the search procedure and sparse sampling. Physically, one doesn't expect to see a sharp cutoff. In addition, one must bear in mind that a turnover in a distribution at a particular separation does not necessarily imply a mass limit based on the lifetime at that separation. Binary systems with small separations will have larger separations on average in time.

In my opinion, the candidate pairs from the Woolley catalog constitute the best sample to date; since the catalog is volume limited, many of the selection difficulties that we found in analyzing the Weistrop sample vanish. On the other hand, we still do not have many pairs. Based on the 9 old main sequence candidate pairs discussed in §5, the best fit model does not give an interesting limit on the dark objects. The results so far are tantalizing and are interesting in and of themselves, but we have fallen short of our goal. Given the paucity of data, I could only guess as to the eventual resolution of this question. We have only begun to determine the stellar distribution in the solar neighborhood systematically and much more data are needed to infer a mass limit with confidence. With the recent resurgence of interest in careful astrometry driven by technical advancements, improved catalogs are only a matter of time.

I would like to acknowledge Ira Wasserman with whom many of the ideas in this talk were conceived. This work has been supported by NSF grants PHY-86-20266 and NJ High Technology Grant 89-240090-2. I gratefully acknowledge travel support provided by NATO which allowed me to attend this meeting.

References

Abt, H. A. 1983. *Ann. Rev. Astron. Astrophys.* **21**, 343.

Abt, H. A. and Levy, S. G. 1976. *Astrophys. J. Suppl.* **30**, 273.

Antonov, V. A. and Latyshev, I. N. 1972. *Soviet Astron.* **15**, 676.

Bahcall, J. N. 1984a. *Astrophys. J.* **276**, 156.

Bahcall, J. N. 1984b. *Astrophys. J.* **276**, 169.

Bahcall, J. N., Hut, P., and Tremaine, S. 1985. *Astrophys. J.* **290**, 15 (BHT).

Bahcall, J. N. and Soniera R. M. 1980. *Astrophys. J. Suppl.* **44**, 73.

Bahcall, J. N. and Soniera R. M. 1981. *Astrophys. J.* **264**, 122 (B&S).

Blitz, L. 1989, private communication

Eggen, O. J. 1965. In *Galactic Structure* ed. A. Blaauw and M. Schmidt (Chicago: University of Chicago Press), p. 111.

Eggen, O. J. 1986. *Astron. J.* **92**, 125.

Eggleton, P.P., Fitchett, M. J., and Tout, C. A. 1986. *Astron. J.* **92**, 125.

Eggleton, P.P., Fitchett, M. J., and Tout, C. A. 1989. *Astrophys. J.* in press.

Gleise, W. 1969. *Catalog of Nearby Stars,* Edition 1969. *Veroff. Astron. Rechen-Inst. Heidelberg,* Nr. 22.

Halbwachs, J. L. 1983. *Astron. Astrophys. Suppl.* **66**, 131.

Halbwachs, J. L. 1986. *Astron. Astrophys.* **168**, 161.

Hoffleit, D. 1982. *The Bright Star Catalog,* 4th edition (New Haven: Yale University Observatory).

Iben, I. and Laughlin, G. 1989. *Astrophys. J.* **341**, 312.

Jeffers, H. D., van den Bos, W. H., and Greeby, F. M. 1963. *Index Catalog of Visual Double Stars,* Lick Observatory.

Latham, D. W., Tonry, J., Bahcall, J. N., Soniera, R. M., and Schechter, P.S. 1984. *Astrophys. J. Lett.* **281**, L41.

Poveda, A., Allen, C. and Parrao, L. 1982. *Astrophys. J.* **258**, 589.

Retterer, J. M., and King, I. R. 1982. *Astrophys. J.* **254**, 214.

Saarinen, S. and Gilmore, G. 1989. *Mon. Not. Roy. Astron. Soc.* **237**, 311.

Torbett, M. V. and Smoluchowski, R. 1984. *Nature* **331**, 641.

Wasserman, I. and Weinberg, M. D. 1987. *Astrophys. J.* **312**, 390 (WW).

Weinberg, M. D., Shapiro, S. L. and Wasserman I. 1985. *Astrophys. J.* **312**, 367 (WSW).

Weinberg, M. D. and Wasserman I. 1988. *Astrophys. J.* **329**, 253.

Weistrop, D. 1972. *Astron. J.* **77**, 366.

Woolley, R., Epps, E. A., Penston, J. J. and Pocock, S. B. 1979. *Catalog of Stars within twenty five parsecs of the Sun,* Roy. Obs. Ann., No. 5.

DARK MATTER IN THE GALACTIC DISK

Gerard Gilmore
Institute of Astronomy
University of Cambridge
Madingley Road
Cambridge CB3 0HA
England

ABSTRACT. The total amount of mass of whatever type distributed on length scales from a few parsecs to thousands of parsecs can be determined to precisions of better than 25% from dynamical studies of globular clusters and of the solar neighbourhood. The baryonic mass distribution on the same scales can be determined from measurement of the mass in the inter-stellar medium and by deduction from the local stellar luminosity function. The total (dynamical) amount of mass within about 1kpc of the Sun is about 30% more than the baryonic mass associated with the Galactic disk. There is no evidence that any of this unidentified mass – which may be baryonic or non-baryonic, we have no direct information either way – is associated with the Galactic disk near the Sun, and there is some model–dependant evidence that none of this mass is associated with the disk. The mass–to–light ratio of the Galactic disk near the Sun of $M/L_{\odot,V} \approx 3$ is however substantially lower than the universal value of ~ 25 required by standard Big Bang nucleosynthetic models.

1. Orders of Magnitude

A convenient measure of the amount of baryonic luminous matter in any place is the mass to light ratio, M/L, expressed in solar units. B-band, V-band and bolometric mass to light ratios are often used, usually without being specifically identified, though for practical purposes the difference between these different passbands is usually smaller than the uncertainty in the determination. In order to set measured M/L ratios in context, it is useful to remember that a Universe with $\Omega = 1$ has $M/L \approx 1500$, using recently measured field galaxy luminosity and spatial distribution data. For comparison, standard homogeneous Big Bang nucleosynthesis suggests that $0.01 \lesssim \Omega_b(H_0/100)^2 \lesssim 0.10$, or equivalently, $5 \lesssim M/L_{baryons} \lesssim 100$. A value of perhaps 20-30 is most consistent with elemental abundance data, though inhomogeneous Big Bang models (Fowler 1990) allow much larger values of M/L.

The total mass of the Milky Way, deduced from the motions of the outer globular clusters and the Milky Way's satellite companions, and estimated from the dynamics and timing argument applied to the Local Group, is $2 - 10 \times 10^{11} \mathcal{M}_{\odot}$. This implies a total M/L $\approx 10 - 50$ associated with the Milky Way. This value is interestingly similar to the values

137

D. Lynden-Bell and G. Gilmore (eds.), Baryonic Dark Matter, 137–157.

noted above as being derived for the Universe as a whole from cosmological arguments. Thus the possibility exists that all the mass in normal galaxies, including that required to explain extended flat rotation curves, may be of similar amount to the total baryonic mass required by standard homogeneous Big Bang nucleosynthesis models. One may therefore reasonably consider the possibility that the two values do in fact agree, and that baryonic matter makes up the dark matter in galaxies.

For elliptical galaxies the available data are much more model-dependent. Dynamical studies suggest that the dynamical $M/L \approx 10$ in their inner regions, though there is as yet no information on the fraction of this mass which is baryonic. It is a common misconception that $M/L \approx 10$ is consistent with a normal old metal-rich stellar population. Measured values for stellar populations in the Milky Way are about 3. Thus mass to light ratios as high as 10 are evidence for either a very different stellar population in big elliptical galaxies from those studied to date in the Local Group, or for the presence of substantial amounts of mass which is not associated with stars or detectable hydrogen in the luminous regions of elliptical galaxies.

The hypothesis that all the mass in galaxies is baryonic can be tested in part by determination of the length scales on which mass is distributed in the Milky Way, and by improving the direct measures of the total amount of mass associated with low mass, high mass to light ratio (baryonic) stars near the Sun.

Reasonably reliable dynamical mass measurements are now available for several globular clusters and for the Galactic disk within ~ 1 kpc of the Sun. The globular cluster results are presented in Table 1, and cover an order of magnitude in metallicity, total cluster mass, and tidal radius. They show a remarkable consistency, however, with all clusters observed having a measured mass to light ratio, in solar visual units, consistent with the value $M/L = 2.5$. The evolutionary effects of dynamical mass segregation mean that this value is somewhat lower than that which would be associated with the primordial stellar mass distribution. This latter effect is however not large – it is not able to allow consistency with the value $M/L = 10$ noted above as being deduced in elliptical galaxies or the value of $20 - 30$ required by nucleosynthetic arguments. A mass to light ratio of about 3 may therefore be considered to be a well established measurement appropriate to an old stellar population. We now consider in turn the directly observed and the dynamically measured mass to light ratio for stars, and for all matter respectively, near the Sun.

TABLE 1

Mass–to–Light Ratios of Globular Clusters

cluster	[Fe/H]	tidal radius (pc)	total mass $\mathcal{M}_\odot \times 10^6$	M/L_V
M92	−2.24	60	0.4	2.4
M3	−1.66	190	0.6	2.2
M13	−1.65	20	0.7	3.6
M2	−1.62	50	0.9	2.7
ω Cen	−1.59	80	3.9	2.9
47 Tuc	−0.71	60	0.7	1.8
MEAN				2.6

2. THE MASS FUNCTION FOR LOW MASS STARS

Discussions of the absolute number of very low mass stars have, like the stars themselves, tended to generate more heat than light. In part this is due to the intrinsic importance of very low mass stars for an understanding of the location and initial mass function of stellar formation – do low mass stars form in the same places as high mass stars? – and for their possible contribution to the dark matter problem, and in part to the rather poor observational and theoretical information which has been available until recently. Salpeter (1955) first showed that the initial mass function can be written approximately as a power-law of the form $\xi(m)dm = 0.013m^{-\alpha}dm$, with $\alpha = 2.35$, for masses in the range $0.4\mathcal{M}_\odot$ to $10\mathcal{M}_\odot$. Very little information on lower mass stars was available at that time.

Miller & Scalo (1979, hereafter MS) later carried out an extensive study of the stellar mass function, showing that the best available representation is by a half Gaussian in $\log_{10} m$ in the mass range $0.1\mathcal{M}_\odot \leq m \leq 60\mathcal{M}_\odot$. The Miller-Scalo mass function flattens at a mass of approximately $0.6\mathcal{M}_\odot$ and can thus be extrapolated to zero mass without divergence. It is important to remember that the stellar mass function derived by Miller & Scalo assumes a distribution of field stars by magnitude, or luminosity function, which increases smoothly with increasing magnitude to $M_V \approx +15$, and a mass-luminosity relation which is roughly linear in $\log_{10} m$ for low masses.

The most detailed and recent study of the stellar mass function is that by Scalo (1986). He derived a mass function for low mass stars that peaks at $m \approx 0.3\mathcal{M}_\odot$ and decreases rapidly for $m < 0.2\mathcal{M}_\odot$. Again this conclusion is sensitive to the assumption that the mass-luminosity relation is roughly a power law for low masses.

In all this work the shape of the derived stellar mass function has reflected the shape of the stellar luminosity function. It is important to remember however that the stellar luminosity function is related to the stellar mass function through the slope of the mass-luminosity relation. Thus the possibility that the mass-luminosity relation can be the origin of features, such as a change of slope or even of a maximum, in the luminosity function, and hence that there may be no such features in the stellar mass function, deserves re-investigation. Such an investigation requires consideration of both the stellar luminosity function and the stellar mass-luminosity relation, and has been carried out recently by Kroupa, Tout & Gilmore (1990).

There have been a variety of recent determinations of the stellar luminosity function for low luminosity field stars near the Sun, following the method of Reid & Gilmore (1982). The most extensive of these are those by Gilmore, Reid & Hewett (1985) and by Hawkins & Bessell (1988). Recently, these have been combined with the results of their own new survey by Stobie, Ishida & Peacock (1989). The stellar luminosity function combined from all these surveys for low mass field stars in the solar neighbourhood is now well established to show a maximum at $M_V \approx 12$ and a subsequent decrease as least as far as $M_V \approx 16.5$. This luminosity function is shown in figure 1, together with a seven knot cubic spline fit to $\psi(M_V)$, where $\psi(M_V)dM_V$ is the number of stars per cubic parsec with magnitudes between M_V and $M_V + dM_V$ required in the analysis here. Note that this spline representation of the luminosity function has no theoretical basis. The analysis requires only that it be a smooth and satisfactory representation of the data.

140

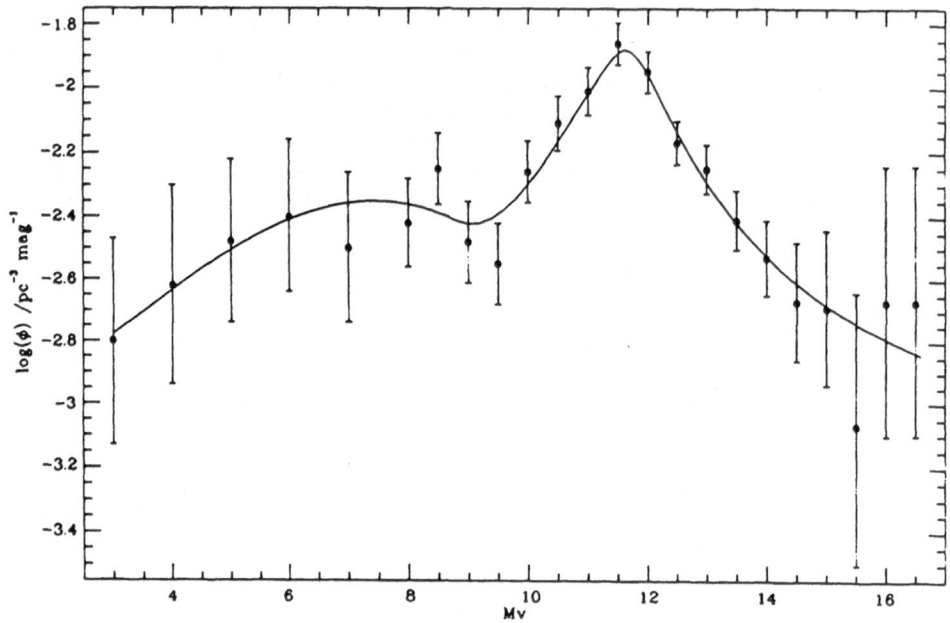

Figure 1. The stellar luminosity function for field stars near the Sun (points) together with
a smooth fit (solid line) adopted for the analysis by Kroupa, Tout & Gilmore (1990). The
luminosity function data are from Stobie, Ishida, & Peacock (1989) for $M_V \geq +8$, and from
Scalo (1986) for more luminous stars.

At lower luminosities than $M_V \approx 16.5$ poor statistics, a shortage of stars with good
parallax distances to calibrate the absolute magnitude-colour relations, considerable uncertainties in the bolometric corrections required for comparison with evolutionary tracks, and
particularly the expectation of very considerable structure in the mass luminosity relation
as a wide variety of molecules become important contributions to the stellar atmospheric
opacity and affect the interior equation of state, all continue to limit one's confidence in
any derived mass function. The consistency of recent studies of the solar-neighbourhood
luminosity function for stars with $M_V \leq 16.5$ however allows us to derive the stellar mass
function for stars with mass $\gtrsim 0.3\mathcal{M}_\odot$, and to estimate the allowed range of mass functions
for lower mass stars.

Essentially we are dealing with the following question: how much of the observed structure in the stellar luminosity function *must* be due to structure in the stellar initial mass
function and how much *could* be due to structure in the stellar mass-luminosity relation?

The form of the low-mass mass-luminosity relation is not well known and we thus test the hypothesis that the maximum in the luminosity function near $M_V \approx 12$ is due to a point of inflection in the mass-luminosity relation and is not caused by a real feature in the mass function. To proceed, we calculate the mass-luminosity relations which converts each of a range of mass functions into the luminosity function shown in figure 1. Each mass-luminosity relations is tested for consistency with available observational constraints to identify the range of mass functions which is consistent with the observed luminosity function. Theoretical mass-luminosity models are calculated to check both the plausibility and the physical cause of the resulting structure in the mass-luminosity relation. Since for present purposes we are primarily interested in high mass to light ratio stellar populations, and not galactic evolution, we restrict the analysis to stars less massive than $0.9 \mathcal{M}_\odot$. These stars have not evolved off the main sequence in the Galactic lifetime, so that one need not model the star formation history of the solar neighbourhood.

2.1. FROM LUMINOSITY FUNCTION TO MASS FUNCTION

By the definition of the luminosity function we have,

$$\psi = \frac{dN}{dM_V} = -\frac{dN}{dm}\frac{dm}{dM_V} = -\xi\frac{dm}{dM_V}, \tag{1}$$

where N is the number of stars per cubic parsec, m the stellar mass, M_V the absolute magnitude and $\xi(m)dm$ is the number of stars per cubic parsec with masses between m and $m + dm$. Equation 1 can be integrated to obtain the relation between mass and absolute magnitude. As a boundary condition at high masses we utilise a zero-age solar model. Standard theoretical models show the absolute visual magnitude of a zero-age Sun to be $M_V = +5.15$. We specify the low-mass boundary condition as mass $m = 0.09 \mathcal{M}_\odot \equiv M_V = 16.5$. This ensures that the mass-luminosity relation is consistent with the observational data points for the lowest mass stars which have reasonably precise mass and luminosity determinations available.

Integration of equation 1 provides

$$\int_{16.5}^{M_V} \psi \, dM_V' = -\int_{0.09}^{m} \xi \, dm', \tag{2}$$

and

$$\int_{16.5}^{5.15} \psi \, dM_V' = -\int_{0.09}^{1} \xi \, dm'. \tag{3}$$

The second integral has the effect of normalizing the mass function, ξ, so that the number of stars observed between $M_V = 16.5$ and $M_V = 5.15$ is always equal to the number of stars with masses between $0.09 \mathcal{M}_\odot$ and $1.0 \mathcal{M}_\odot$. To compute the mass-absolute magnitude relation we first apply this normalization and then compare the values of the two integrals in equation (2) over the range of interest.

2.2. THE MASS–LUMINOSITY RELATION

The constraints on the adopted mass-luminosity relation are the observations of binary stars compiled by Popper (1980), and shown in figure 2. These data, and all recent stellar models for stars of very low mass, show there to be a change in slope of the mass-absolute magnitude relation near $m \approx 0.7 \mathcal{M}_\odot$, $M_V \approx 7$. This is due to the increasing importance of H^- as an opacity source, and leads to a flattening of the observed stellar luminosity function below $M_V \approx 7$, a feature which has sometimes been called the "Wielen Dip".

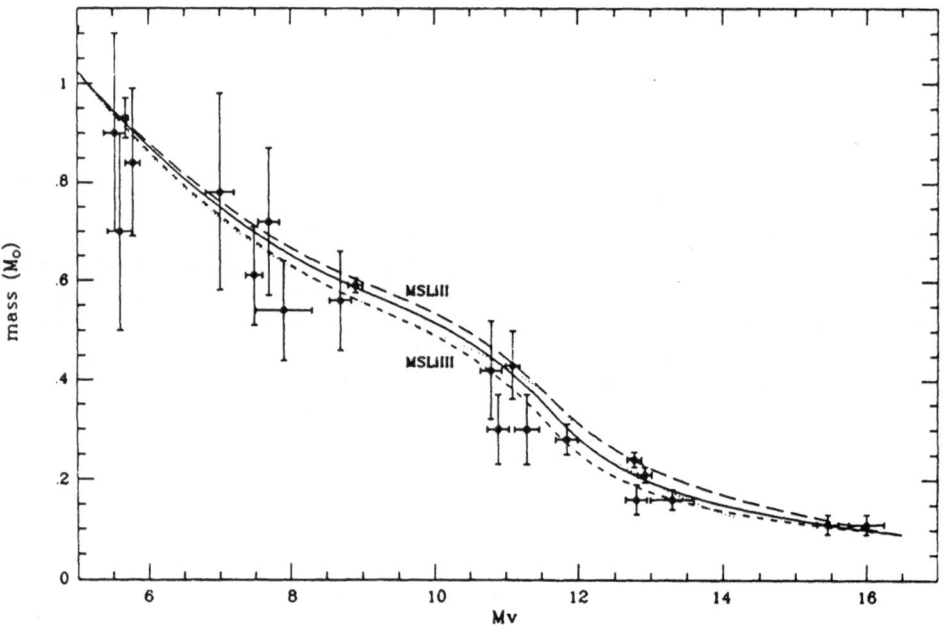

Figure 2. The mass-absolute visual magnitude relations required to transform between the observed luminosity function shown in figure 1 and the Miller & Scalo (1979) mass function, with three different linear extrapolations to very low masses. The three extrapolations are a mass function which is constant below $0.36\mathcal{M}_\odot$ (solid line), and the maximally decreasing (long dashes;MSli II) and maximally increasing (short dashes;MSli III) mass function extrapolations allowed by the observational constraints on the mass-luminosity relation. A theoretical model from the analysis by Kroupa, Tout & Gilmore (1990) is also shown for comparison (dotted line). The points are the observational data for all eclipsing binaries with good masses compiled by Popper (1980).

There is some evidence for a point of inflection near $0.33\mathcal{M}_\odot$. This second change in the mass–luminosity relation is due to two effects. The mean molecular weight of hydrogen molecules is twice that of hydrogen atoms, so that the formation of hydrogen molecules leads to a reduction in pressure, with consequent contraction, core heating, and increasing luminosity. Additionally, there is a major change in the interior structure of a star near $0.33\mathcal{M}_\odot$, with more massive stars having radiative cores and lower mass stars being completely convective. This further enhances the contraction and increases the luminosity. This effect is not a new feature of recent models, but has been established for many years. Copeland, Jensen & Jørgensen (1970) first noted the existence of this point of inflection, attributing it to the lowering of the adiabatic gradient in the H_2 dissociation zone, but not commenting on the implications for the stellar initial mass function.

2.3. THE STELLAR MASS FUNCTION

The identification of changes in the gradient of the mass-luminosity relation has important consequences for the shape of the stellar mass function. Since the mass-luminosity relation is not required to be (nearly) linear at low masses, features in the mass-luminosity relation can be the explanation of features in the observed luminosity function. Thus a maximum in the observed luminosity function need not correspond to a maximum in the stellar mass function.

We proceed by allowing any level of structure in the mass-luminosity relation which is not inconsistent with the direct observational constraints. We then specify a wide range of smooth mass functions, and determine which can be made consistent with the observed luminosity function by adoption of an allowed mass-luminosity relation. The final step is to calculate the required mass-absolute magnitude relations and compare them with the data using a χ^2 test. The mass functions considered by Kroupa etal (1990) included power laws, a gaussian, and the function derived by Miller & Scalo (1979). A range of linear extensions to the Miller-Scalo mass function with various positive and negative slopes was tried, to estimate the allowed range of smooth extrapolations of the mass function consistent with available data. The mass–absolute visual magnitude relation which maps the Miller–Scalo mass function to the luminosity function of figure 1 is shown in figure 2, as is a theoretical curve described by Kroupa, Tout & Gilmore (1990).

Mass functions defined by a Gaussian, by the Miller-Scalo function, and by the Miller-Scalo function with a flat extension to low masses cannot be distinguished within a 60% confidence limit. Power-law mass function lies well outside the 0.1% confidence level, while the Salpeter mass function is completely ruled out, in agreement with the results of Scalo (1986). A single power-law is in fact an adequate description of the mass function for masses above $\sim 0.4\mathcal{M}_\odot$, but deviates substantially, in predicting too many stars, at lower masses. Below about $0.4\mathcal{M}_\odot$ the mass function can again be described well by a power law, but this time with a flat slope.

The range of mass functions which is consistent with both the observational constraints on the mass–absolute visual magnitude relation and with the observed stellar luminosity function for field stars near the Sun is shown in figure 3. We emphasise that a smooth mass function is not *required* by available data, any more than is a mass function with structure. The important result is that a smooth mass function is *consistent* with all available data.

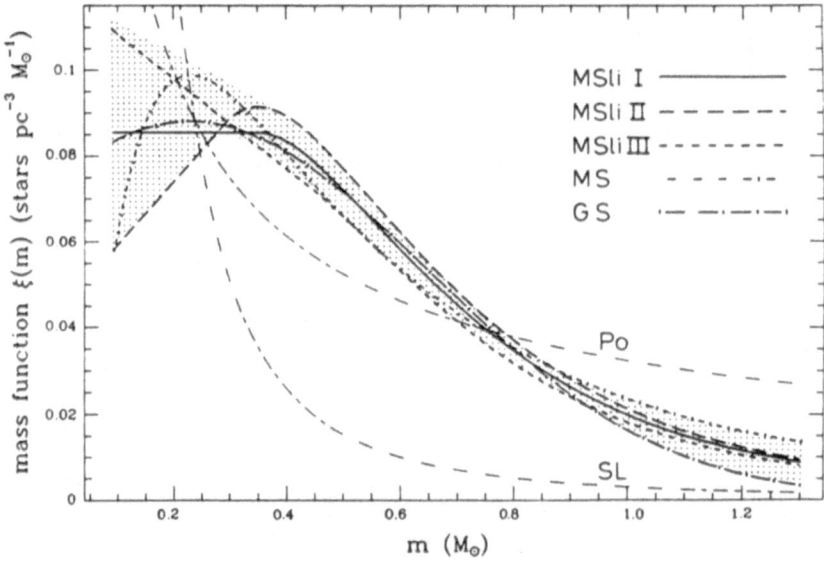

Figure 3. The shaded area represents the range of mass functions which are consistent at the 95% level with the observed stellar luminosity function in figure 1 and with the observational constraints on the mass–absolute visual magnitude relation shown in figure 2.

It is straightforward to calculate the mass density in stars in the solar neighbourhood by integrating the extrapolated mass functions to $m = 0$ from $m = 0.35 \mathcal{M}_\odot$. All the mass functions shown to be adequate in figure 3 are convergent. The resulting mass densities range from $0.005 \mathcal{M}_\odot \mathrm{pc}^{-3}$ to $0.008 \mathcal{M}_\odot \mathrm{pc}^{-3}$. Either the Salpeter (1955) mass function or that due to Miller & Scalo (1979) can be used to calculate the mass in stars more massive than $0.35 \mathcal{M}_\odot$, which is $0.05 \mathcal{M}_\odot \mathrm{pc}^{-3}$.

It is important to note that all the mass functions consistent with both the observed luminosity function and the mass-luminosity relation are convergent when extrapolated to low masses. The total amount of mass contained in a smooth extrapolation of the stellar mass function to zero mass from $0.35 \mathcal{M}_\odot$ is less than about 20 per cent of the mass contained in the stellar mass function for higher mass stars. Over the range from $\sim 0.2 \mathcal{M}_\odot$ to $\sim 1.0 \mathcal{M}_\odot$ the mass function for field stars near the Sun is smooth. The several features evident in the stellar *luminosity* relation over this range are all consistent with being due to structure in the mass-luminosity relation, caused by the effects of varying opacity sources on the stellar equation of state.

Thus there remains no robust evidence to support the existence of either a dynamically significant population of very low mass stars, or of any intrinsic structure in the stellar initial mass function. This result has considerable implications for models of galactic evolution involving bimodal star formation processes, and provides no support for the concept of brown dwarfs as the location of baryonic dark matter in galaxies.

With the mass functions shown in figure 3 we can now calculate the mass to light ratio of the stellar component of the Galactic disk near the Sun. The total mass in stars more massive than $0.35 \mathcal{M}_\odot$ is $\sim 0.05 \mathcal{M}_\odot \mathrm{pc}^{-3}$, while the mass in lower mass objects, *assuming a smooth linear extrapolation of the stellar mass function to zero mass*, is $\lesssim 0.008 \mathcal{M}_\odot \mathrm{pc}^{-3}$. A further $\sim 0.005 \mathcal{M}_\odot \mathrm{pc}^{-3}$ is in white dwarfs. The corresponding luminosity of the local disk has been measured from star counts to be $j = 0.063 \pm 0.006 \mathcal{L}_\odot, \mathrm{vpc}^{-3}$. That is, the mass to light ratio of the local disk, in solar visual units, is $M/L \approx 1$. Very similar values are measured for open clusters. For the *old disk* stars only, the corresponding value is $M/L \approx 2$. This value is very similar to the value of 2.5 measured dynamically for globular clusters, and significantly smaller than the dynamical M/L value of about 10 deduced for elliptical galaxies. We now consider how this value compares to the local dynamical mass. The remainder of this paper is based on the analysis of the mass distribution in the Galactic disk of Kuijken & Gilmore (1989a,b,c), and follows the discussion in Gilmore, Wyse & Kuijken (1989).

3. THE MASS DISTRIBUTION IN THE GALACTIC DISK

The distribution of the total amount of mass of all types in the Galactic disk is character-ized by two numbers, its local *volume* density ρ_o and its total *surface* density $\Sigma(\infty)$. Both these dynamical quantities are derived from a measurement of the vertical Galactic force field, $\mathcal{K}_z(z)$. Although $\Sigma(\infty)$ and ρ_o are different measures of the distribution of mass in the Galactic disk near the Sun, they are related. Of the two, the most commonly discussed is the local *volume* mass density — *ie* the amount of mass per unit volume near the Sun, which for practical purposes is the same as the volume mass density at the Galactic plane. This quantity has units of $\mathcal{M}_\odot \mathrm{pc}^{-3}$, and its local value is often called the 'Oort limit' in honour of the early attempt at its measurement by Oort (1932). The contribution of *iden-tified* material to the dynamically determined Oort limit may be determined by summing all local observed matter — an observationally difficult task, which leads to considerable uncertainties. The uncertainties arise in part due to difficulties in detecting very low lu-minosity stars, even very near the Sun, in part from uncertainties in the binary fraction among low mass stars, in part from uncertainties in the stellar mass–luminosity relation, but mostly from uncertainties in determining the volume density of the interstellar medium (ISM). This latter uncertainty is exacerbated since the physically important quantity for dynamical purposes is the mean volume density of the patchily distributed ISM at the so-lar galactocentric distance. The best available determination of the local mass density in identified material is $\sim 0.1 \mathcal{M}_\odot \mathrm{pc}^{-3}$, with a very-poorly defined uncertainty of perhaps as much as 25% in this value. Comparison of this value with that determined from dynamical analyses is required to test for the existence of dark matter associated with the Galactic disk.

The second measure of the distribution of mass in the solar vicinity is the integral surface mass density. This quantity has units of $\mathcal{M}_\odot pc^{-2}$, and is the total amount of disk mass in a column perpendicular to the Galactic plane. It is this quantity which is required for the interpretation of rotation curves and the large-scale distribution of mass in galaxies. Recent determinations of this surface mass density lead to values near $50\mathcal{M}_\odot pc^{-2}$. As an indication of the global dynamical significance of this mass density, the contribution of the disk potential generated by some known local surface mass density to the local circular velocity, assuming an exponential disk with the Sun 2.5 radial scale lengths from the Galactic centre, is

$$V_{c,disk} \sim 150 \left(\frac{\Sigma_{local}}{60\mathcal{M}_\odot pc^{-2}} \right)^{\frac{1}{2}} km\,s^{-1}. \qquad 4$$

The local circular velocity is $\sim 220\,km\,s^{-1}$. The contribution of the potential due to a given enclosed mass $\mathcal{M}$ to the circular velocity is approximately $V_c^2 = G\mathcal{M}/r$, so that contributions to the observed local circular velocity from the various mass components generating the Galactic potential add in quadrature. Thus the Galactic disk provides only about 50% of the total Galactic potential at the solar galactocentric distance. The nature of the mass which generates the other half of the potential remains unknown. One might hope to restrict the range of possibilities for its nature by identifying the smallest characteristic scale on which this dark mass is distributed. This is the basis of attempts to determine in detail the mass distribution near the Sun.

If one knew both the local *volume* mass density and the integral *surface* mass density of the Galactic disk, one could immediately constrain the scale height of any contribution to the local volume mass density whose existence was proven from a precise measurement of the Oort limit, but whose nature was not identified. For example, one might suspect that some fraction of the local volume mass density was unidentified (*ie* a local 'dark mass' problem), but also determine a surface density which is effectively fully explained by observed mass. Then the unidentified contribution to the local volume density would have to have a small scale height, in order that its integral contribution to the surface density be small. In view of the very small scale height on which it must be distributed, it would then be plausible to deduce that any 'local' dark mass unidentified in the volume mass density near the Sun was not the same 'dark' mass which dominates the extended outer parts of galaxies.

3.1. MEASUREMENT OF THE GALACTIC POTENTIAL

Determination of the volume mass density and the integral surface mass density near the Sun require similar observational data, namely distances and velocities for a suitable sample of tracer stars, but rather different analyses.

All determinations of the mass distribution in the Galactic disk require a solution of the collisionless Boltzmann equation. In view of the inconvenience of general solutions of this equation derived from real data, in practise one utilises its vertical velocity moment, the vertical Jeans' equation:

$$\mathcal{K}_z = \frac{1}{\nu}\frac{\partial}{\partial z}(\nu\sigma_{zz}^2) + \frac{1}{R\nu}\frac{\partial}{\partial R}(R\nu\sigma_{Rz}^2) \qquad 5$$

where $\nu(R,z)$ is the space density of the stars, and $\vec{\sigma}_{ij}(R,z) = \langle v_i v_j \rangle - \langle v_i \rangle \langle v_j \rangle$ their velocity dispersion tensor.

The first term on the right hand side of equation 5 is dominant, and contains a logarithmic derivative of the stellar space density $\nu(z)$, and a derivative of the vertical velocity dispersion, σ_{zz}. The part of this term containing the derivative of the space density dominates for tracer stellar populations which are strongly concentrated to the Galactic plane. This point is not often appreciated adequately, but means that most determinations of the local mass density have been limited by their determination of the stellar density distribution, and not by the precision or number of the velocity data available. This point is discussed further by Crézé $etal$ (1989).

The second term in the Jeans' equation 5 describes the tilt of the stellar velocity ellipsoid away from the local cylindrical–polar coordinate system in which velocity dispersions are measured. One therefore needs the R-gradients of σ_{Rz} and of ν. There are no general analytical solutions for this term, as it depends on the unknown "third integral" of the motion, and has been ignored until recently. Numerical calculation of orbits in realistic Galactic potentials, however, shows that one may however derive a realistic upper limit on the importance of the second term in the Jeans' equation by considering velocity ellipsoids which are oriented towards the Galactic centre. In this case, if the disk of the Galaxy is self-gravitating, radially exponential and has a constant vertical scale height, as is seen in external disk galaxies, vertical balance implies (for disk surface density μ) $\sigma_{zz}^2 \propto \mu$, and hence

$$\sigma_{zz}^2 \propto \mu \propto \nu \propto e^{-R/h_R}. \qquad 6$$

Thus we obtain

$$\frac{1}{R\nu}\frac{\partial}{\partial R}(R\nu\sigma_{Rz}^2) = 2(\alpha^2 - 1)\sigma_{zz}^2 \left(\frac{\alpha^2 z^3}{(\alpha^2 z^2 + R^2)^2} - \frac{Rz}{h_R(\alpha^2 z^2 + R^2)} \right) \qquad 7$$

as the tilting term for a radially exponential population of constant vertical scale height with a velocity ellipsoid of constant axis ratio $\sigma_{RR} : \sigma_{zz} = \alpha$ which points at the Galactic centre (Kuijken & Gilmore 1989a). Since this term is proportional to σ_{zz}, inserting it into the Jeans' equation (equation 5) gives a linear equation in σ_{zz}, from which one can deduce $\mathcal{K}_z$. The tilt term is clearly unimportant near the plane, as $z \Rightarrow 0$, and may legitimately be ignored in analyses restricted to local determinations of ρ_o. As one attempts to derive the gravitational field at increasingly larger distance from the Galactic disk, however, the low mass of the disk, with its corollary that the Galactic potential is dominated by the extended dark mass distribution, means that the limiting feature of analyses becomes one's lack of knowledge of the large–scale distribution of matter, and not the sampling errors in available data.

3.2. FROM GRAVITY TO MASS

Given a measurement of the gravitational field $\vec{\mathcal{K}}(R, z)$ in an axisymmetric galaxy, the total density ρ of gravitating matter follows from Poisson's equation:

$$\nabla \cdot \vec{\mathcal{K}} = -4\pi G \rho. \qquad 8$$

In the case of a disk galaxy we can express the R-gradient in $\nabla \cdot \vec{\mathcal{K}}$ in terms of the observed circular velocity at the Sun, v_c, or in terms of the Oort constants of Galactic rotation A and B

$$
\begin{aligned}
\rho &= -\frac{1}{4\pi G}\left\{\frac{\partial \mathcal{K}_z}{\partial z} + \frac{1}{R}\frac{\partial}{\partial R}\left(R\mathcal{K}_R\right)\right\} \\
&= -\frac{1}{4\pi G}\left\{\frac{\partial \mathcal{K}_z}{\partial z} - \frac{1}{R}\frac{\partial\left(v_c^2\right)}{\partial R}\right\} \\
&= -\frac{1}{4\pi G}\left\{\frac{\partial \mathcal{K}_z}{\partial z} + 2\left(A^2 - B^2\right)\right\}.
\end{aligned}
\qquad 9
$$

For a disk galaxy with an approximately flat rotation curve the second term is small within a few kpc of the disk plane (Kuijken & Gilmore 1989a provide a more exac' calculation; for an exactly flat rotation curve $A^2 - B^2 \equiv 0$ at $z = 0$), so we can integrate in z to obtain the total column density $\Sigma(z)$ between heights $-z$ and z relative to the disk plane $z = 0$:

$$\Sigma(z) = \int_{-|z|}^{|z|} \rho(z)dz = \frac{|\mathcal{K}_z|}{2\pi G} - \frac{\left(A^2 - B^2\right)}{\pi G}|z|. \qquad 10$$

The physical interpretation of a determination of the Galactic $\mathcal{K}_z(z)$ force law can now be seen from inspection of equations 5, 9, and 10. In effect, one measures the pressure–gravity balance of the *collisionless* stellar 'fluid'. The hydrodynamic analogy following from the description of the collisionless Boltzmann equation as the equation of stellar hydrodynamics is particularly appropriate here. The dominant first term on the right hand side of equation 5 contains a logarithmic derivative of the stellar spatial density $\nu(z)$, and the stellar velocity dispersion σ_{zz}. The spatial density term plays the role of a scale height, the velocity dispersion is analogous to a temperature, and the product $\nu\sigma_{zz}$ is a pressure.

Some complexity is introduced since the stellar fluid is collisionless. This means that non-diagonal terms in the velocity dispersion tensor exist [the second term on the right hand side of equation 5] in consequence of the fact that pressure is not isotropic, and we have no equation of state to close the series of moment equations. It also means that one must measure the temperature-density balance locally by point–by–point sampling of many stars, imposing some inconvenience on the observational techniques, but conversely allowing the possibility of more sophisticated analyses.

It is evident from the equations above that determinations of the local volume mass density ρ_o depend on the square of any distance scale errors in the tracer population, since they are derived from the second derivative of the (log of the) stellar space density distribution, while determinations of the surface mass density are linearly proportional to the distance scale, being based on the first derivative. Since the stellar space density is itself a derivative from the basic star count data, it is evident that determinations of $\mathcal{K}_z(z)$

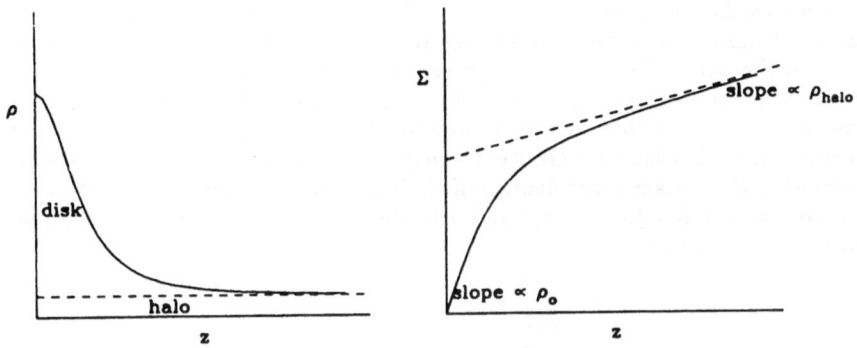

Figure 4. The volume density (a, left) and surface mass density, or equivalently $\mathcal{K}_z(z)$, (b, right) as a function of vertical distance z from the disk plane in a simple disk-halo model.

and particularly of ρ_o are very sensitive to sampling noise (and systematic distance errors) in the data.

A further illustration of the physical significance of determinations of the $\mathcal{K}_z(z)$ law can be noted by considering the distribution of mass in a simple disk-halo galaxy. The total mass density along a slice perpendicular to the disk for a generic disk-halo system is shown schematically in figure 4. (By 'halo' in this context we mean any mass component which is not distributed like the dominant mass in the old disk. Nothing is implied as regards the nature of this mass; it includes the luminous 'spheroid' or 'bulge' as well as any roughly spherical distribution of dark matter.) At low z, the density is mostly due to the disk component, while further away from the plane the density of the halo, which at large galactocentric radii is essentially constant over a few disk scale heights, dominates. The surface density, or equivalently $|\mathcal{K}_z|$, for such a system is also shown in figure 4. This force law has the following generic features:

♡ At low z, smaller than the scale heights of the dominant disk components, $|\mathcal{K}_z|$ rises almost linearly, with slope $4\pi G\rho_o$.

♡ At large z, beyond most of the mass of the disk, $|\mathcal{K}_z|$ is again linear, but with a much reduced slope, equal to $4\pi G\rho_{\mathrm{halo}}$.

♡ The extrapolation of this latter linear portion back to $z = 0$ has an intercept of $2\pi G\Sigma_{\mathrm{disk}}$.

An accurately measured $\mathcal{K}_z$-profile over a wide range in z thus yields the disk surface density, as well as the volume density of the halo and that of the disk at $z = 0$.

3.3. DETERMINATION OF THE SURFACE MASS DENSITY

Because high-energy stars are present at all heights above the Galactic plane, measurements of the potential very close to the plane still require knowledge of the high-energy tail of the distribution function. Therefore either the tail of the velocity distribution at low z, or the density *and* potential at high z, are required to measure the potential at low z, and hence to deduce the local volume density of matter ρ_o. The phase-space distribution function we discuss below, however, depends on the density only at points farther from the plane than the height at which data are being analysed. It is possible to capitalise on this insensitivity to the detailed shape of the potential (equivalently, the detailed mass distribution) near the plane to derive the potential at large distances from the plane from high-z data alone. Since a measurement of $\mathcal{K}_z$ at any height relates directly to the total surface density integrated to that height, this is extremely useful, allowing us to obtain meaningful determinations of the surface mass density of the Galactic disk from high-z data alone.

Given a distribution function $f_z(E_z)$ and a potential $\psi(z)$, we can calculate the density $\nu(z)$, which is just a moment of f_z:

$$\nu(z) = \int_{-\infty}^{\infty} f_z(z, v_z) dv_z$$
$$= 2 \int_{\psi(z)}^{\infty} \frac{f_z(E_z)}{\sqrt{2(E_z - \psi(z))}} dE_z. \qquad 11$$

Reparameterizing the z-height in terms of the potential ψ, we have

$$\nu(\psi) = 2 \int_{\psi}^{\infty} \frac{f_z(E_z)}{\sqrt{2(E_z - \psi)}} dE_z. \qquad 12$$

This equation is an Abel transform, which can be inverted to give (see *eg* Binney & Tremaine 1987):

$$f_z(E_z) = \frac{1}{\pi} \int_{E_z}^{\infty} \frac{-d\nu/d\psi}{\sqrt{2(\psi - E_z)}} d\psi, \qquad 13$$

so that there is a unique relation between $\nu(\psi)$ and $f_z(E_z)$. Because of this equivalence of $\nu(\psi)$ and $f_z(E_z)$, there is a triangular mathematical relationship between the three functions $\psi(z)$, $\nu(z)$ and $f_z(E_z)$: any one of them can be deduced from the other two.

The important feature of equation 13 is that $f_z(E_z)$ depends on the density only at points where the potential exceeds E_z, *ie* beyond the point $z = \psi^{-1}(E_z)$. It is this property which allows the derivation of $\mathcal{K}_z(z)$ at large z independently of the poorly known distribution of mass very near the plane.

When starting from a set of (z, v_z) data for the tracer population, only the first of the three quantities $\nu(z)$, $f_z(E_z)$ and $\psi(z)$ is known, as one needs the potential to be able to convert $f_z(z, v_z)$ into $f_z(E_z)$. Therefore, before being able to make an inversion such as that given in equation 13 we have to make some assumption about the form of the potential, or about $f_z(E_z)$. Assuming that the tracer is isothermal ($f_z \propto e^{-E_z/\sigma_{z_*}^2}$) is an example of the latter, leading to a trivial inversion.

An analysis technique based on equation 13 has been devised by Kuijken & Gilmore (1989a,b) for the determination of $\mathcal{K}_z(z)$, and $\Sigma(z)$. The essential feature of that analysis is that one avoids the assumption of isothermality, by instead postulating a range of potentials $\psi(z)$, and for each of them calculating $f_z(z, v_z)$ from $\nu(z)$. The range of model distribution functions can then be compared to the observed distribution function of velocity–distance data, and used to select the best-fitting model potential. One may then derive the integral surface mass density of the Galactic disk.

This technique has been developed by Kuijken & Gilmore (1989a,b), and applied to a new set of stellar data they obtained, which extends to 2kpc towards the South Galactic pole. From these data, they determine a dynamical total surface mass density $\Sigma_\infty = 46 \pm 9 \mathcal{M}_\odot \mathrm{pc}^{-2}$. These same authors integrate the local observed volume mass density in stars through their derived potential, and add the directly observed mass in the inter-stellar medium, to derive an observed integral surface mass density of $\Sigma_{\infty,obs} = 48 \pm 8 \mathcal{M}_\odot \mathrm{pc}^{-2}$. There is thus evidence that there is no significant unidentified mass associated with the Galactic disk.

3.4. DETERMINATION OF THE VOLUME MASS DENSITY

Almost all determinations of the local volume mass density ρ_o model the distribution function of the tracer stars as one or more *isothermal* components. An isothermal component by definition has constant velocity dispersion, independent of z. Then one can easily see that the velocity distribution has to be gaussian: by equation 5, if σ_{zz} is independent of z,

$$\nu(z) \propto e^{-\psi(z)/\sigma_{zz}^2},\qquad 14$$

and hence the Abel inversion of equation 12 yields

$$f_z(E_z) \propto e^{-E_z/\sigma_{zz}^2} \propto e^{-\frac{1}{2}v_z^2/\sigma_{zz}^2}.\qquad 15$$

This method is especially simple if the tracer appears to be a single isothermal component and the analysis is restricted to stars near the plane, as then the Jeans' equation simplifies to

$$K_z = \sigma_{zz}^2 \frac{\partial}{\partial z}(\ln\nu(z))\qquad 16$$

and only the density gradient of the tracer population is needed.

An almost invariable finding in early determinations of the volume mass density was that a *maximum* was found in the $K_z(z)$ law a few hundred pc from the plane. Such a result is physically impossible, corresponding to a layer of negative mass. The solution to this inconvenient situation was found by Oort (1960), who imposed consistency with the Poisson equation 5 on his solution of equation 16. In effect, Oort assumed that uncertainties in the data rather than in the assumptions underlying the analysis are at fault, and that the theoretically reasonable solution which is most consistent with the (defective) data will produce an answer which is close to the 'true' answer. Assignment of a realistic uncertainty to the resulting answer is of course somewhat problematic in this case.

More recently, Bahcall (1984a,b,c) has extended this idea, and has improved the theoretical methods with which to determine the local volume density of matter ρ_o, by deriving a joint solution of the Poisson and collisionless Boltzmann equations. This solution requires the derived $K_z(z)$ law to be physically possible, which is a highly desirable and quite reasonable constraint. The analytical techniques developed by Bahcall (1984a,b,c) represent a considerable improvement over those applied previously, and for the first time allow a derivation of ρ_o which is limited by the quality of the available observational data, rather than by the approximate nature of the analysis.

In this technique, the solar neighbourhood is divided into different isothermal components with local volume density $\rho_{i,0}$ and spatial density distribution ν_1, each of which responds to the potential ψ *via* its velocity dispersion:

$$\nu_i = \nu_{i,0}\, e^{-\psi/\sigma^2_{z,i}}. \tag{17}$$

Self-consistency of the potential and the total matter density in these components then requires that Poisson's equation be satisfied, *ie*

$$4\pi G\rho = 4\pi G \sum_i \rho_{i,0}\, e^{-\psi/\sigma^2_{z,i}} = \frac{d^2\psi}{dz^2}. \tag{18}$$

Here the density includes a constant 'effective halo' density, due to the halo mass and the radial gradients of the global gravitational field of the Galaxy. With the boundary conditions $\psi = \psi' = 0$ at $z = 0$, equation 18 can easily be integrated forwards to obtain $\psi(z)$. A variety of dark matter components can be added in, using the same prescription as for the visible components, and the resulting potentials calculated.

A critical problem is to analyse the gas and stars into isothermal components; it may be possible to do this to reasonable precision very near the plane, but at higher z the calculated potential becomes increasingly sensitive to the precise $z = 0$ velocity dispersions and densities. The modelling of the gas is also a problem, as it accounts for about one-half of the locally identified volume density, but its precise density and spatial distribution are poorly known.

Reliable deconvolution of non-isothermal populations into isothermal components is neither trivial nor robust, nor a well-defined concept in principle. Most importantly, there is the question of how physically meaningful it really is to separate a stellar sample into discrete isothermal subsamples. Given continual star formation and the continual but poorly understood diffusion processes thought to be active in heating the disk, and the consequent uncertainty in knowing the expected shape of the velocity distribution function in even a coeval stellar population, this can at best be only an *ad hoc* approximation to the true distribution. A very large number of discrete isothermals is clearly necessary for a close approximation to a real galactic disk. Second, such deconvolutions are far from unique, leading to non-unique force laws.

At distance z_j, a superposition of N_{iso} isothermal components, each with velocity dispersion $\sigma^2_{zz,i}$ and density ν_i at $z = 0$ has density

$$\nu(z_j) = \sum_{i=1}^{N_{iso}} \nu_i e^{-\psi(z_j)/\sigma^2_{z,i}}. \tag{19}$$

This emphasises that at different distances from the plane the relative densities of the 'isothermal' components are not the same, but that those of highest velocity dispersion dominate increasingly at higher z. If we were able to specify the velocity dispersions and the spatial density normalisation at $z = 0, \nu_i(0)$ of a sufficiently large number of components, regardless of the fact that such 'components' have no physical meaning in the sense of being astrophysically identifiable stellar tracer populations, this would allow a direct solution for the potential from the density profile. However in practice we have to derive the number density of the high–velocity dispersion components at $z = 0$ either from the sparsely populated wings of the distribution function, or from high−z velocity data, as only at greater heights are these components sufficiently in evidence to allow reliable determination of their density. However, for $z \neq 0$, the relative densities of the isothermal components depend on the potential as well as the ν_i: thus we are faced with a self-consistency problem involving both the density profile and the potential far from the plane even if we wish to solve for the potential only near the plane. This an irreducible consequence of Poisson's equation (equation 8 above), which shows the *local* volume mass density to depend on the second derivative of the potential at the Sun. One cannot derive a second derivative from purely local measurements.

Bahcall (1984a,b) used his algorithm to reanalyse the available F dwarf and K giant high Galactic latitude data with new models which are self-consistent in the sense that the matter which generates the gravitational field itself responds to the field in a manner described by the collisionless Boltzmann equation. Bahcall found that:

(i) the gravitational field due to the $0.10\mathcal{M}_\odot\text{pc}^{-3}$ of stars and gas that are identified in the solar neighbourhood is inconsistent with the gravitational fields derived from the data;

(ii) depending on its scale height, a further 0.06–$0.14\mathcal{M}_\odot\text{pc}^{-3}$ of unidentified matter is required. This unidentified matter is not part of a very extended halo, though that must also exist and have a local volume mass density of $\sim 0.01\mathcal{M}_\odot\text{pc}^{-3}$ so that there is sufficient mass in the Galaxy to generate the potential required to explain the local circular velocity. Hence this result implies significant amounts of disk-like, dissipational dark matter in the solar neighbourhood.

If the determinations of both the local volume mass density and the local surface mass density described above are correct, then very severe constraints on the nature of the dark mass follow. The specific limit on the spatial distribution of the dark mass suspected to exist in the local volume, is that the dark mass must be distributed with an effective scale height 2H, such that $\Sigma = 2\rho_0\text{H}$, where

$$\rho_\text{dark} < \frac{10\mathcal{M}_\odot\text{pc}^{-2}}{2\text{H}} \qquad\qquad 20$$

Large amounts of local missing mass thus require a very small scale height for consistency with the measured low value for the surface mass density. For example, if one really believed that $0.085\mathcal{M}_\odot\text{pc}^{-3}$ were unidentified near the Sun, that matter must be distributed with an effective scale height of less than 60pc. This is less than the scale height of the cold ISM.

3.5. UNCERTAINTIES IN THE LOCAL VOLUME MASS DENSITY

In view of the important consequences of the existence of large amounts of dissipational dark mass, it is of interest to examine the uncertainties in the determination of the Oort limit. The sensitivity of determinations of the local volume mass density ρ_o to uncertain data lies in the modelling of the stellar velocity distribution near the Galactic plane, and in the determination of the stellar density distribution with distance from this plane. Both F dwarf and K giant tracer samples have been analysed to determine ρ_o, with both producing a result of $\rho_o \sim 0.20\mathcal{M}_\odot\text{pc}^{-3}$, where the identified mass provides $\rho_{o,\text{obs}} = 0.10\mathcal{M}_\odot\text{pc}^{-3}$ (Bahcall 1984c).

The effect of *random* errors on determinations of the Oort limit can be investigated using Monte Carlo simulations of the data acquisition and analysis. Random errors can produce

154

both random and systematic effects on a dynamically-determined quantity, through effects similar to Malmquist bias. It is therefore important to understand the effects of such errors. Analyses of this type have been undertaken, but unfortunately disagree in their conclusions. Gilden & Bahcall (1985) conclude that random errors produce an unbiased uncertainty of $\sim 12\%$ in ρ_o, while Bienaymé, Robin & Crézé (1987) and Crézé, Robin & Bienaymé (1989) conclude that random errors produce an uncertainty of $\sim 50\%$, and also produce a bias towards an erroneous detection of unidentified mass. The difference in these results is due to different techniques for handling observational errors in the simulations, suggesting that the appropriate uncertainty to apply to determinations of the Oort limit is is not yet well quantified, but that it might be very large.

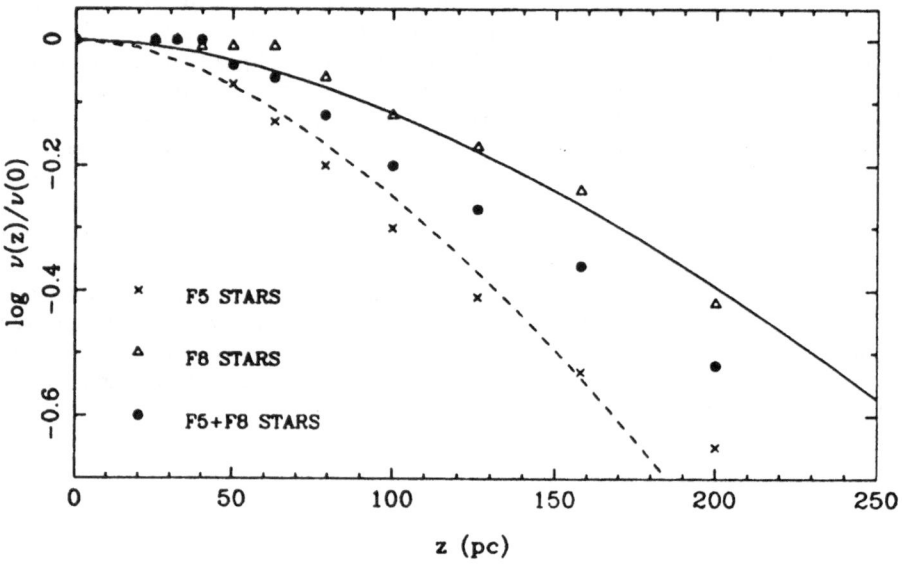

Figure 5. The Hill *et al.* (1979) F star samples. The difference between the density profiles of the F5 and the F8 samples is evident. The curves show separate model fits calculated using the algorithm devised by Bahcall (1984a) to the F5 and the F8 subsets of the data defined by Hill *et al.* Only the averaged sample (solid points) was analysed by Bahcall (1984b). The models shown have local volume mass densities: $\rho_o = 0.11 \mathcal{M}_\odot \mathrm{pc}^{-3}$, *ie* with no missing mass (solid line), and $\rho_o = 0.29 \mathcal{M}_\odot \mathrm{pc}^{-3}$ (dashed line).

An alternative, and perhaps more objective, method for the determination of an error bar is to do the experiment with two different but supposedly similar samples of stars, and compare the resulting answers. Fortunately for this purpose, the F star sample analysed is the sum of two sub-samples (F5 and F8, Hill *et al.* 1979), with no evidence for a difference between their velocity distributions (Adamson *et al.* 1988). For steady-state stellar populations, two tracer populations with the same kinematics in the same gravitational potential must follow the same spatial density distribution. For the F5 and F8 samples the data shown in Figure 5 show that this is not the case. One or both of the data or the assumptions underlying the modelling of the F star kinematics is thus clearly in error. The amplitude of the resulting uncertainty can be found by deducing ρ_o from each of the three F star samples, F5, F5+F8 and F8, using the algorithm derived by Bahcall (1984a). The resulting values of ρ_o are $0.29\mathcal{M}_\odot\mathrm{pc}^{-3}$, $0.185\mathcal{M}_\odot\mathrm{pc}^{-3}$ (reproducing the result derived by Bahcall (1984b) exactly), and $0.11\mathcal{M}_\odot\mathrm{pc}^{-3}$ respectively. Thus one may deduce that there is twice as much mass missing as observed in the local volume density, just as much missing as observed, or no missing mass at all, depending on which sample of stars one chooses to analyse. Clearly, the available F star data are not capable of providing any evidence either for or against the concept of missing mass near the Sun. These results are in good agreement with the simulations of Crézé *etal* (1989).

UPGREN K GIANT SAMPLE

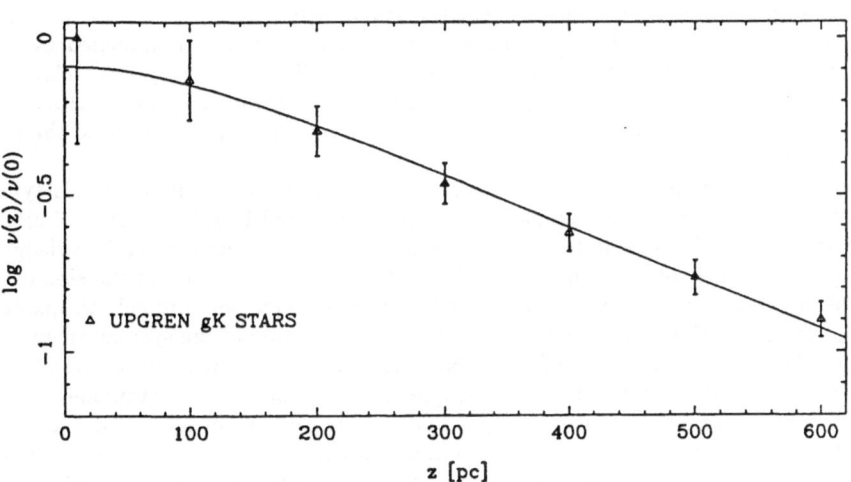

Figure 6. Weighted fit to the Upgren (1962) K giant data, using the velocity distribution measured by Hill (1960). The model shown contains no dark matter in the Galactic disk, and has $\rho_o = 0.10\mathcal{M}_\odot\mathrm{pc}^{-3}$.

The sample of K giants which has been analysed previously has been shown to have a velocity distribution which is consistent with a single isothermal, with a velocity dispersion of $\sim 20\,\mathrm{km\,s}^{-1}$ (Bahcall 1984c). Thus, unlike the F stars, in this model the K giants consist entirely of old disk stars, with neither young disk nor thick disk star representatives. Since stars of a wide range of masses become K giants, including the present F dwarfs, this model is inherently implausible. Remember also that Oort (1960) found it necessary to model the same K giant sample as the sum of three isothermal components, with (crudely) one component each for the young disk stars, the old disk stars, and the high velocity halo stars. A further complication follows from a feature of Bahcall's analysis, which assigns high weight to the density profile near the plane, where the number of stars counted is smallest. Reanalysis of published data including weighting of the density data by its Poisson noise and using the detailed fit to the local K giant velocity data derived by Hill (1960), leads to a value of $\rho_o = 0.10\mathcal{M}_\odot\mathrm{pc}^{-3}$ (Figure 6; *cf.* Kuijken & Gilmore 1989c). The previously derived value from the same data using the same analysis technique was $\rho_o = 0.21 \mathcal{M}_\odot \mathrm{pc}^{-3}$ (Bahcall 1984c).

The fundamental reason for the difficulty with analysing data by deconvolving the tracer sample into a few isothermal populations has been outlined by Fuchs & Wielen (1989). Basically, it is physically impossible for such a thing as a real "isothermal" stellar population to exist, since stars are formed and evolve dynamically as a continuum. Thus any "isothermal" stellar group must include a range of ages and kinematics. Determination of the local volume mass density is much more than linearly sensitive to deviations from isothermality. Hence undetectable deviations of a tracer sample from the idealised assumptions of the analysis produce significant errors in the resulting volume mass density. Fuchs & Wielen (1989) have calculated the amplitude of this effect. They show that the known rate of kinematic diffusion for stars near the Sun operating for the age range corresponding to the observed F star sample means that the apparently isothermal F star sample, which has an observed velocity dispersion of $\sim 11.5\mathrm{km/s}$, can be better modelled as having a range of velocity dispersions from $10\mathrm{km/s}$ to $13\mathrm{km/s}$. This apparently small change in the model results in a change in the derived local volume mass density from $0.19\mathcal{M}_\odot \mathrm{pc}^{-3}$ to $0.10\mathcal{M}_\odot \mathrm{pc}^{-3}$. Thus the results of such an analysis are extremely sensitive to the precision of the adopted kinematic model.

Determination of the appropriate uncertainties to apply to analyses of this type is extremely complex, as many of the parameters are correlated in subtle ways. Similarly, the amplitude of possible systematic errors, due for example to our limited knowledge of the gradient in the local rotation curve and of the shape and orientation of the stellar velocity ellipsoid at large distances from the Galactic plane, is extremely difficult to quantify. As with the determination of ρ_o, the most reliable determination of the systematic uncertainty will come from the comparison of the results described here with the results of a similar measurement based on an independent sample of stars, when such is available. Neither the random nor the systematic uncertainties in either the amount of identified mass or in the amount of dynamical mass are well determined, though both types of uncertainty are considerable. It is also true that absence of evidence is not evidence of absence. Occam's razor must be invoked, so that the minimal conclusion required by available data and analyses is clearly that there is no robust evidence for the existence of any dark mass associated with the Galactic disk. There is no significantly detected population of dark baryons associated with the Galactic disk.

The resulting mass to light ratio of the Galactic disk near the Sun is about 3, with the total mass being roughly two-thirds stars and stellar remnants, and the remaining one-third the inter-stellar medium. All of it is baryonic, but the baryonic mass associated with the Galactic disk is small compared to the total Galactic mass, whatever type of matter that is, and is also small compared with the number of baryons expected in the Universe from Big Bang nucleosynthetic arguments.

References

Adamson, A.J., Hill, G., Fisher, W., Hilditch, R.W. & Sinclair, C.D., 1988. *Mon. Not. Roy. Astron. Soc.* **230**, 273.

Bahcall, J.N., 1984a. *Astrophys. J.* **276**, 156.

Bahcall, J.N., 1984b. *Astrophys. J.* **276**, 169.

Bahcall, J.N., 1984c. *Astrophys. J.* **287**, 926.

Binney, J. & Tremaine, S., 1987. *Galactic Dynamics* (Princeton U.P., Princeton, 1987.)

Carlberg, R.G. & Innanen, K.A., 1987. *Astrophys. J.* **94**, 666.

Crézé, M., Robin, A. & Bienaymé, O., 1989. *Astron. Astrophys.* **211**, 1.

Fowler, W. 1990 *Baryonic Dark Matter* eds D. Lynden Bell & G. Gilmore, p257 (Kluwer, Dordrecht)

Fuchs, B. & Wielen, R. 1989. In Preparation.

Gilmore, G., Reid, N. & Hewett, P., 1985. *Mon. Not. Roy. Astron. Soc.* **213**, 257.

Hawkins, M. R. S. & Bessell, M. S., 1988. *Mon. Not. Roy. Astron. Soc.* **234**, 177.

Gilden, D.L. & Bahcall, J.N., 1985. *Astrophys. J.* **296**, 240.

Hill, E.R., 1960. *Bull. Astron. Inst. Netherlands* **15**, 1.

Hill, G., Hilditch, R.W. & Barnes, J.V., 1979. *Mon. Not. Roy. Astron. Soc.* **186**, 813.

Kroupa, P., Tout, C.A. & Gilmore, G., 1990. *Mon. Not. R. astr. Soc.*, in press

Kuijken, K. & Gilmore, G., 1989a. *Mon. Not. Roy. Astron. Soc.* **239**, 571.

Kuijken, K. & Gilmore, G., 1989b. *Mon. Not. Roy. Astron. Soc.* **239**, 605.

Kuijken, K. & Gilmore, G., 1989c. *Mon. Not. Roy. Astron. Soc.* **239**, 651.

Miller, G. E. & Scalo, J. M., 1979. *Astrophys. J. Suppl.* **41**, 513.

Oort, J.H., 1932. *Bull. Astron. Inst. Netherlands* **6**, 249.

Oort, J.H., 1960. *Bull. Astron. Inst. Netherlands* **15**, 45.

Popper, D. M., 1980. *Ann. Rev. Astron. Astrophys.* **18**, 115.

Reid, I.N. & Gilmore, G. 1982. *Mon. Not. Roy. Astron. Soc.* **201**, 73.

Salpeter, E. E., 1955. *Astrophys. J.* **121**, 161.

Scalo, J. M., 1986. *Fundamentals of Cosmic Physics*, **11**, 1.

Stobie, R. S., Ishida, K. & Peacock, J.A., 1989. *Mon. Not. Roy. Astron. Soc.* **238**, 709.

Upgren, A.R.Jr., 1962. *Astrophys. J.* **67**, 37.

SYSTEMATIC PROPERTIES OF ROTATION CURVES AND DARK MATTER

Stefano Casertano and Tjeerd S. van Albada
Kapteyn Astronomical Institute
University of Groningen
9700 AV Groningen
Netherlands

ABSTRACT. All galaxies for which extended rotation curves are available contain a significant amount of dark matter. The luminous matter is probably self-gravitating in the inner parts of the galaxy, while dark matter dominates beyond the optical radius. The mass-to-light ratio of the luminous material, corrected for variations in the present rate of star formation, is remarkably constant and in agreement with classical population synthesis models. Rotation curves are not featureless; rather, their shape correlates with their luminosity and size, with a preference for rising rotation curves for dwarf galaxies, and declining curves for bright compact galaxies. The latter contain less dark matter than the typical average galaxy, the former more. The variations in dark-to-luminous mass ratio and their correlation with galaxy properties provide material for models of galaxy formation, and may indicate that at least part of the dark material is baryonic.

1. Introduction: General properties of rotation curves

1.1. FLATNESS AND THE NEED FOR DARK MATTER

Let us start with a brief discussion of the main evidence for dark matter around spiral galaxies.

In the 1970's 21-cm line observations of neutral hydrogen in galaxies showed that the circular velocity outside the optical radius is more or less constant. This was an unexpected result because it was customary to think in terms of stars as the main contributors to the mass. Since most of the light comes from the region within the optical radius, the circular velocity was expected to decline at large radii.

The best-known example is perhaps that of NGC 3198, shown in Fig. 1. The upper panel illustrates the distribution of stellar light. A look at the lower panel shows that the circular velocity measured from the 21-cm line of neutral hydrogen (dots with error bars) is more or less constant with radius out to two optical radii. The continuous curve represents the rotation curve calculated from the distribution of light given in the upper panel, assuming a constant mass-to-light ratio, and the neutral hydrogen gas. This predicted rotation curve shows the expected Keplerian decline beyond the optical radius. Inside the peak it agrees with the observed rotation curve.

D. Lynden-Bell and G. Gilmore (eds.), Baryonic Dark Matter, 159–177.
© 1990 *Kluwer Academic Publishers.*

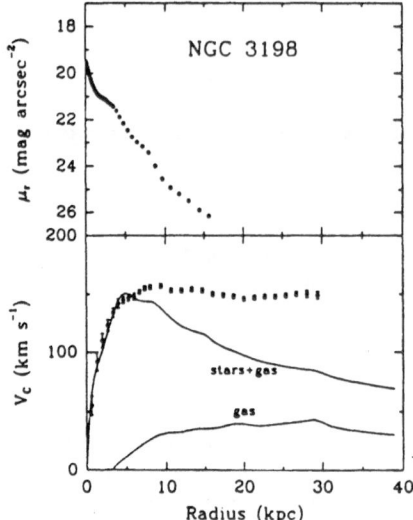

Figure 1. Light profile and rotation curve of NGC 3198. Top: luminosity profile in the r-band from Kent(1987a). Bottom: observed rotation curve (dots with error bars), and rotation curves predicted from the light and neutral hydrogen distributions. (Figure from Begeman 1987)

Because one does not know the mass-to-light ratio of the luminous material accurately, one can only predict the shape of the rotation curve, not its amplitude. In Fig. 1 the M/L value of the luminous matter has been scaled to the maximum value compatible with the observed rotation curve. In this way the discrepancy between the observed and predicted rotation curves in the inner region is minimized.

The discrepancy between observed and predicted rotation curves in the outer region shows that there must be additional matter beyond the optical radius. In other words, spiral galaxies are surrounded by a dark halo. Alternatively, Newton's law does not hold on the scale of galaxies. For a review of this possibility see Sanders (1990).

The amount of mass in the dark halo is large. For NGC 3198, the circular velocity at the outermost point is about 150 km/sec. The maximum predicted circular velocity at that point is about half of this. Thus, in NGC 3198, there is about 4 times as much matter inside the last measured point than can be accounted for by stars and gas. In §4 we address this point more quantitatively.

The same result is obtained in all galaxies for which an extended HI rotation curve is available, about 20 cases. We will come to this point again later. An extreme case is that of NGC 2841 (Begeman 1987). There the neutral hydrogen extends to 16 scalelengths of the stellar disk, and the rotation curve is more or less flat out to the last measured point, while the Keplerian decline should start at about 3 scalelengths.

Not all rotation curves are quite as flat, however. Two examples are NGC 2903 (Begeman 1987) and NGC 2683 (Casertano and van Gorkom 1989), which both show some velocity decline outside the optical radius. This point will be discussed further in §§2 and 3.

1.2. UNCERTAINTIES

The evidence on missing mass results from a comparison of observed rotation curves with models based on the luminosity distribution. Doubts have been raised in the past on the reliability of both light profiles and rotation curves. Most of these doubts have disappeared in view of the improvement of the observations and of the analysis methods. Let us review these points in somewhat closer detail.

1.2.1. *Light profiles.*

Light profiles are now mostly measured by CCD photometry. This eases the problems of non-linearity and non-uniformity of the detectors traditionally encountered with photographic plates. Uniformity of the detector is especially important, as Binney (1986) pointed out, when the surface brightness of the galaxy is a small fraction of the sky brightness. Kent (1987b) has stressed that CCD photometry is reliable at least down to about one percent of the level of the night sky. This level is reached roughly at 5 disk scalelengths. The discrepancy between observed and predicted rotation curves starts well within this point, at about 3 disk scalelengths. Therefore light profiles are sufficiently well-known for the purpose of establishing the existence of dark matter.

There is no proof, however, that light profiles continue their fall-off beyond five scale lengths. Lake (this conference) has raised the possibility that a faint component exists with low surface brightness and a large mass-to-light ratio, compatible with low-luminosity stars. This component, as yet undetected, could make up a large fraction of the dark matter inferred from the rotation curves.

1.2.2. *Rotation curves.*

Deriving a 21-cm rotation curve is a complex procedure, in which several assumptions are made in order to interpret the observed velocities. Let us review the standard procedure.

The starting point is the velocity field of the galaxy, a map of the gas velocity projected on the plane of the sky. In the best cases, this contains about 1000 independent positions on the sky. The velocity values represent the peak velocity of HI along the line of sight, and they are accurate to about 5 km/s (1σ).

To obtain the rotation curve from the velocity field the galaxy is divided into a number of rings. At this point the assumption is made that the hydrogen clouds in each ring move about the center of the galaxy in circular orbits. The spatial orientation of each ring is left free. From a least-squares solution of all data points in a ring, one then obtains the spatial orientation of the ring (inclination and position angle), and its circular velocity. Thus, the inclinations used to determine the circular velocity come directly from the velocity field, and not from the shape of the distribution of hydrogen. But in general there is good agreement between inclinations derived from the velocity field and from the shape of the HI distribution.

This procedure is usually done separately for the two halves of the galaxy, yielding inclination, position angle and velocity for the two sides. The values for NGC 3198 are plotted in Fig. 2. For both halves the inclination changes from about 70 o to 76 o in the outer parts, and the rotation curves obtained for the two halves agree to within 5 km/s. This symmetry is a requirement for a good rotation curve, and we use the difference between the two halves to estimate the uncertainty in the final result. The formal error in the fit is typically about $2\,\mathrm{km\,s^{-1}}$. This does not include systematic errors due to the fitting

162

procedure, which could be somewhat larger. by at least a factor of two. (Systematic errors could well behave similarly for the two halves of the galaxy. and thus the symmetry of the system does not provide useful constraints.)

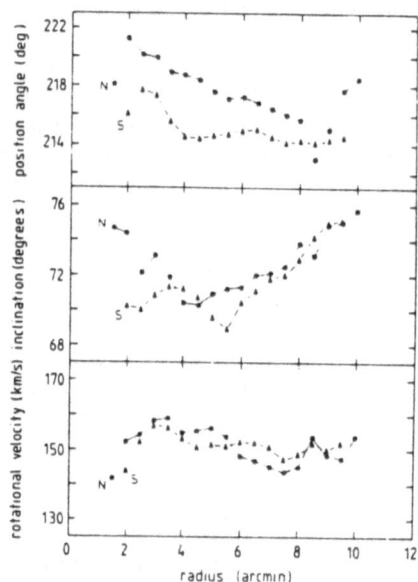

Figure 2. Position angle, inclination and rotational velocity for the two halves of NGC 3198.

Clearly, the crucial part of this procedure is the assumption that the individual HI clouds move in circular orbits about the center. This assumption can be tested to some extent by calculating a model velocity field from the adopted rotation curve and comparing this with the observed velocity field. The residual velocity fields show large scale patterns with amplitudes of the order of 10 km/s. This shows that deviations from circular motion do occur. The hope is that by averaging data points along a ring, motions that are faster than circular and those that are slower than circular average out.

An important remaining problem is that one can not differentiate between circular streamlines in a given plane, and elliptical streamlines in some other plane. For each individual galaxy this is an additional, unknown source of uncertainty, not included in the present error bars. To our knowledge, no detailed studies have been made of the possible bias that this effect may produce on rotation curves. However, for a sample of galaxies most of the bias should average out. For some galaxies the outer parts may be more edge-on than the inner parts, but for others it should be the other way around. The rather good agreement between kinematic and isophotal inclinations also indicates that the effect is probably not large.

In conclusion, while the accuracy of rotation curves and light profiles may be challenged at the level of a few percent, there is no reason to believe that something could be seriously wrong with either in such a way as to falsify the conclusions about dark matter.

2. Deviations from flatness

Rotation curves are often described as 'flat' to emphasize that their shape does not resemble the Keplerian-like decline expected on the basis of the light distribution (van Albada and Sancisi 1986, Sancisi and van Albada 1987, Bahcall and Casertano 1985). A closer look, however, shows that there *is* structure in rotation curves. This structure provides crucial information on the properties of dark and luminous mass. We start with a qualitative description of the deviations from flatness.

2.1. LOW LUMINOSITY GALAXIES

Rotation curves of low luminosity galaxies tend to continue rising beyond their optical radius. Notable examples are DDO 154 (Carignan and Freeman 1988) and NGC 1560 (Broeils, this conference).

One aspect is basically different for these late–type dwarfs: they contain more gas than stars. This means that the amplitude of the predicted rotation curve is determined by the amount of HI in the system, which can be calculated by adopting a value for the distance to the galaxy. Doubling the distance means multiplying the amount of gas by a factor of 4. This leads to an increase in the amplitude of the rotation curve of a factor of $\sqrt{2}$.

Using generally accepted distances and the observed amount of gas, the predicted amount of dark matter in DDO 154 exceeds that of ordinary matter by a factor of 10, and M/L is about 80. But it is striking that the shape of the predicted rotation curve agrees rather well with the observed rotation curve. Thus, if one could increase the amount of gas in some way by a factor of order 10, keeping the same spatial distribution, no dark halo would be needed at all!

2.2. BUMPS AND WIGGLES

With few exceptions, the shapes of predicted and observed rotation curves are very similar over a large fraction of the visible part of galaxies for which mass models have been made (Kent 1986; cf. §4). In several cases, even relatively small features in the observed rotation curves are reproduced by the rotation curves calculated from the light profile. No objective study of these small scale features and of their significance has been made yet.

2.3. VELOCITY DROP AT THE EDGE OF THE STELLAR DISK

In some galaxies, the stellar disk appears to end sharply at 4–5 scale lengths. Several cases of 'disk truncation' have been reported by van der Kruit and Searle (1981a, b, 1982)

164

in edge-on galaxies. The truncation is often accompanied by a break in the rotation curve, occurring near, but sometimes not precisely at, the location of the optical truncation. The photometric and velocity features are most likely related (Casertano 1983).

An example is NGC 5033 (Fig. 3). In this case the observed rotation curve shows a break close to the truncation radius of the disk, but somewhat sharper than predicted by the combination of a disk represented by the light and a halo. (This break is seen clearly on only one side of the galaxy; the HI emission on the other side is weak.) The results for this galaxy illustrate again that features in the observed rotation curve agree with predictions from the light profile, in agreement with the idea that luminous matter is self-gravitating.

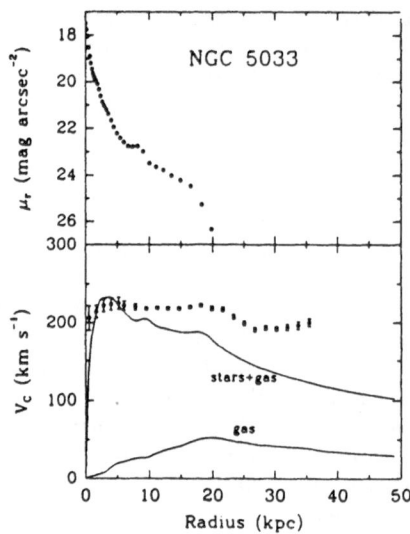

Figure 3. Light profile and rotation curve of NGC 5033, a galaxy with a truncated disk. Top: luminosity profile in the r-band (Kent 1986). Bottom: observed rotation curve (dots with error bars) and rotation curve predicted from the distributions of light and neutral hydrogen (Begeman 1987).

More generally, rotation curves seem to drop somewhat near the edge of the visible disk, whether or not an actual truncation feature can be identified. Salucci and Frenk (1989) find this to be a measurable effect in all but dwarf galaxies. The typical drop is about $20 \, \mathrm{km \, s^{-1}}$, and occurs near 4–5 exponential scale lengths from the center of the galaxy. The drop in the rotation curve near the edge has also been used to estimate the density of the disk (Casertano 1983).

2.4. GALAXIES WITH DECLINING ROTATION CURVES

In a number of galaxies, a velocity decline larger than 20–$30 \, \mathrm{km \, s^{-1}}$ is observed, mostly beyond the optical radius. Two puzzling cases are the edge-on spirals NGC 891 and NGC 5907 (Sancisi and van Albada 1987); in both galaxies a substantial, almost Keplerian drop-off in velocity is present, but it can only be measured on one side of the galaxy because of lack of HI on the other side. Sancisi and van Albada (1987) point out that, without confirmation from both sides of the galaxy, the apparent velocity decrease can still be reconciled with an intrinsically flat rotation curve and an asymmetric distribution of neutral hydrogen.

More recently, Casertano and van Gorkom (1989) have found two galaxies, NGC 2683 and NGC 3521, in which a large velocity decline can be measured *on both sides* of the disk. The rotation curve for NGC 2683 is shown in Fig. 4. For this galaxy there is no question of projection effects, and the velocity decline is real. One is forced to conclude that some galaxies have declining rotation curves, and perhaps contain substantially less dark matter than others. However, even these galaxies need *some* dark matter to explain their rotation curve.

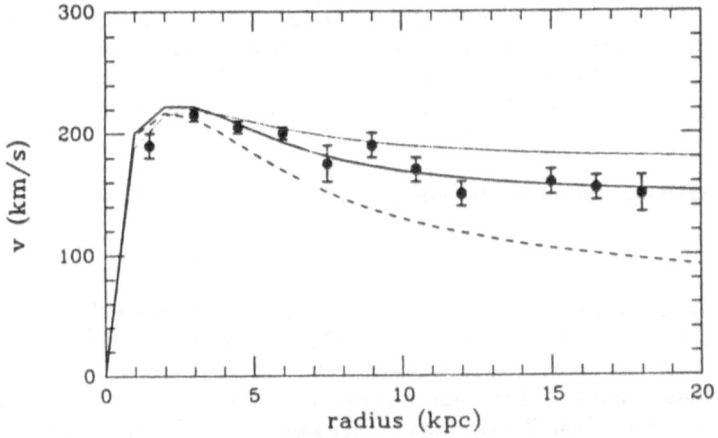

Figure 4. NGC 2683: an example of a galaxy with a declining rotation curve. The dots represent the observed rotation velocity, averaged between the two sides of the galaxy. The curves represent different mass models: no dark matter (dashed), maximum disk (solid), and half maximum disk (dotted). Only the maximum-disk model agrees with the observations. See also §4.3.

2.5. MASS TYPES

Burstein and Rubin (1985) have suggested that the total mass distribution of spiral galaxies, as derived from their rotation curves, follows one of three different patterns, called 'mass types'. Furthermore, they have claimed that mass types do not correlate well with the shape of the light distribution, in disagreement with the idea that luminous matter is gravitationally important (the 'maximum-disk' hypothesis, see §4). The most recent assessment of the merit of 'mass types' (Forbes and Whitmore 1988), using objective statistical tests, indicates however that mass distribution and Hubble type *do* correlate, removing this apparent problem for the maximum-disk hypothesis. We believe that detailed mass models for galaxies with accurate velocity *and* photometric data remain the most powerful method of addressing the relationship between mass and light.

2.6. SHAPE *vs.* ENVIRONMENT

Rubin *et al.* (1988) and Whitmore *et al.* (1988) have studied the shape of the inner part of the rotation curve, well inside the optical radius, for galaxies at different distances from the center of rich clusters. They confirm a correlation, already indicated by Burstein *et al.* (1986), between shape of the rotation curve and position of the galaxy in the cluster; namely, rotation curves of galaxies deep inside clusters appear to drop, while galaxies far from the cluster center have flat rotation curves. Notice that this drop is not the same as the velocity decline discussed in §§2.2 and 2.3 above, for it would take place at smaller galactocentric distances. They claim that the velocity drop is inconsistent with a constant mass-to-light ratio inside the luminous region, although the evidence for this statement is qualitative—based on approximate mass-to-light ratios rather than on a full modelling of the rotation curve. If confirmed, this interpretation of the velocity drop would indicate that dark mass is important at all radii, and its amount decreases faster for galaxies in highly populated regions, perhaps due to a larger merging/stripping activity there.

3. Correlations between rotation curve shape and galaxy properties

A systematic pattern emerges from the description of rotation curve shapes. Low luminosity galaxies have rising rotation curves. Many bright galaxies show a slight drop in velocity beyond the optical radius. A few galaxies have *large* velocity declines; Casertano and van Gorkom (1989) point out that these galaxies are not particularly bright, but they seem to be more *compact*, *ie* of smaller size, than other galaxies of the same total luminosity.

In order to lend substance to this pattern, we have put together the available information on extended rotation curves. We have included all HI rotation curves available to us that extend beyond at least 1.2 optical radii, and fulfil two additional conditions: 1) projection effects are probably negligible, or they have been corrected for; and 2) HI is observed on both sides of the galaxy, and the rotation curves are in substantial agreement. Our sample consists of 23 galaxies with extended rotation curves. Note that not all rotation curves are of the same quality; some have errors smaller than 5% , for others the uncertainties could be as large as 10%. In the framework of this statistical presentation, we feel justified to include also the less accurate curves, even though somewhat uncertain when considered by themselves.[1] Even though individual measurements of velocity decrease (and increase) could be marred by difficulties, they can still provide useful information on general trends in the shape of rotation curves.

Rotation curves are plotted in Fig. 5. The galaxies have been divided in four categories, on the basis of the maximum rotation velocity v_{max} and of the scale length h: dwarfs, with $v_{max} < 100 \, km \, s^{-1}$; intermediate, with $100 < v_{max} < 180 \, km \, s^{-1}$; bright, with $180 <$

---

[1] One exception is M81, which was included by Kent (1987a), but which we chose to leave out since its rotation curve, as determined by Visser (1980), falls on one side but remains flat on the other. With the rotation curve used by Kent (1987a), M81 would have fitted perfectly into our proposed correlations.

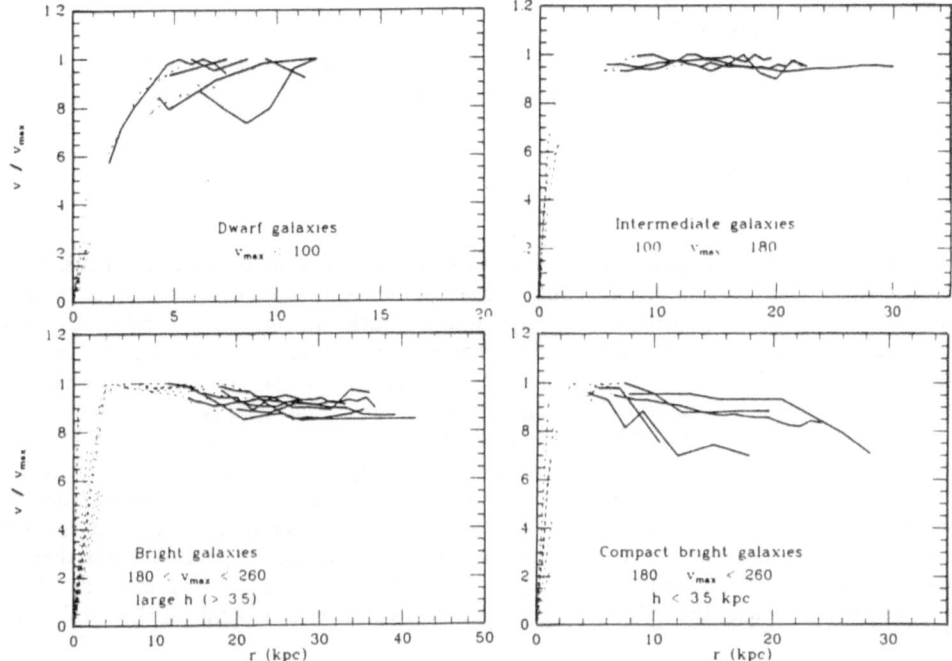

Figure 5. A comparison of 23 extended rotation curves. Four panels include: dwarfs, top left; intermediate, top right; bright, bottom left; bright compact, bottom right. Rotation curves have been rescaled to the same maximum velocity; they are drawn dashed inside three optical scale lengths, solid outside. References for individual rotation curves are: Begeman 1987, 1989 (NGC 2403, NGC 2903, NGC 6503, NGC 5033, NGC 2841, NGC 7331, NGC 5371); Bosma 1981 (NGC 5055); Bosma *et al.* 1977 (NGC 4736); Bottema *et al.* 1987 (NGC 4013); Carignan 1985 (NGC 3109); Carignan and Freeman 1985 (NGC 247, NGC 300); Carignan and Freeman 1988 (DDO 154); Carignan *et al.* 1988 (UGC 2259); Newton and Emerson 1977 (NGC 224); Shostak 1973 (NGC 4236); van Albada 1980 (NGC 4258); van Albada *et al.* 1985 (NGC 3198); van Albada and Sancisi 1986 (NGC 4565, NGC 5907); Casertano and van Gorkom 1989 (NGC 2683, NGC 3521).

$v_{max} < 300 \, \text{km s}^{-1}$ and $h > 3.5 \, \text{kpc}$; and bright compact, with $180 < v_{max} < 300 \, \text{km s}^{-1}$ and $h < 3.5 \, \text{kpc}$. (We use the maximum rotation velocity instead of the total luminosity because it is distance-independent and is not affected by internal or galactic absorption.) Each panel in Fig. 5 refers to one of these classes; rotation curves have been rescaled to the same peak velocity, but the horizontal scale has been left unchanged. Since we are interested in the properties of the dark matter, we concentrate on the shape of the rotation curves in the outer parts; the region inside three scale lengths is affected by bulges and by the actual light distribution of the galaxy, and is less useful to constrain the properties of dark matter. Rotation curves are drawn with dashed lines inside three scale lengths.

Just looking at the four panels of Fig. 5 immediately conveys the difference in rotation curve shapes. Rotation curves in Panel 1 (dwarfs) all rise more or less to their last measured point. Panel 2 (intermediate luminosity) contains the flattest rotation curves known. In

Panel 3 (bright galaxies) rotation curves are rather flat, but most drop somewhat in the outer parts. Galaxies in Panel 4 (bright compact) all show a significant drop in rotation velocity, from the $30\,\mathrm{km\,s^{-1}}$ of NGC 2903 to the $60 + \mathrm{km\,s^{-1}}$ of NGC 2683. Even though some of the rotation curves collected in Fig. 5 may be doubtful, the correlation between shape, luminosity and size of these galaxies is striking.

3.1. CONSEQUENCES

In terms of standard disk-halo models, the different shapes of rotation curves have a straightforward interpretation. Declining rotation curves mean less dark matter inside the optical radius; rising rotation curves require more dark matter. The correlation between rotation curve shape and scale length of the galaxy can be understood in terms of variations in the initial angular momentum parameter λ: smaller λ means a larger collapse factor for the luminous matter (Fall and Efstathiou 1980), and a final galaxy which is both more compact and more dominated by luminous matter.

The *stronger* correlation between rotation curve shape and total luminosity or rotation velocity of the galaxy is less easily understood. If the constituents of dark and luminous matter were initially well-mixed, there would be no reason for faint galaxies to have proportionally more dark matter than luminous ones. One tantalizing possibility is that dark matter might be baryonic material that somehow differentiated in low-luminosity objects, such as brown dwarfs; the efficiency of this process could well depend on the total mass of the galaxy.

4. Distribution of dark and luminous matter:
   The maximum–disk hypothesis.

4.1. DEGENERACY OF DISK-HALO MODELS

The amount of dark matter, and its distribution, can be 'measured' by means of a two-component mass model for the galaxy: a disk component in which the distribution of matter follows that of the light, and a spherically symmetric halo following a simple density law. The mass-to-light ratio (M/L) of the disk is assumed constant. When a bulge is present, its mass is also included, again with constant M/L. Disk and bulge need not have the same mass-to-light ratio.

The obvious problem is that the M/L value of the stellar disk is not known *a priori*. One prescription is to use the maximum value of M/L compatible with the observed rotation curve. This is referred to as the 'maximum-disk' hypothesis (van Albada and Sancisi 1986, Sancisi and van Albada 1987). The maximum-disk model for NGC 3198, mentioned earlier in §1, is shown in Fig. 6.

The main property of a maximum-disk model is that the stars and gas determine the circular velocity in the inner region, while the dark matter determines the circular velocity in the outermost region. In the region in between dark and luminous matter conspire to produce an approximately flat rotation curve, through the combination of a declining rotation curve for the light with a rising one for the halo.

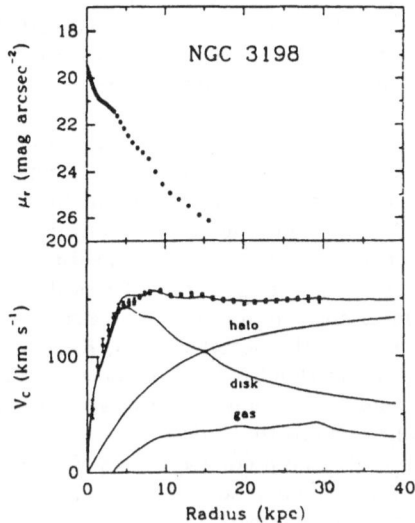

Figure 6. Maximum-disk model for NGC 3198. Top: luminosity profile observed by Kent. Bottom: observed rotation curve (dots) and the rotation curves of the individual mass components and their sum (Begeman 1987)

At first sight there would be no conspiracy if one lowers the contribution of luminous material to the rotation curve. Disk-halo decompositions with disks with smaller M/L than of the maximum disk also give good fits to the observed rotation curve (van Albada and Sancisi 1986, Fig. 5). In fact, a study of χ^2 shows that all these fits are equally good.

In the model fits with low disk mass the dark material is the main contributor to the gravitational field. If this case represents reality, the M/L inside the optical radius would be a measure of *dark* mass over disk luminosity. Yet in §5 we show that the intrinsic dispersion in the mass-to-light ratios is quite small. Therefore, in all cases there must be a close coupling between dark and luminous material inside the optical radius, whatever the contribution of the luminous disk to the rotation curve.

Accepting the maximum-disk hypothesis removes the freedom associated with the choice of an M/L value and defines the disk-halo models uniquely. Such models have now been calculated for a number of galaxies: for a synopsis of their properties see Begeman (1987) and Kent (1987a).

4.2. DARK MATTER INSIDE THE OPTICAL RADIUS

Optical rotation curves have been collected by Rubin and co-workers (Rubin *et al.* 1980, 1982, 1985) ; Burstein *et al.* 1982) for a sample of about 60 galaxies. These rotation curves have high spatial resolution and, because of this, are often better suited than HI curves to study the mass distribution inside the optical radius. Many of these rotation curves appear quite flat (e.g. UGC 2885), much like the rotation curves obtained from HI *outside* the optical radius. The optical rotation curves typically extend to 0.8 R_{25}.

The general flatness of these optical rotation curves has in the past often been attributed to dark matter, using the argument that constant circular velocity holds for a spherical system with mass proportional to radius, and that therefore dark matter must be present. This would contrast the maximum-disk hypothesis.

This conclusion that dark matter dominates inside the optical radius would be premature: perhaps the distribution of luminous mass *is* consistent with a nearly constant circular velocity over a large range in radii. To see whether the shape of the rotation curve of a galaxy is consistent with the distribution of light, one needs surface photometry.

Kent obtained CCD photometry for most of the galaxies from Rubin *et al.* (1985). For each he determined light profiles along the major axis, which he used to calculate a model rotation curve by assuming that the local mass-to-light ratio is constant with radius. Comparing these calculated rotation curves with the observed ones, Kent finds that in nearly all cases there is very good agreement inside about 80% of the optical radius. Outside that radius, the predicted rotation curves tend to fall below the observed ones, indicating a need for dark matter. This conclusion applies also to UGC 2885.

For a few Sb and Sc galaxies the observed and predicted curves disagree in the innermost region. The most likely cause for this is that there is something wrong with the light profile, for example obscuration by dust in edge-on galaxies.

Discrepancies for Sa galaxies are more frequent. For these Kent argues that the emission line velocities in the region of the bulge may not represent the true circular velocity.

A quantitative comparison between rotation curve and light distribution can also be achieved by parameterizing their shapes. Kent (1987b) does this by defining a *velocity* and a *luminosity* concentration index. The velocity concentration index is based on where the velocity reaches half its value at the optical radius; the luminosity concentration index on the ratio of the radii enclosing 80% and 20% of the total luminosity. These two indices correlate nicely with each other, as they should if the luminous matter is dynamically important.

The main result from Kent's work is that, for the large majority of spiral galaxies, there is very good agreement between the observed distribution of light and the distribution of matter calculated from the rotation curve, provided one allows to vary the mass-to-light ratio from galaxy to galaxy. A typical value of M/L that Kent finds is about 2 (corrected to face-on). Such a value agrees with the local M/L in the solar neighborhood.

Value of M/L around 2 can also be understood on the basis of stellar population analyses. Therefore, the conclusion must be that there is no dynamical evidence for unseen material in the inner region of spiral galaxies, inside say $(2/3) \times R_{25}$. 2/3 of the optical radius. Dark matter is only needed beyond this radius.

4.3. RECENT ARGUMENTS REGARDING THE MAXIMUM DISK HYPOTHESIS

The maximum-disk hypothesis is capable of producing good disk-halo decompositions for all extended rotation curves. It is supported by the correlation between luminosity distribution and shape of the rotation curve. The maximum-disk hypothesis is also strengthened by the small intrinsic scatter in mass-to-light ratios corrected for variations in the present rate of star formation (see §5).

Additional support comes from the modelling of galaxies with declining rotation curves. For NGC 2683, for example, the velocity decline is nicely fitted by a maximum disk model in which the luminous matter accounts for most of the gravitational field inside the optical radius (solid line in Fig. 4). Less dark matter is needed than in other galaxies; hence the decrease in velocity. If the amount of luminous matter was reduced by a factor two, the decline could no longer be fitted with a halo with constant asymptotic velocity (dashed line in Fig. 4). In that case one must introduce an *ad hoc* explanation, such as the existence of a boundary to the halo, for which there is no independent evidence.

Athanassoula, Bosma and Papaioannou (1987) have used the well-known argument that a massive dark halo suppresses spiral structure. Using Toomre's swing amplification theory they have worked this out quantitatively for a large sample of galaxies with known rotation curves and light profiles. They also use the argument that some halo is needed to suppress $m = 1$ modes. From this they find that the true disk must lie in between the maximum disk and 0.5 times the maximum disk.

A possible problem is represented by the *shape* of the spiral structure in some of these galaxies. Sellwood (this conference) estimates the disk surface density from the pitch angle of the spiral arms. His best models for NGC 3198 and NGC 2403 have about half the luminous mass of a maximum-disk model.

On the basis of a statistical study of the shape of rotation curves, Salucci and Frenk (1989) and Salucci (this conference) have disputed the applicability of the maximum-disk model to dwarf galaxies. However, dwarfs are the class of galaxies for which both rotation and luminosity information is least satisfactory. The case for dwarfs should be evaluated again when better information becomes available.

In conclusion, most recent findings add to the evidence in favor of the maximum disk hypothesis reviewed by van Albada and Sancisi (1986). There is at present no strong reason to believe that dark matter might be dominant in the inner regions of galaxies.

5. Mass-to-light ratios

5.1 GLOBAL M/L AND STAR FORMATION RATE

The dispersion in global mass-to-light ratios inside the optical radius among galaxies is large. At a given type, the range is nearly a factor of 4 (Kent 1986). If there is only luminous material inside the optical radius, this M/L is the true value for the stellar population. The large variation among galaxies must thus be explained in terms of some measurable property of the stellar populations. Failure to do so would be an argument against the maximum-disk hypothesis.

An explanation in terms of stellar population differences can indeed be given (van Albada, in preparation). The critical variable is the *present rate of star formation*. The evidence comes from a comparison of M/L with the 100 μm flux measured by IRAS for the Sb and Sc galaxies in the sample of Rubin *et al.* studied by Kent. The 100 μm luminosity L_{100} is an indicator of the star formation activity over the past 10^8 years or so, and L_{100}/M is therefore a measure of the present star formation rate per unit mass (e.g. Soifer *et al.* 1987). In Fig. 7 the mass-to-light ratio is plotted against L_{100}/M. The correlation is quite evident. Galaxies with a high star formation rate have low M/L because of the enhanced luminosity.

This interpretation is supported by the models with monotonically declining functions for the star formation rate (SFR) calculated by Tinsley (1981). We assume that L_{100}/M is proportional to the current SFR per unit mass in these models. Thus, the position of Tinsley's models in Fig. 7 contains an arbitrary scaling factor for the abscissa. The models have been shifted by a small amount in the vertical direction for clarity. The trend shown by the models (solid line in Fig. 7) agrees nicely with the observations.

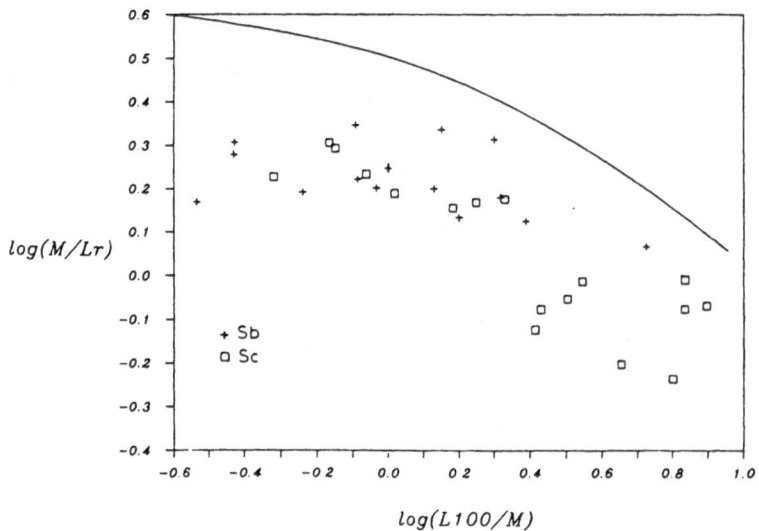

Figure 7. Mass-to-light ratio as a function of L_{100}/M for Sb and Sc galaxies from Rubin *et al.* M represents the mass of the maximum disk. A low $100\,\mu$m flux implies a low present star formation rate, and thus a decreased luminosity. Different symbols represent different galaxy types. The curve connects models calculated by Tinsley (1981); see text.

The figure also shows that there is no intrinsic difference in stellar population between Sb and Sc galaxies. The average Sc galaxy has a higher present rate of star formation and therefore more young, bright stars; this fully explains its lower M/L.

The total range in M/L at fixed star formation rate is about a factor of 1.8, compared to a total range of a factor of 4. The remaining spread is compatible with observational uncertainties. Thus, any cosmic dispersion in M/L at fixed present rate of star formation must be substantially smaller than 20% (1 sigma).

5.2 LOCAL VARIATIONS IN MASS TO LIGHT RATIO; OR, CAN ALL MASS INSIDE THE OPTICAL BOUNDARY BE STELLAR?

The total amount of matter inside the last point of extended HI rotation curves is, for some galaxies, at least four times the mass we can see. There is no doubt that the mass distribution extends further out than the light distribution. But is it possible that all mass inside the optical radius can be accounted for by stars? In other words, how must the local M/L change to explain the rotation curve inside the optical radius?

Again, consider the test case of NGC 3198. Kent (1987b, Fig. 1) compares the observed surface brightness (Wevers 1984, Kent 1987a) with the surface mass density calculated from the rotation curve under the assumption that all matter lies in a disk (Binney 1986). The ratio between the two would give the *local M/L*, at least as far out as the luminosity measurements are reliable.

Inside about 2 scalelengths the derived M/L is constant, as we have seen before. This region corresponds with the rising part of the rotation curve, up to and including the peak. Further out M/L increases. At about 6 scalelengths the two curves differ by about 3 magnitudes, which corresponds to a factor of 16, and the local value of M/L is about 60. Beyond 6 scalelengths the surface brightness drops below 1% of the sky level, which is too faint for a reliable measurement. Bosma and van der Kruit (1979) quote values of up to several hundred on the basis of photographic photometry.

The rise in M/L required to have all mass inside the optical radius in stars is modest, and could be explained with a gradual variation in the stellar population with radius.

One might ask how much M/L would vary with radius if also the dark matter *outside* the optical radius was in a disk. The problem is that only upper limits can be given, since we don't know how much light there is. (Extrapolating the exponential law leads to artificially high values of the M/L.) Lake (this conference) addresses this problem in greater detail, and finds that a stellar population similar to that of the galactic spheroid is compatible with the rotation data and the available upper limits on the surface brightness at large galactocentric distances.

5.3. COMMENTS

Two important points can be made about M/L inside the optical radius:

1) Variations in the global mass-to-light ratio are due to different present rates of star formation. At constant rate of star formation, the rms spread in M/L is less than 20%, compatible with observational errors.

2) M/L ratios are constant inside about 2.5 disk scalelengths. This follows from the work of Kent (1986, 1987a, 1987b).

Both points reinforce the opinion that any non-luminous material that may be present inside 2.5 disk scalelengths is a close relative of stars and gas, and also resides in the disk. In other words there is no evidence for an independent spherical component of dark matter dominating the gravitational field in the inner parts.

A question that remains open is whether the increase in local M/L from 4 to 60, between 2.5 and 5 disk scale lengths, could be due to a decrease in star formation rate with radius, in the same way that galaxies with a high M/L have low present star formation rates. The increased amount of neutral hydrogen present in this region seems to argue against this idea, but little is known about the efficiency with which stars form as a function of density.

6. Shape of dark haloes

Rotation curves do not, by themselves, provide evidence about the three-dimensional distribution of the dark matter. A very flattened disk-like distribution would do just as well as the more popular spherical halo to explain the mass discrepancy. Let us review briefly some of the arguments that have been used in the past to set constraints on the shape of the dark matter component.

From a dynamical point of view, dark haloes are necessary to stabilize the disk against lopsided $m = 1$ and bar-like $m = 2$ distortions. This fact has been used in the past to provide support for a nearly spherical shape for the dark mass. However, a hot disk-like distribution can also be effective in stabilizing the disk (Athanassoula and Sellwood 1986); the criterium for stability appears to be a value of the Toomre stability parameter, $Q \gtrsim 2.2$. For a typical maximum-disk model, this implies that the radial velocity dispersion of the dark matter is of the same order as the rotation velocity. This can be achieved with a very thin distribution.

Nearly spherical haloes have also been invoked to stabilize warps (Tubbs and Sanders 1979, Petrou 1980). More recently, Sparke and Casertano (1988) have found that moderately flattened haloes can explain some characteristics of observed warps. The mechanism they suggest for the persistence of warps is rather insensitive to exactly how much the halo is flattened, and would work with axial ratios between 0.6 and 0.9 (in density). Haloes flatter than 0.6 have not been explored in detail.

Polar ring dynamics has also been used to explore the shape of the halo, by comparing the orbital speed of material in equatorial and polar orbits. On this basis, Schweizer *et al.* (1983) suggested that the dark matter distribution should be very nearly spheroidal in one polar ring galaxy, and Whitmore *et al.* (1987) made similar claims for two more cases. However, the more detailed analysis of one of these galaxies by Sparke and Sackett (1989; also Sackett, this conference) does not confirm these results. They find that the axis ratio of the halo can be as small as 0.4, with a preferred value of 0.6, although a nearly spherical halo cannot be ruled out altogether. Thus the evidence from polar rings must at the moment be regarded as inconclusive.

The HI disk flares in the outer parts of several galaxies, including our own. If the dark matter was concentrated near the galactic plane, it would confine HI more effectively, and no flaring would be observed. Van der Kruit (1981) used this argument in his quantitative study of the three-dimensional HI distribution in NGC 891. He found that a spherical halo model was better than a model with all the mass in the disk. Later, Athanassoula and Bosma (1987) have disputed the accuracy with which this method can discriminate between models. Better data have since been obtained of a number of edge-on galaxies, but we are not aware of any conclusive evidence in either direction.

From a study of the local mass density in the solar neighbourhood, the so-called Oort limit, Bahcall (1984a, b) has argued that there *must* be some dark matter in the disk, with a scale height *not larger* than 700 pc. The disk dark matter could be sufficient to bring up the rotation curve of the disk to the measured value. Bahcall's result has been recently disputed by Kuijken and Gilmore (1989; see also Gilmore 1990), who used a new and larger set of data to derive a value for the dynamical local density in agreement with what is observed. The two results are marginally compatible from a statistical point of view, but lead to completely different views of where the dark matter is.

To sum up, the available evidence is rather weak. A disk as thin as the observed young stellar disk can be ruled out from considerations of stability, the thickness of HI disks, and perhaps polar ring results; a perfectly spherical halo would reopen the question of warp persistence. But anything between an axis ratio of 0.2 and 0.9 is consistent with all firm constraints available at the moment. It is to be hoped that the work on the Oort limit and on polar rings will improve the constraints on the shape of dark haloes.

7. Conclusions

Recent results on rotation curves confirm earlier conclusions on the ubiquitous presence of dark matter in the outer regions of spiral galaxies. Two new elements are presented here. The first is an interpretation of the large variations in mass-to-light ratios between galaxies in terms of their present rate of star formation, estimated from their far-infrared flux. Galaxies actively forming stars appear brighter, and thus have smaller mass-to-light ratios. The small residual scatter in mass-to-light ratios, after correction for this effect, reinforces the belief that the luminous matter is essentially self-gravitating in the inner parts.

The other element is the correlation between the dark matter content of a galaxy and its total luminosity. Bright galaxies, especially if small, contain proportionally less dark matter than faint ones. This correlation, neither explained nor predicted by present models of galaxy formation, may prove a powerful constraint on the properties of dark matter.

References

Athanassoula, E. & Bosma, A. 1987, in *Large Scale Structures of the Universe*, IAU Symp. *130*, eds. J. Audouze, M.-C. Pelletan and A. Szalay (Dordrecht: Kluwer), p. 391.

Athanassoula, E., Bosma, A., & Papaioannu, S. 1987, *Astron. Astrophys.* **179**, 23.

Athanassoula, E. & Sellwood, J. A. 1986, *Mon. Not. R. astron. Soc.* **221**, 213.

Bahcall, J. N. 1984a, *Astrophys. J.* **276**, 169.

Bahcall, J. N. 1984b, *Astrophys. J.* **287**, 926.

Bahcall, J. N. & Casertano, S. 1985, *Astrophys. J. (Letters)* **293**, L7.

Barnes, J. E. 1987, in *Nearly Normal Galaxies: From the Planck Time to the Present*, ed. S. M. Faber (New York: Springer), p. 154.

Begeman, K. 1987, *PhD Thesis*, University of Groningen.

Begeman, K. 1989, to be submitted.

Binney, J. 1986, *Phil. Trans. R. Soc. Lond.* A **320**, 465.

Blumenthal, G. R., Faber, S. M., Flores, R., & Primack, J. R. 1986, *Astrophys. J.* **301**, 27.

Bosma, A. 1981, *Astron. J.* **86**, 1791.

Bosma, A. 1981b, *Astron. J.* **86**, 1825.

Bosma, A., van der Hulst, J. H., & Sullivan, W. T., III 1977, *Astron. Astrophys.* **57**, 373.

Bosma, A. & van der Kruit, P. C. 1979, *Astron. Astrophys.* **79**, 281.

Bottema, R., Shostak, G. S., & van der Kruit, P. C. 1987, *Nature* **328**, 401.

Broeils, A. H., this conference.

Burstein, D., & Rubin, V. C. 1985, *Astrophys. J.* **297**, 423.

Burstein, D., Rubin, V. C., Ford, W. K. Jr., & Thonnard, N. 1982, *Astrophys. J.* **253**, 70.

Burstein, D., Rubin, V. C., Ford, W. K., Jr, & Whitmore, B. C. 1986, *Astrophys. J. (Letters)* **305**, L11.

Carignan, C. 1985, *Astrophys. J.* **299**, 59.

Carignan, C. 1987, in *Dark Matter in the Universe*, *IAU Symp. 117*, eds. J. Kormendy and G. R. Knapp (Dordrecht: Kluwer), p. 135.

Carignan, C., this conference.

Carignan, C. & Freeman, K. C. 1985, *Astrophys. J.* **294**, 494.

Carignan, C. & Freeman, K. C. 1988, *Astrophys. J. (Letters)* **332**, L33.

Carignan, C., Sancisi, R. & van Albada, T. S. 1988, *Astron. J.* **95**, 37.

Casertano, S. 1983, *Mon. Not. R. astron. Soc.* **203**, 735.

Casertano, S. & van Gorkom, J. H. 1989, in preparation.

Fall, S. M. & Efstathiou, G. 1980, *Mon. Not. R. astron. Soc.* **193**, 189.

Forbes, D. A. & Whitmore, B. C. 1989, *Astrophys. J.* **339**, 657.

Gilmore, G., 1990, in *Baryonic Dark Matter* eds D. Lynden-Bell &G. Gilmore (Kluwer: Dordrecht) p137.

Lake, G., this conference.

Lake, G. & Feinswog, L. 1989, *Astron. J.* **98**, 166.

Kent, S. M. 1984, *Astrophys. J. Suppl.* **56**, 105.

Kent, S. M. 1985, *Astrophys. J. Suppl.* **59**, 115.

Kent, S. M. 1986, *Astron. J.* **91**, 1301.

Kent, S. M. 1987a, *Astron. J.* **93**, 816.

Kent, S. M. 1987b, in *Nearly Normal Galaxies: From the Planck Time to the Present*, ed. S. M. Faber (New York: Springer), p. 81.

Kuijken, K., & Gilmore, G. 1989, *Mon. Not. R. astron. Soc.* **239**, 651.

Newton, K., & Emerson, D. T. 1977, *Mon. Not. R. astron. Soc.* **181**, 573.

Petrou, M. 1980, *Mon. Not. R. astron. Soc.* **191**, 767.

Rubin, V. C. 1987, in *Dark Matter in the Universe*, *IAU Symp. 117*, eds. J. Kormendy and G. R. Knapp (Dordrecht: Kluwer), p. 51.

Rubin, V. C., Ford, W. K. Jr., & Thonnard, N. 1980, *Astrophys. J.* **238**, 471.

Rubin, V. C., Ford, W. K. Jr., Thonnard, N., & Burstein, D. 1982, *Astrophys. J.* **261**, 439.

Rubin, V. C., Burstein, D., Ford, W. K. Jr., & Thonnard, N. 1985, *Astrophys. J.* **289**, 81.

Rubin, V. C., Whitmore, B. C., & Ford, W. K., Jr. 1988, *Astrophys. J.* **333**, 522.

Sackett, P. D., this conference.

Salucci, P., this conference.

Salucci, P. & Frenk, C. S. 1989, *Mon. Not. R. astron. Soc.* **237**, 247.

Sancisi, R. & van Albada, T. S. 1987, in *Dark Matter in the Universe, IAU Symp. 117*, eds. J. Kormendy and G. R. Knapp (Dordrecht: Kluwer), p. 67.

Sanders, R. H. 1990, *Astron. Astrophys. Rev. , in press.*

Schweizer, F., Whitmore, B. C., & Rubin, V. C. 1983, *Astron. J.* **88**, 909.

Sellwood, J. A., this conference.

Shostak, G. S. 1973, *Astron. Astrophys.* **24**, 411.

Soifer, B. T., Houck, J. R., & Neugebauer, G. 1987, *Ann. Rev. Astron. Astrophys.* **25**, 187.

Sparke, L. S. & and Casertano, S. 1988, *Mon. Not. R. astron. Soc.* **234**, 873.

Sparke, L. S. and Sackett, P. D. 1989, preprint.

Tinsley, B. M. 1981, *Mon. Not. R. astron. Soc.* **194**, 63.

Tubbs, A. D. & Sanders, R. H. 1979, *Astrophys. J.* **230**, 736.

van Albada, G. D. 1980, *Astron. Astrophys.* **90**, 123.

van Albada, T. S., Bahcall, J. N., Begeman, K., & Sancisi, R. 1985, *Astrophys. J.* **295**, 305.

van Albada, T. S. & Sancisi, R. 1986, *Phil. Trans. R. Soc. Lond.* A **320**, 447.

van der Kruit, P. C. 1981, *Astron. Astrophys.* **99**, 298.

van der Kruit, P. C. & Searle, L. 1981a, *Astron. Astrophys.* **95**, 105.

van der Kruit, P. C. & Searle, L. 1981b, *Astron. Astrophys.* **95**, 116.

van der Kruit, P. C. & Searle, L. 1982, *Astron. Astrophys.* **110**, 61.

Visser, H. C. D. 1980, *Astron. Astrophys.* **88**, 149.

Wevers, B. M. H. R. 1984, *PhD Thesis*, University of Groningen.

Whitmore, B. C., Forbes, D. A., & Rubin, V. C. 1988, *Astrophys. J.* **333**, 542.

Whitmore, B. C., McElroy, D. B., & Schweizer, F. 1987, *Astrophys. J.* **314**, 439.

ARE THERE MASSIVE BLACK HOLES IN GALACTIC NUCLEI?

Martin J. Rees
Institute of Astronomy
University of Cambridge
Madingley Road
Cambridge CB3 0HA
England

ABSTRACT. Most of the talks at this conference deal with dark matter in the outlying parts of galaxies. This paper, in contrast, discusses the dark objects, probably massive black holes, which lurk at the *centres* of such normal galaxies as M31, and perhaps even our own. This subject is approached historically by considering the cosmic history of the quasar population. The evidence for massive black holes in the centres of some nearby galaxies is reviewed. Some interesting observational consequences (*eg* flares from tidally disrupted stars) are proposed and discussed.

The Quasar Population, and Expected Remnants

Investigations of quasars and active galaxies have established the luminosity function for such objects at different redshifts. Quasars were much more common at a redshift z = 2 than at the present epoch, in the sense that there were then, when the universe was 2 or 3 billion years old, almost 1000 times more luminous quasars per comoving volume than there are today. At still higher redshifts, corresponding to still earlier times, quasars seem to thin out, though the details are less clear beyond z $\simeq$ 4. A current estimate of the evolution in the quasar density with redshift is shown on the left–hand side of Figure 1. This same data can, following Schmidt (1989), be presented much more dramatically in the manner shown on the right–hand side, where time and comoving density are plotted linearly rather than on logarithmic scales. Clearly quasar activity was sharply peaked at a particular cosmic epoch. It is an anti–anthropic irony that the most exciting time to have been an astronomer was when the universe was 2 billion years old, before the Earth had formed.

The reasons for the sharp rise and subsequent fall in quasar activity, presumably related in some way to galaxy formation and evolution, are not my subject today. However, one can, from quasar statistics, draw the important inference that about 10^7 solar rest masses of radiation were emitted by quasars, for every bright galaxy present in the universe today. Quasars have generated radiation amounting to about $3000 \mathcal{M}_\odot c^2$ per cubic comoving Mpc (Soltan 1982, Phinney 1983). Many features of quasars remain enigmatic: active galactic nuclei display many phenomena on various scales and different wavebands, and it is hard to fit them into a single pattern. On the other hand, there is a much stronger consensus on

179

D. Lynden-Bell and G. Gilmore (eds.), Baryonic Dark Matter, 179–194.
© 1990 *Kluwer Academic Publishers.*

180

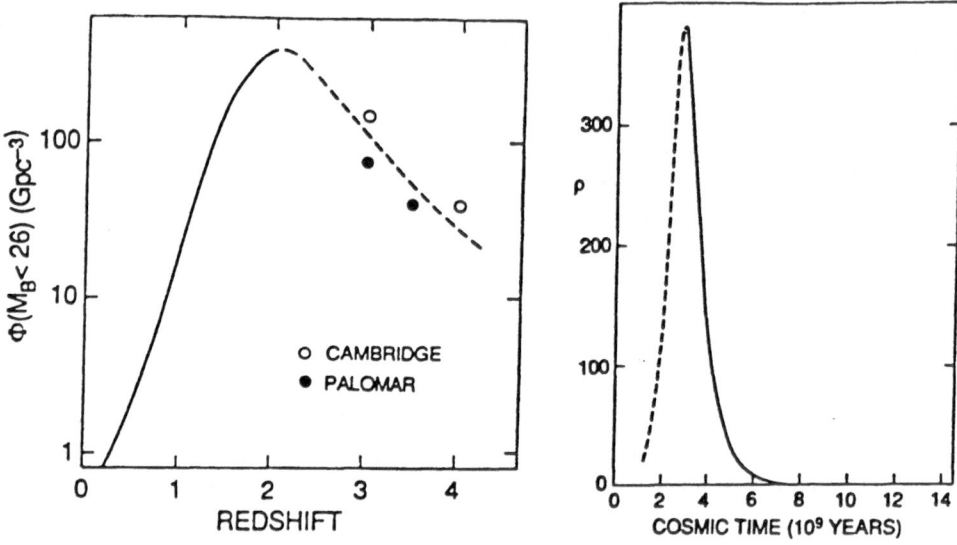

Figure 1. The comoving density of powerful quasars is given on the left as a function of redshift z. (Beyond $z \simeq 3$ there is a fall-off, but the quantitative details are uncertain. The filled and open circles correspond to the results from two different surveys.) The same data are, following Schmidt (1989), replotted on the right on a linear scale, where the horizontal axis is cosmic time, assuming an Einstein-de Sitter cosmology with $H_0 = 50$ km s^{-1} Mpc^{-1}. This plot dramatizes the relative brevity of the 'quasar era', when the Universe was 2–3 billion years old.

what a *dead* quasar should be. Given that quasars derive their energy primarily via gravitation, there seems no way of evading the conclusion that a substantial fraction of the mass involved must eventually collapse to a massive black hole. Even if we optimistically assign an efficiency of 10% to the overall energy generation in quasars, we must then conclude that their black hole remnants have a total mass amounting to an *average of* $10^8 \mathcal{M}_\odot$ per galaxy.

Even at the epoch of peak quasar activity, the comoving density of quasars is only 1 or 2 per cent of the present galactic density. One might therefore surmise that quasar remnants would be expected in only 1 or 2 per cent of galaxies, and that each would weigh around $10^{10} \mathcal{M}_\odot$. However, we must remember that Figure 1 delineates the evolution of the quasar *population*, which decays on a timescale of $t_{Evo} \simeq 2$ billion years. While this may relate directly to the lifecycle of a typical quasar, there is the alternative possibility that individual quasars have much briefer lives, so that many generations flare and fade during the 2 billion year period of peak quasar activity.

The contrasting implications of two different hypotheses about typical quasar lifetimes are displayed in Table 1. The masses scale with luminosity, given in units of $10^{47} \, \text{erg s}^{-1}$, and inversely with the efficiency, given in units of 0.1. In practice, of course, it is likely to be an oversimplification to characterise quasars by a single typical lifetime – the highest luminosities may well correspond to shorter lifetimes, for instance. We also have to consider how lower level AGNs, such as Seyfert nuclei, fit into the scheme.

Table 1

One generation of quasars	~ 50 generations of quasars
$t_Q \simeq t_{Evo}$	$t_Q \simeq 4 \times 10^7 \, \text{yr} \simeq 0.02 \, t_{Evo}$
$M = 2.5 \times 10^{10} \varepsilon_{0.1}^{-1} \, L_{47} \, \mathcal{M}_\odot$	$M = 5 \times 10^8 \varepsilon_{0.1}^{-1} \, L_{47} \, \mathcal{M}_\odot$
$L \ll L_{Ed}$	$L \simeq L_{Ed} \, \varepsilon_{0.1}$
Broad–line regions gravitationally bound	Broad–line region *not* gravitationally bound
Very massive remnants in ~2% of galaxies	~ $10^8 \, \mathcal{M}_\odot$ remnants in most bright galaxies

The table does, however, suffice to indicate that, if typical quasar lifetimes were a few times 10^7 years (a value which is supported by many theoretical models) one would expect dead remnants, massive black holes now starved of fuel, to lurk in the nuclei of most galaxies, including nearby ones.

Effects of Central Mass on Surrounding Stars

A massive black hole will inevitably affect the orbits of stars passing close to it, and evidence for just such effects in the centres of several nearby galaxies has recently been reported by several observers. To appreciate the nature of this evidence, it is helpful to define a few characteristic length scales. These are illustrated and defined in Figure 2 and its caption. In numerical terms, r_h and r_c are approximately given by

$$r_h \simeq \frac{GM_h}{\sigma_c^2} \simeq 10^6 r_g \left(\frac{\sigma_c}{300 \, \text{km s}^{-1}} \right)^{-2} \tag{1}$$

$$r_c = \frac{GM_h}{v_*^2} = r_g \left(\frac{c}{v_*} \right)^2 \simeq 10^5 r_g \left(\frac{v_*}{1000 \, \text{km s}^{-1}} \right)^{-2} \tag{2}$$

The tidal radius r_T, defined as the radius within which a star gets tidally disrupted, obviously has a value depending on the type of star being considered. For solar–type stars it is approximately

$$r_T \simeq \left(\frac{M_h}{m_*} \right)^{\frac{1}{3}} r_* \simeq 3 \times 10^{13} \, (M_h/10^8 \, \mathcal{M}_\odot)^{\frac{1}{3}} \, (r_*/r_\odot) (m_*/r_\odot)^{-\frac{1}{3}} \tag{3}$$

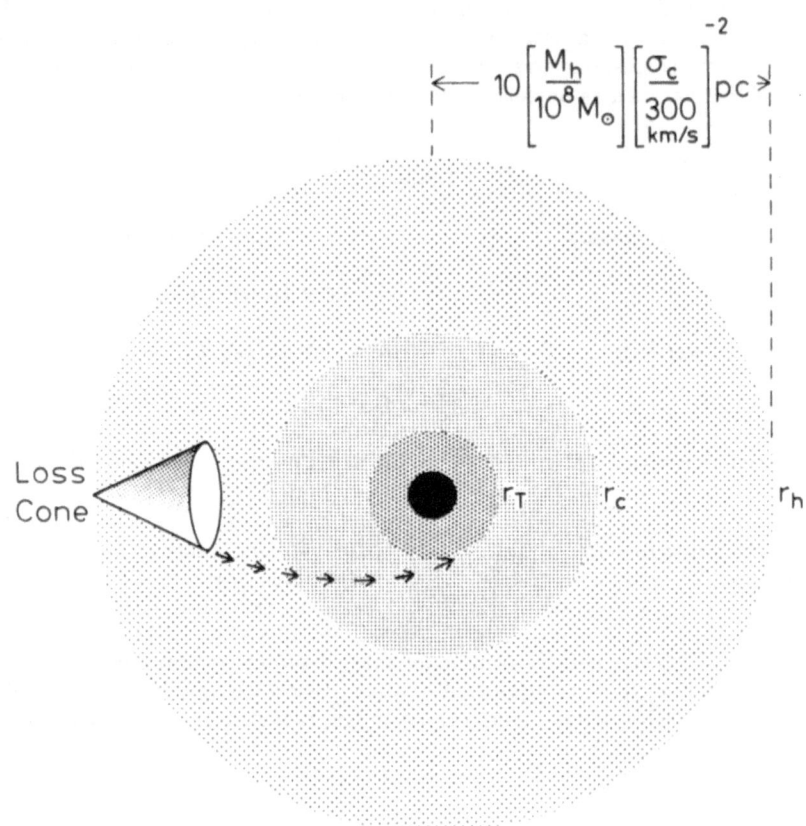

Figure 2. This diagram depicts, not to scale, various characteristic radii around a massive black hole in a stellar system. If the velocity dispersion in the core of the galaxy is σ_c the hole influences the stellar motions within a radius r_h (equation 1). Within r_c (equation 2) stars would be moving so fast that they would be more likely to experience (generally disruptive) physical collisions with each other rather than to undergo two-body encounters of the kind that can be treated by point-mass approximations. r_c is the radius where the escape velocity from the hole is comparable with the escape velocity v_* from the surface of a star. Tidal disruptions occur within a radius r_T (equation 3). To be disrupted, a star must cross the sphere at $r \simeq r_h$ on a nearby radial 'loss cone' orbit.

Note that this is always much smaller than r_c, and is indeed inside the gravitational radius $r_g = 1.5 \times 10^{13} \, (M_h/10^8 \, \mathcal{M}_\odot)$ cm for black hole masses exceeding $10^8 \, \mathcal{M}_\odot$, for solar-type stars. If $M_h << 10^8 \, \mathcal{M}_\odot$ it is, however, sufficiently far outside r_g that the disruption can be approximated as a Newtonian process and it makes little difference whether the hole is described by a Schwarzschild or a Kerr metric. Tidal disruption is, of course, a complicated process and the details depend on the density profile within the star. Giants are subject to

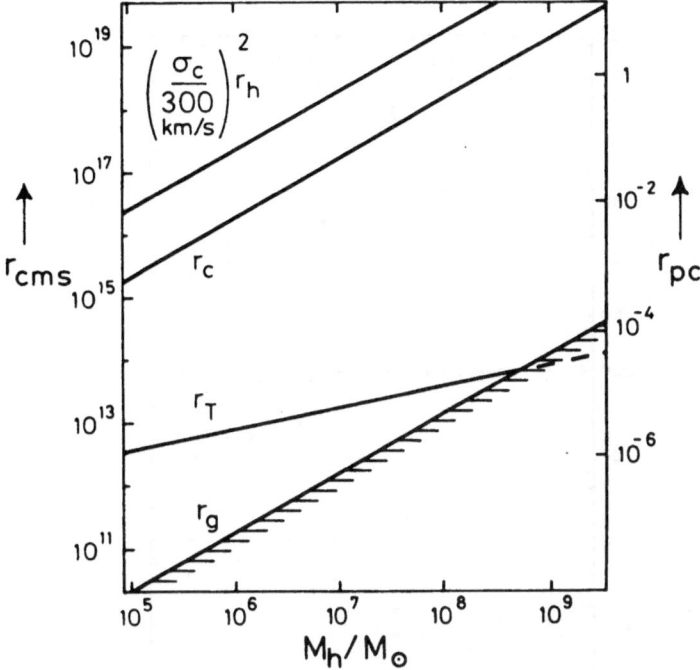

Figure 3. The various radii depicted in Fig. 2 are here plotted, on a logarithmic scale, as a function of the hole mass M_h. The radii r_c and r_T are plotted for a solar-type star with $v_* \simeq 1000 \, \text{km s}^{-1}$. Note that a hole of $\gg 10^8 \, \mathcal{M}_\odot$ can swallow solar-type stars (though not, of course, giants) without first disrupting them. When $r_T \simeq r_g$, tidal disruption effects would be restricted to the general–relativistic domain where the hole's tidal effects cannot be adequately modelled by an r^{-3} Newtonian approximation.

tidal effects even for hole masses $\gtrsim 10^{10} \, \mathcal{M}_\odot$.

At the distances of even the nearest galaxies, we would expect the gravitational effect of the central black hole to be discernible only within an angular distance of a few arcseconds of the galactic nucleus. The first indications of such effects, dating from the late 1970s, related to the giant elliptical M87, where a mass of several times $10^9 \, \mathcal{M}_\odot$ was claimed (Sargent *et al.* 1977, 1978). Of course, at the M87 distance of around 20 Mpc, a more moderate mass would not have been discernible. This evidence remains rather controversial, partly because of uncertainty about the contribution to the central light from nonthermal processes. Recently, somewhat less ambiguous evidence has arisen for M31 and some other nearby galaxies (Tonry 1984, 1987; Kormendy 1988; Dressler and Richstone 1988). Data for M31 are shown in Figure 4. There is evidence not only for an increase in the velocity dispersion near the centre, but also that the rotation curve indicates a sharper concentration in the central mass than the corresponding central peak in the luminosity. One can directly infer that the mass–to–light ratio in the centre exceeds 35 solar units.

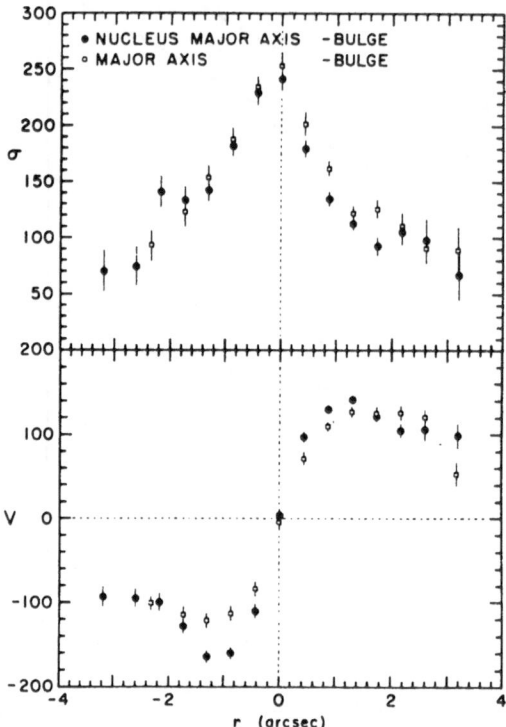

Figure 4. Data from Kormendy (1988) showing the velocity dispersion (upper panel) and rotation (lower panel) for stars in the nucleus of M31, the Andromeda Galaxy. Observations were made along the apparent major axis of the bulge and along the (misaligned) axis of the flattened nuclear star distribution; the plotted velocities were obtained after subtracting the spectral contribution from the stars in the bulge.

This limit does not self-evidently require a black hole. For instance, one might envisage a dense star cluster, consisting of stars with a different mass function from those in the body of the galaxy. The half mass relaxation time for such a cluster, with mass M, half-mass radius r, composed of stars each of mass m_*, is approximately

$$t_{rel} \simeq 10^9 \left(\frac{m_*}{m_\odot}\right)^{-1} \left(\frac{M}{10^8 \, \mathcal{M}_\odot}\right)^{-\frac{1}{2}} (r/1\,\mathrm{pc})^{\frac{3}{2}} \, \mathrm{yrs} \tag{4}$$

This implies that a cluster of 10^8 neutron stars, within a radius of 1 parsec, could marginally survive for around 10^{10} years. If the dimensions could be shown (for instance, by HST observations) to be much less than 1 parsec, then this option could be ruled out. Moreover, for a cluster containing a range of different masses, evolution would take place in much less than 10 relaxation times, because the heavier stars would segregate towards the centre (Murphy and Cohn, 1988; Quinlan and Shapiro 1989). At first sight one might infer from this equation that a cluster of lower mass stars, correspondingly more numerous so as to produce the same total mass, would survive for longer. However, for very low mass stars, physical encounters would become more important than two–body 'coulomb' deflections in determining the evolution time for the cluster.

Apart from the data on Andromeda, there is evidence for a central mass of $5 \times 10^6 \, \mathcal{M}_\odot$ in M32 (Tonry 1984, 1987), and for $10^9 \, \mathcal{M}_\odot$ in NGC 4594 the Sombrero galaxy (Jarvis and Dubath 1988; Kormendy 1989). Dressler and Richstone (1988) conjecture that the masses may be proportional to the total mass in the bulge. Quasar remnants would then be found in ellipticals, the smaller holes in disc galaxies being relics of lower–level activity such as is manifested by Seyfert galaxies.

Although one cannot yet exclude alternative interpretations of the apparent dark central mass concentration – which could conceivably involve a dense cluster, or be an artefact of anisotropic velocities for the visible stars – massive black holes certainly seem the most natural inference from this body of recent evidence. (For a recent assessment, see Goodman and Lee (1989), or Binney and Petit (1989)).

One would like some independent corroboration of the black hole hypothesis, or, conversely, some way of ruling it out. *Stellar disruption* potentially offers such a test. The inner part of a galaxy could in principle be swept completely clean of gas. On the other hand, we directly observe a concentration of stars near the putative hole. As stellar orbits diffuse in phase space, it therefore seems inevitable that some may wander sufficiently close to the hole that they suffer tidal disruption. It is therefore of interest to estimate the rate for this process, and to explore the observational manifestations of a tidally–disrupted star.

Tidal Disruption of Stars

If the star density in the galactic nucleus, just outside r_h, is N^*, and the velocities are isotropic, then the rate of disruption is

$$\sim 10^{-3} \left(\frac{M_h}{10^7 \, \mathcal{M}_\odot} \right)^{\frac{4}{3}} \left(\frac{N^*}{10^5 \, \text{pc}^{-3}} \right) \left(\frac{\sigma_c}{300 \, \text{km s}^{-1}} \right)^{-1} \text{yr}^{-1} \tag{5}$$

The actual rate could be lower than this approximate expression because radial loss cone orbits get depleted. Or it could be higher, because stars accumulate on orbits between r_h and r_c, and the density of stars in this cusp bound to the hole may exceed N^*. In fact, neither of these countervailing effects is likely to be of great importance for holes less massive than around $10^8 \, \mathcal{M}_\odot$ (Bahcall and Wolf, 1976; Frank and Rees, 1976; Lightman and Shapiro 1976). It may seem, however, that even the modest rate of stellar disruptions given above could have conspicuous consequences. If one supposed that the debris from a disrupted star were all swallowed by the hole, with efficiency 0.1 $\varepsilon_{0.1}$, and that the mean disruption rate led to steady accretion, then the resultant luminosity would be

$$6 \times 10^{42} \left[\frac{\text{disruption rate}}{10^{-3} \, \text{yr}^{-1}} \right] \varepsilon_{0.1} \, \text{erg s}^{-1} \tag{6}$$

This predicted luminosity does not seem to be observed in M31. Do we therefore have to abandon the black hole interpretation of stellar motions in this and other galactic nuclei? The answer is probably not, because of uncertainty about three things.

i) The fraction of the debris which is swallowed, rather than expelled.

ii) The radiative efficiency.

iii) The timescale for accretion or expulsion of debris. Maybe we should expect bright flares with short duty cycles, rather than a steady luminosity?

The energy required to tear the star apart (*ie* the star's self–binding energy) is supplied at the expense of the orbital *kinetic* energy (which, at $r \simeq r_T$, is larger by $\sim (M_h/m_*)^{\frac{2}{3}}$). Unless there were some explosive energy input, which would be expected only if the star passed several times closer than r_T (Carter and Luminet 1982), the debris would be *on average* bound to the hole unless the star were initially on a hyperbolic orbit with asymptotic velocity $> v_*$, which is $\sim 1000 \, \text{km s}^{-1}$ for solar–type stars.

Several effects would, however, impart orbital energies *spread widely about* this mean to gas from different parts of the disrupting star; this spread crucially influences what we would actually observe. The dominant such effect is the following. While falling inwards towards the hole, the star would develop a quadrupole distortion which attains an amplitude of order unity by the time of disruption at $r \sim r_T$. The resultant gravitational torque would 'spin it up' to a good fraction of its corotation angular velocity by the time it gets disrupted: it would consequently, by that stage, be spinning at close to its break–up angular velocity. The parts on the 'outside track' *furthest* from the hole would therefore have an *extra* velocity, over and above the orbital velocity $v_{orb} \simeq (2GM/r_T)^{\frac{1}{2}} \simeq c\,(r_g/r_T)^{\frac{1}{2}}$, of order $v_* = (m_*/M_h)^{\frac{1}{3}} v_{orb}$; those *closest* to the hole would have a comparable velocity *deficit*. Moreover, the slower-moving gas on the 'inside track' is deeper in the potential well by an amount $\sim (GM_h/r_T)(r_*/r_T) \simeq (Gm_*/r_*)$. There would consequently be a spread of order $v_{orb}\Delta v$, where $\Delta v \simeq v_*$, in the energies of different bits of debris. (See Figure 5 and its caption.) Other processes during the flyby – for instance, impulses from shocks or nuclear energy released during the drastic compression and distortion of the stellar material (*eg* Luminet, 1987) – could further enhance Δv.

Even though the *mean* specific binding energy of the debris to the hole would be positive, and comparable with the self-binding energy (Gm_*/r_*) of the original star, the *spread about this mean* is larger by $(M_h/m_*)^{\frac{1}{3}}$ – a factor which is $\gtrsim 100$ for hole masses in the range relevant to galactic nuclei. (Lacy *et al.* 1982, Rees 1985, 1988.) Whenever a solar–type star passes within the tidal disruption radius r_T, some of the debris would be flung out on hyperbolic orbits with escape velocities up to $10^4 \, \text{km s}^{-1}$. When $M_h \gtrsim 10^6 \, \mathcal{M}_\odot$, the *bound* debris would be on orbits with characteristic specific binding energy $\gtrsim 10^{-3}c^2$ rather than $\sim 10^{-5}c^2$. The actual mass-fractions in a particular energy range, and the precise high and low energy cut–offs, would depend on the details. The bound orbits are very eccentric; for solar–type stars the orbital major axis of even the most tightly–bound debris is $\sim 10^3 M_6^{\frac{1}{3}} r_g$, and the period is only $0.03 M_6^{\frac{1}{2}}$ years. For $M_6 \lesssim 100$ this is certainly small compared with the interval between successive disruptions (equation (5)). Unless it takes many orbital periods to swallow the bound gas, the debris from each star would be digested separately – in contrast to Hills' (1975) 'debris cloud' model of quasars, where the disruptions were postulated to be frequent enough to generate (*c.f.* equation (5)) a quasar-level luminosity, but the orbital periods of the debris $\left(\propto M_h^{\frac{1}{2}}\right)$ are longer. The fate of bound debris, where the quantitative details depend on viscosity, relativistic precession effects, etc., is discussed more fully elsewhere (Rees 1988).

When individual stars are being captured at the modest rate expected in relatively

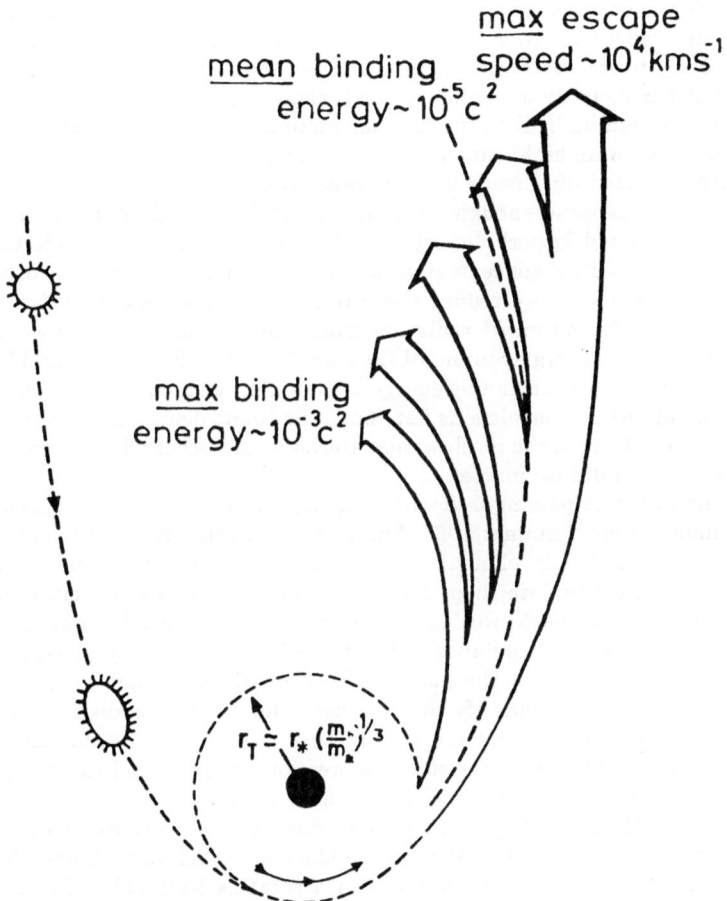

Figure 5. A solar–type star approaching a massive black hole on a parabolic orbit with pericentre distance r_T is distorted and spun up during infall, and then tidally disrupted. The average specific binding energy of the debris to the hole is $\sim 10^{-5}c^2$ (of order the self–binding energy of the original star). However, the *spread* in this energy [of order $v\Delta v$, where $v \simeq c(r_T/r_g)^{-1/2}$ and $\Delta v \simeq (Gm_*/r_*)^{1/2}$] is $\sim 10^{-3}c^2$ for hole masses $M_h \gtrsim 10^6 \, \mathcal{M}_\odot$. Almost half the debris would therefore escape on hyperbolic orbits with speeds up to $\sim 10^4$ km s^{-1}, the most tightly bound debris would traverse an elliptical orbit with major axis $\sim 10^3 r_g$ before returning to $r \simeq r_T$. Radiation from this debris, much of which may swirl down into the hole, creates a conspicuous 'flare'.

quiescent nuclei, the bulk of the debris from each would be swallowed or expelled *rapidly* compared with the interval between successive stellar captures – the only conspicuous luminosity being a flare (predominantly of thermal UV or X–ray emission) with $L = L_E$, fading within a few years (see Figure 6 and caption). The infalling debris forms a torus at $r \simeq r_T$, within which radiation pressure is dominant; its subsequent flow, controlled by viscosity, is likely to lead to accretion on a timescale which is a modest multiple of the rotation period at $r \simeq r_T$ (only a few hours). It is clear from the behaviour of $\dot{m}$ in Figure 6 that most debris would be 'fed' to the hole far more rapidly than it could be accepted if the radiative efficiency were high; much of the bound debris must then either escape in a radiatively–driven directed outflow or be swallowed inefficiently.

When a star passes close enough to be disrupted in a single flyby, almost half the debris escapes on unbound hyperbolic orbits. The kinetic energy of the ejecta is $\sim 10^{51}$ ergs. The material would be concentrated in a cone or 'fan' close to the orbital plane. Adiabatic cooling would severely reduce the internal radiative content before the debris became translucent. The 'prompt' radiation from the unbound debris would therefore release less than the initial energy content of the star (just as a supernova would be optically inconspicuous were it not for continuing energy injection in the months after the explosion). There would therefore be no conspicuous flare until the *bound* debris fell back onto the hole. The main observable effects of the outflowing material would occur when it was decelerated by running into external diffuse matter.

The behaviour of stars passing *well within* the tidal radius r_T exhibits special features, (Carter and Luminet 1982, Luminet 1987, Luminet and Carter 1986, Luminet and Marck 1985). Such stars are not only elongated along the orbital direction but are even more severely compressed into a prolate shape (*ie* a 'pancake' aligned in the orbital plane). This compression is halted by a shock, raising the matter (which then rebounds perpendicular to the orbital plane) to a higher adiabat. Also, there is the possibility of explosive energy release. The resultant spread in the energy of the debris may then exceed $v_* v_{orb}$, but one needs a detailed model to quantify this because velocities perpendicular to the orbital plane yield only a second–order contribution, and are therefore less important than those in the plane. The orbits for which extreme compression occurs would enter regions where relativistic effects were more important than for those with $r_{min} = r_T$.

For hole masses $M_h \gtrsim 10^8 \, \mathcal{M}_\odot$, solar–type stars cannot be disrupted without entering the strongly relativistic domain. The form of the black hole (Schwarzschild or Kerr?) then has an important quantitative effect, as does (for a rotating Kerr hole) the orientation of the stellar orbit relative to the hole's spin axis. Stars on counter–rotating orbits are more readily captured, with the result that Kerr holes would spin down if they gained mass primarily from stellar capture. When the hole mass is $>> 10^8 \, \mathcal{M}_\odot$ most main–sequence stars would be swallowed whole (*ie* $r_T \not> r_g$), and only giants would generate debris outside the hole.

Observability of Flares from Disrupted Stars

The most distinctive consequence of a $10^6 - 10^8 \, \mathcal{M}_\odot$ black hole's presence would be transient flares whenever bound debris from a star was swallowed, the luminosities being as high as $L_E \simeq 10^{44} M_6 \, \mathrm{erg \, s^{-1}}$. In a given object, these flares would have a duty cycle of order 10^{-3}

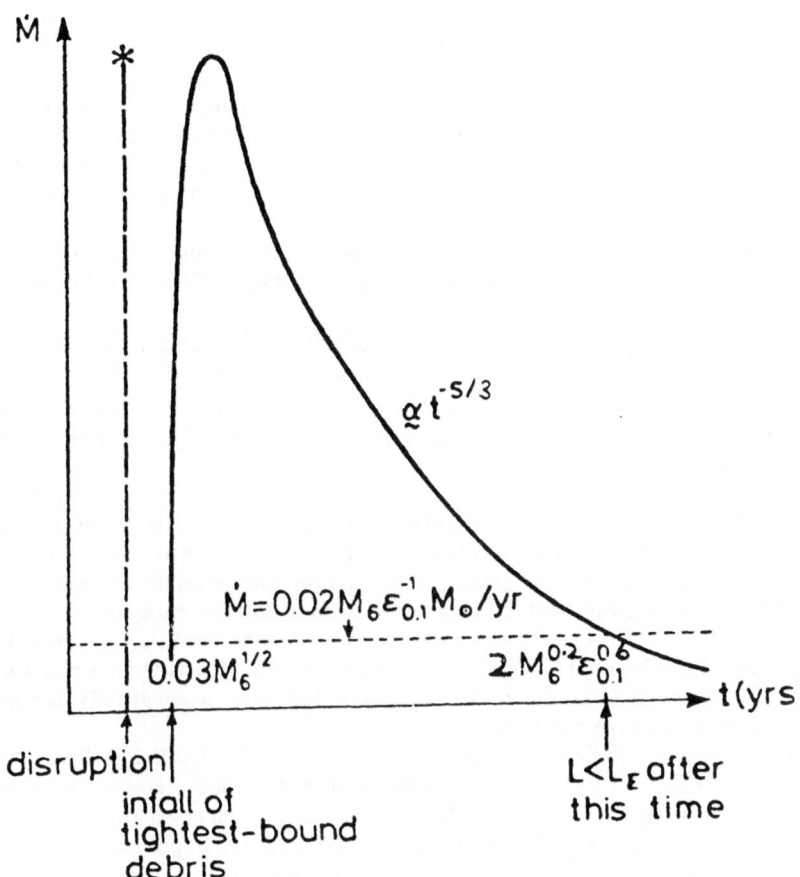

Figure 6. When a disruptive flyby occurs, the bound debris starts to rain down on to the hole after a timelag $\sim 0.03(M_h/10^6 \mathcal{M}_\odot)^{1/2}$ years. This diagram shows schematically the subsequent behaviour of $\dot{m}$. The peak infall rate is highly 'super-Eddington', but a thermal luminosity $\sim L_E$ can only be maintained for $\lesssim 1$ year. Thereafter the 'flare' would fade. At late times, the infall rate declines as $t^{-5/3}$ (the power law $-5/2$ given by Rees (1988) results from a stupid error). This is probably the dominant effect controlling the decline in luminosity after each flare.

at peak luminosity. The rise time and the peak bolometric luminosity can be predicted with some confidence. However, the effective surface temperature (and thus also the fraction of the luminosity that emerges in the visible band) is harder to predict – this depends on the size of the effective photosphere that shrouds the hole, particularly when $\dot{m}$ is high. After each luminosity peak, there would be a decline as the 'dregs' were swallowed (see Figure 6), but the median luminosity would be far below that which would result from *steady efficient* accretion of the mass supply implied by equation (6). Therefore we would not yet expect to have detected such a flare. On the other hand, a sufficiently large sample of such galaxies should reveal some members of the ensemble in a flaring state. Such objects could be searched for out to large distances: they would last rather longer than supernovae and would differ from typical AGNs through the lack of any extended structure (emission line or radio components). The central location of the phenomenon, however, militates against its detection in supernova searches, which are notoriously incomplete in the inner high–surface–brightness regions of galaxies. If $10^6 \mathcal{M}_\odot$ holes were prevalent in small or even dwarf galaxies, the nearest such flares, in any given year, may be no further away than the Virgo Cluster.

The ejected matter, though less spectacular than the accretion–powered flares, could nevertheless have more sustained cumulative effects. There are 2 ways in which matter can be expelled:

i) When a solar–type star is disrupted in a single flyby, the 'spray' of ejected debris on hyperbolic orbits (see Figure 5) moves outward at $\sim 10^4 M_6^{\frac{1}{3}}$ km s^{-1}. The kinetic energy output would be $\sim 10^{51} M_6^{\frac{2}{3}}$ erg s^{-1} – for $M_6 \gtrsim 1$ (*ie* in all cases of interest here) this exceeds the energy of a supernova. Stars that penetrate well inside r_T would cause still more violent ejection.

ii) The bound debris falls back to $r \simeq r_T$, where it can acquire centrifugal support. If energy is dissipated at a supercritical rate as this material swirls closer to the hole, some gas may then be ejected in a radiation–driven wind, or even in double jets aligned with the angular momentum of the debris. It is in principle possible for only a fraction $\varepsilon^{-1}(r_g/r_T)$ to be actually swallowed, ε being the efficiency, all the remainder being ejected. The characteristic speeds here – of order the escape velocity from r_T – would be even higher than those of the type (i) outflow discussed above. The debris from a star that does not approach close enough to be destroyed in a single flyby but is eventually disrupted by tidal dissipation could behave similarly.

Several aspects of the stellar disruption phenomenon call for detailed stellar–dynamical and hydrodynamic calculations (*c.f.* Nolthenius and Katz, 1982; Evans and Kochanek 1989). Also, relativistic precession effects have an important effect on the flow at $r \lesssim r_T$. Obviously, precise modelling should allow for a realistic stellar population. The preliminary estimates presented here are nevertheless already relevant to the searches, with high sensitivity and spatial resolution, for mass concentrations in the centres of nearby galaxies, and to the properties of AGNs in general.

Mergers and Binary Black Holes

We have seen that there may be black holes in most galaxies. It is also implied by the data in Figure 1 that most of these had already formed by the time the universe was 2 or 3 billion years old. There have certainly been a substantial number of galactic mergers since that time. Indeed, according to some models for galaxy formation, the majority of large galaxies today will have experienced at least one merger since that time. This raises the question of what happens during a merger if *each* of the galaxies involved harbours a central black hole.

Dynamical friction, operating on a timescale inversely proportional to the mass of the holes, quickly causes them to settle towards the centre of the merged galaxy. The holes eventually approach close enough to become a binary, where each is influenced more by the gravitational pull of the other than by the galaxies overall gravitational field. The black hole binary continues to tighten, as it transfers energy to stars, until it gets sufficiently close for gravitational radiation, operating on a timescale proportional to the fourth power of separation, to bring about eventual coalescence. The shape of the curve in Figure 7 shows that a binary may spend a long time at a separation of order 10^4 or $10^5 r_g$, with orbital velocity in the range $1000 - 3000$ km s^{-1}. For its orbit to shrink by a factor of 2 through this range, the binary must interact with $\sim (M_h/m_*)$ stars, each of which is imparted a velocity of order the orbital speed, and therefore probably ejected from the galaxy. The time taken for the binary to shrink can be straightforwardly calculated if the stellar orbits are isotropic, but in practice the binary may expel the stars on nearly radial orbits in the galaxy faster than these can be replenished, in which case loss cone effects will reduce the rate of orbital shrinkage. Perhaps interaction with gas, which could be associated with an epoch of nuclear activity, is needed to bring the binary close enough for eventual coalescence.

If black holes in galaxies are brought together as a result of galactic mergers, the interesting question arises of whether the resultant holes remain in the merged galaxies. There may be a recoil due to emission of net *linear* momentum by gravitational waves in the final coalescence (Redmount and Rees 1989). If the holes have unequal masses, a preferred longitude in the orbital plane is determined by the orbital phase at which the final plunge occurs. For spinning holes there may be a rocket effect perpendicular to the orbital plane, since the spins break the mirror symmetry with respect to the orbital plane. The expected velocities arising from these processes are several hundred kilometres per second. These estimates are based on extrapolating a weak field argument almost beyond the limits of its validity, but a better estimate must await full general relativistic numerical calculations in three dimensions without special symmetry. If a third hole drifts in before the binary has merged, one can get a Newtonian 'sling shot'. Ejection then occurs with speeds of order the orbital velocity of the binary, in other words up to 3000 km s^{-1}.

Our Own Galactic Centre

If there are indeed black holes in most nearby galaxies, our own Galaxy would be under-privileged if there was not also one in its own centre. There has for a number of years been dynamical evidence for a concentrated central mass of order $3 \times 10^6 \mathcal{M}_\odot$. This dynamical evidence comes, however, primarily from the motions of gas streams rather than of stars. Since the gas could be subject to nongravitational forces, and therefore need not follow ballistic trajectories, there is some ambiguity in this interpretation. The 'substantial but not fully convincing' evidence for a central mass has been reviewed by Genzel and Townes (1987).

A unique compact radio source appears to lie at the dynamical centre of the galaxy (Lo 1986). Its proper motion indicates that it is moving at < 40 km s^{-1} relative to the centre, and it can be naturally, though not uniquely, interpreted in terms of a model involving low level accretion onto a black hole (Rees 1982). The variable electron–positron annihilation gamma–ray flux may be related to this object. There remains a certain ambiguity because this source does not lie at the centre of the pattern delineated by the peculiar arm–like

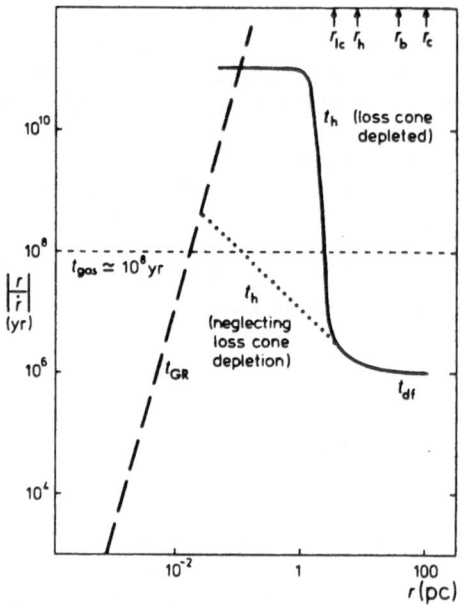

Figure 7. The timescales involved in the approach, and eventual coalescence, of a supermassive binary (after Begelman, Blandford and Rees, 1980). The core parameters chosen are those that might be appropriate to a giant elliptical: $\sigma_c \sim 300\,\mathrm{km\ s^{-1}}$, $N^* \sim 2 \times 10^3$ and $m_* \sim 1$. The members of the binary are taken to have masses $M_8 = 1$ and $m_8 = 0.3$. For this system the dynamical friction time is $t_{df} \sim 10^6$ yr. Within r_h the evolution timescale would be $(r_h/r)\,t_{df}$ if loss cone depletion could be ignored; however, unless collective effects permit replenishment of the loss cone on much less than the ordinary stellar relaxation time, the evolution within r_{l_c} proceeds on a very much longer timescale. Influx of gas into the system at a rate $\dot{m} \sim 1\,\mathcal{M}_\odot\,\mathrm{yr^{-1}}$ would however yield a timescale $t_{gas} \simeq 10^8$ yr. Gravitational radiation would take over as the dominant mechanism within $1.3 \times 10^{-2}(t_{gas}/10^8\,\mathrm{yr})^{1/4}$ pc, and the binary would then evolve towards coalescence. The recoil in the final burst may be enough to eject the newly–merged hole from the core of the galaxy.

gas features in the central 2 parsecs. Nor is it located symmetrically with respect to the infrared sources which make up the components of IRS 16.

An energy input is required into the gas filling the volume in which the gas streams are embedded. The mean inferred input seems higher than the present input. One can speculate that this, and the gas streams, could be due to the last few tidally–disrupted stars. Equation (5) suggests that 1 star may be swallowed every $\sim 10^4$ years. Unless there has been such an event recently, we would expect that the current luminosity will be below the mean.

Let me mention finally two more speculative, but potentially more definitive, tests of whether there is indeed a massive black hole in our galactic centre.

First, one can ask whether a single star could be tidally captured into a circular orbit at $r = r_T$ without disruption. Such a star would have an orbital period of order 1 hour, but a remarkable velocity exceeding 0.1c. It is rather unlikely that a star could get into such an orbit without disruption, because it would have to radiate many times its own internal binding energy, and the time needed to do this would be many Kelvin times. Since the energy would be dissipated much more quickly than this, it seems almost unavoidable that the star would get disrupted. However, a related possibility, suggested by Hills (1988), is that the stars in the central parsec of our Galaxy may include a population of close binaries. If such a binary got sufficiently near to the hole, it would be disrupted, one component being left in an eccentric bound orbit with period less than a year (which may then circularise), its companion being expelled at $\sim 1000\,\mathrm{km\ s^{-1}}$. Such hypervelocity stars, if they were detectable, would be compelling testimony that there is indeed a massive central hole.

References

Bahcall, J.N. and Wolf, R.A. 1976. *Astrophys. J.* **209**, 214.

Begelman, M.C., Blandford, R.D. and Rees, M.J. 1980. *Nature* **287**, 307.

Binney J. and Petit, J.-M. 1989, in *'Dynamics of Dense Stellar Systems'*, ed. D. Merritt (CUP) p 43.

Carter, B. and Luminet J.P. 1982. *Nature* **296**, 211.

Dressler, A. and Richstone, D.O. 1988. *Astrophys. J.* **324**, 701.

Evans, C.R. and Kochanek, C.S. 1989. *Astrophys. J. Lett.* in press

Frank, J. and Rees, M.J. 1976. *Mon. Not. Roy. Astron. Soc.* **176**, 633.

Genzel, R. and Townes, C.H. 1987. *Ann. Rev. Astron. Astrophys.* **25**, 377.

Goodman, J. and Lee, H.M. 1989. *Astrophys. J.* **337**, 84.

Hills, J.G. 1975. *Nature* **254**, 295.

Hills, J.G. 1988. *Nature* **331**, 687.

Jarvis, B.J. and Dubath, P. 1988. *Astron. Astrophys.* **201**, L33.

Kormendy, J. 1988. *Astrophys. J.* **325**, 128.

Kormendy, J. 1989. *Astrophys. J.* in press

Lacy, J.H., Townes, C.H. and Hollenbach, D.J. 1982. *Astrophys. J.* **262**, 120.

Lightman, A.L. and Shapiro, S.L. 1976. *Astrophys. J.* **211**, 244.

Lo, K.Y. 1986. *Science* **233**, 1394.

Luminet, J.P. 1987 in *'Gravitation in Astrophysics'*, eds. B. Carter and J. Hartle, (Reidel Dordrecht), p 215.

Luminet, J.P. and Carter, B. 1986. *Astrophys. J.* **61**, 219.

Luminet, J.P. and Marck, J.A. 1985. *Mon. Not. Roy. Astron. Soc.* **212**, 56.

Murphy, B.W. and Cohn, H.N. 1988. *Mon. Not. Roy. Astron. Soc.* **232**, 835.

Nolthenius, R.A. and Katz, J.I. 1982. *Astrophys. J.* **263**, 377.

Phinney, E.S. 1983, Cambridge PhD Thesis

Quinlan, G.D. and Shapiro, S.L. 1989. *Astrophys. J.* in press

Redmount, I. and Rees, M.J. 1989. *Comments on Astrophys.* in press

Rees, M.J. 1982 in 'The Galactic Center', eds. G. Riegler and R.D. Blandford, (AIP, New York) p 166.

Rees, M.J. 1985 in *'The Milky Way Galaxy'*, eds. H. van Woerden *et al.* (Reidel, Dordrecht) p 379.

Rees, M.J. 1988. *Nature* **333**, 523.

Sargent, W.L.W., Schechter, P.L., Boksenberg, A. and Shortridge, K. 1977. *Astrophys. J.* **212**, 326.

Sargent, W.L.W., Young, P.J., Boksenberg, A., Shortridge, K., Lynds, C.R. and Hartwick, F.D.A. 1978. *Astrophys. J.* **221**, 731.

Schmidt, M. 1989. *Highlights of Astronomy* **8**, 31.

Soltan, A. 1982. *Mon. Not. Roy. Astron. Soc.* **200**, 115.

Tonry, J.L. 1984. *Astrophys. J. Lett.* **283**, L27.

Tonry, J.L. 1987. *Astrophys. J.* **322**, 632.

THE FORMATION OF DARK MATTER IN COOLING FLOWS

A.C. Fabian
Institute of Astronomy
University of Cambridge
Madingley Road
Cambridge CB3 OHA
U.K.

ABSTRACT. X-ray images and spectra of clusters of galaxies show strong evidence for cooling flows. In many clusters, the hot gas in the core is cooling at rates of $\sim 100\,\mathcal{M}_\odot\,\mathrm{yr}^{-1}$ and greater. The cooled product is not yet observed but is proposed to be some form of low-mass star (perhaps brown dwarf or even jupiter-mass objects). Cooling flows must form baryonic dark matter. A scenario for low-mass star formation in a cooling flow is presented here. The main point is that no molecular clouds, which are responsible for the formation of massive stars in our own Galaxy, can exist at large radii in a cooling flow. Clouds formed at large radii in a central cluster galaxy are broken into smaller pieces by their motion relative to the intracluster gas and produce only low-mass stars. Some larger clouds, and therefore some more massive stars, can form in the central few kpc of a flow. The wider implications of the formation of dark matter in cooling flows are discussed in the context of the formation of massive galaxies.

1. Introduction

Enormous quantities of diffuse hot gas are observed in clusters and groups of galaxies. The thermal pressure of the gas necessary to support it against collapse in the large gravitational field of the cluster requires that the sound speed of the gas is similar to the typical velocity of a cluster galaxy (the velocity dispersion of the cluster). This is generally in the range 500 to 1200 km s^{-1}, implying that the gas temperature is $\sim 10^7 - 10^8$ K. The main energy loss of gas at such high temperatures is bremsstrahlung, which is the origin of the diffuse X-radiation from clusters of galaxies and our principal source of information on their intracluster medium (ICM). Further indirect evidence for the gas is found in 'head-tail' radio sources and from theories of the propagation of double-lobe radio sources. Reviews of the properties of the ICM are given by Sarazin (1986, 1988) and Fabian (1988b, on which parts of this paper is based).

Most of the *observed* intracluster gas has an electron density, n_e, in the range of $10^{-4} - 10^{-2}\,\mathrm{cm}^{-3}$ and a temperature $T \sim 10^7 - 10^8$ K, and is contained within a radius of 1 to 2 Mpc. The total mass of gas in a rich cluster is $\sim 10^{14}\,\mathcal{M}_\odot$ and its bremsstrahlung

195

D. Lynden-Bell and G. Gilmore (eds.), Baryonic Dark Matter, 195–221.
© 1990 *Kluwer Academic Publishers.*

luminosity is $\sim 10^{43} - 10^{46} \, \text{erg s}^{-1}$. An emission line due to highly-ionized iron is observed in all clusters that are bright enough for detection to be possible (see e.g. Rothenflug & Arnaud 1986; Edge 1989) and shows that the gas has ~ 0.3 times solar abundance in iron. The work of Canizares *et al.* (1979; 1982) and Mushotzky *et al.* (1981) on cooling regions in clusters shows O, Ne, Si and S to be also present at abundances close to solar (although O may be super-solar).

The origin of the gas is uncertain. Being metal-enriched, it cannot all be primordial. It is generally assumed that much of the gas was processed through an early-population of stars which then blasted it into intracluster space in supernova explosions. It could also have been stripped from young galaxies during the formation of the cluster. Whichever of these is correct, gas shares the kinetic energy of the galaxies, which is ultimately gravitational in origin, and it has a sound speed similar to the galaxy motions. The presence of so much gas, comparable to all the mass in observable stars and at least 10 per cent of the virial mass of the cluster, suggests that galaxy formation is an inefficient business as far as gas is concerned.

2. Cooling Flows

The intracluster gas is, of course, densest in the core of a cluster and its cooling time, t_{cool}, due to the emission of X-rays such as those observed, is shortest there. A cooling flow is formed when t_{cool} is less than the age of the system, $t_a (\sim H^{-1})$.

In the cases considered here, t_{cool} exceeds the gravitational free-fall time, t_{grav}, within the cluster (except perhaps in some very small region at the centre), so,

$$t_a > t_{cool} > t_{grav}. \tag{1}$$

The flow takes place because the gas density has to rise to support the weight of the overlying gas. It is essentially pressure-driven.

If that is not immediately clear, consider the gaseous atmosphere trapped in the gravitational potential well of the cluster or galaxy to be divided into two parts at the radius, r_{cool}, where $t_{cool} = t_a$. The gas pressure at r_{cool} is determined by the weight of the overlying gas, in which cooling is not important. Within r_{cool}, cooling is tending to reduce the gas temperature and so the gas density must rise in order to maintain the pressure at r_{cool}. The only way for the density to rise (ignoring matter sources within r_{cool}, which is a safe assumption in a cluster of galaxies) is for the gas to flow inward. This is the cooling flow.

If the initial gas temperature exceeds the virial temperature of the central galaxy (which is generally the case for rich clusters but not for poor ones or individual galaxies) then the gas continues to cool as it flows in. However, when the temperature has dropped to the virial temperature of the central galaxy, the gas heats up as it flows further in due to the release of gravitational energy. The gas temperature can eventually drop catastrophically in the core of the galaxy if its gravitational potential flattens there. The net result is that the gas within r_{cool} radiates its thermal energy plus the PdV work and gravitational energy released in the flow.

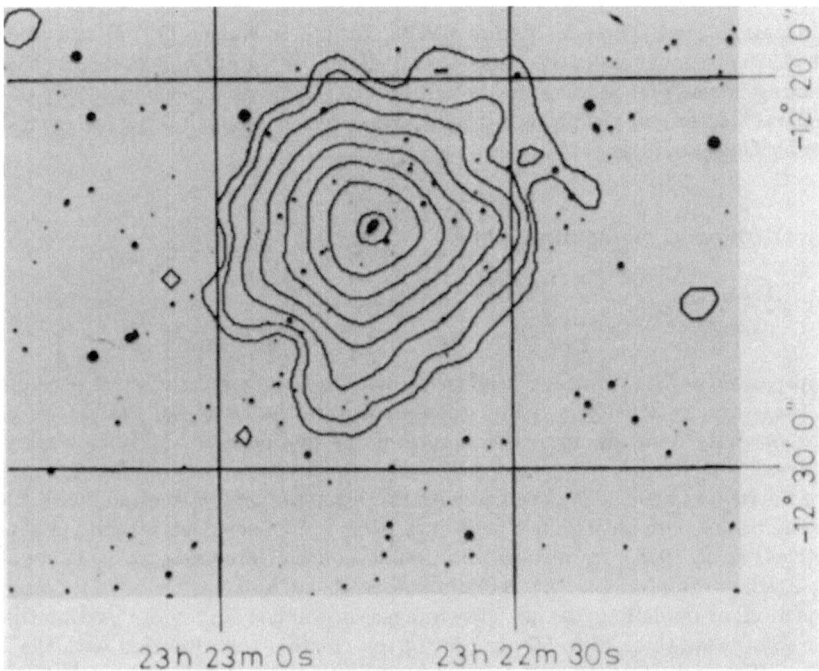

Figure 1. IPC X-ray surface brightness contours of A2197 superimposed on an optical image of the cluster (from Crawford *et al.* 1989). Note that the contours peak onto the central cluster galaxy. The mass deposition rate is about 400 $\mathcal{M}_\odot$ yr^{-1}.

This is how an idealized, homogeneous cooling flow, in which the gas has a unique temperature and density at each radius, will behave. Observations of real cooling flows shows that they are inhomogeneous and must consist of a mixture of temperatures and densities at each radius. The homogeneous flow is, however, still a fair approximation of the mean flow.

General reviews of cooling flows have been made by Fabian, Nulsen & Canizares (1984) and Sarazin (1986, 1988) and some other points of view may be found in the Proceedings of a NATO Workshop (Fabian 1988a). As explained above, the cooling flow mechanism is very simple, although the details of its operation are not. The primary evidence for them is in the X-ray observations. There is no evidence at other wavelengths for the large mass deposition inferred from the X-ray data. I discuss this point more fully later, but it should be stressed that large amounts of distributed low-mass star formation at other wavelengths need not be detectable if the gas is initially at X-ray emitting temperatures. This is, perhaps, the crux of the controversial aspect of cooling flows. They are difficult to prove or disprove in wavebands other than the X-ray. The X-ray evidence is, for me, sufficiently compelling that the existence of large cooling flows is a reasonable and straightforward conclusion.

It was Uhuru observations of clusters that first showed the mean cooling time of the gas in the cores of clusters to be close to a Hubble time (Lea *et al.* 1973). X-ray measurements from the Copernicus satellite showed that the core emission in the Perseus and Centaurus clusters was highly peaked (Fabian *et al.* 1974; Mitchell *et al.* 1975). These, and theoretical considerations, led Cowie & Binney (1977), Fabian & Nulsen (1977) and Mathews & Bregman (1978) to independently consider the effects of significant cooling of the central gas, ie cooling flows. The process was noted by Silk (1976) as a mechanism for the formation of central cluster galaxies from intracluster gas at early epochs and for general galaxy formation by Gold & Hoyle (1959).

3. X-ray Evidence For Cooling Flows

3.1. X-RAY IMAGES

A sharply-peaked X-ray surface brightness distribution is indicative of a cooling flow. It shows that the gas density is rising steeply towards the centre of the cluster or group since the emissivity depends upon the square of the gas density and only weakly on the temperature.

Most of the images have been obtained with the *Einstein Observatory* and with EXOSAT, although the peaks were anticipated with data from the Copernicus satellite (Fabian *et al.* 1974; Mitchell *et al.* 1975), from rocket-borne telescopes (Gorenstein *et al.* 1977) and with the modulation collimators on SAS 3 (Helmken *et al.* 1978).

Deprojection, or modelling, of the X-ray images shows that $t_{cool} < H_0^{-1}$ within the central 100kpc or so of more than 30 to 50 per cent of the clusters well-detected with the Einstein Observatory (Stewart *et al.* 1984b; Arnaud 1988). Additional data from EXOSAT (Edge 1989) now show that more than two-thirds of the 50 X-ray brightest clusters in the Sky (see list in Lahav *et al.* 1989) have cooling flows (Pesce *et al.* 1989; Edge *et al.* 1989). Many of the remaining clusters in this sample have not been imaged so their status is undefined. Since the luminosity associated with the flow does not dominate the total X-ray emission, this high fraction is not a simple consequence of the clusters being X-ray bright. It is due to the data on them generally being of the best quality (ie many X-ray counts detected and the core well-resolved). Whether H_0^{-1} should be used for t_a is debatable, but inspection of the results shows that reducing t_a by 2, say, does not much change the fraction of clusters which contain cooling flows. The overall picture is that the prime criterion for a cooling flow, $t_{cool} < 10^{10}$ yr, is satisfied in a large fraction (> 80 per cent) of clusters. It is also satisfied in a number of poor clusters and groups (Schwartz, Schwarz & Tucker 1980; Canizares, Stewart & Fabian 1983; Singh, Westergaard & Schnopper 1986). Cooling flows must be both common and long-lived, in order that such a high fraction of peaked clusters is observed.

The mass deposition rate, $\dot{M}$, due to cooling (ie the accretion rate, although this is a poor term since most of the gas does not much change its radius) can be estimated from the X-ray images by using the luminosity associated with the cooling region (ie L_{cool} within r_{cool}) and assuming that it is all due to the radiation of the thermal energy of the gas, plus the PdV work done.

$$L_{cool} = \frac{5}{2} \frac{\dot{M}}{\mu m} kT, \tag{2}$$

where T is the temperature of the gas at r_{cool}. Values of $\dot{M} = 50 - 100 \, \mathcal{M}_\odot$ yr^{-1} are fairly typical for cluster cooling flows. (L_{cool} is similar to the excess luminosity measured by Jones

& Forman 1984.) Some clusters show $\dot{M} \sim 500\,\mathcal{M}_\odot\,\text{yr}^{-1}$ (eg PKS0745, A1795, A2597 and Hydra A). The main uncertainties in the determination of $\dot{M}$ are the gravitational potential within the cluster core and t_a. Assuming $t_a \sim 10^{10}\,\text{yr}$, the estimates of $\dot{M}$ are probably accurate to within a factor of 2 (Arnaud 1988).

Since we often measure a surface brightness profile for the cluster core (where the X-ray emission is well-resolved), we have $L_{cool}(r)$ which can be turned into $\dot{M}(r)$, the mass deposition rate within radius r. Generally,

$$\dot{M}(r) \propto r. \qquad (3)$$

The surface brightness profiles are less peaked than they would be if all the gas were to flow to the centre. This means that the gas must be inhomogeneous, so that some of the gas cools out of the flow at large radii and some continues to flow in. The actual computation of $\dot{M}(r)$ is in detail complicated, since we need to take into account how the gas cools and any gravitational work done, but since plain cooling dominates in clusters, a simple analysis gives a fair approximation to the profile (see Fabian, Arnaud & Thomas 1986; Thomas, Fabian & Nulsen 1987; White & Sarazin 1987abc).

3.2. X-RAY SPECTRA

Key evidence that the gas does actually cool is given by moderate to high resolution spectra of the cluster cores. Canizares et $al.$ (1979; 1982), Canizares (1981), Mushotzky et $al.$ (1981) and Lea et $al.$ (1982) used the Focal Plane Crystal Spectrometer (FPCS) and the Solid State Spectrometer (SSS) on the $Einstein$ $Observatory$ to show that there are low temperature components in the Perseus and Virgo clusters, consistent with the existence of cooling flows. Detailed examination of the line fluxes and of the emission measures of the cooler gas by Canizares, Markert & Donahue (1988) and Mushotzky & Szymkowiak (1988) shows that, in the case of the Perseus cluster, the gas loses at least 90 per cent of its thermal energy and that the mass deposition rates are in agreement with those obtained from the images. Good agreement is obtained also in several other clusters. The SSS results show that the emission measures vary with temperature in the manner expected from a cooling gas. The $importance$ of $these$ $data$ $cannot$ be $overemphasized$ $since$ $they$ $show$ $that$ the gas $does$ $cool.$ Any 'alternative interpretation' of the images must confront this spectroscopic evidence successfully.

The cooling time of the gas in the Perseus cluster which emits the FeXVII line ($T <$ 5×10^6 K) is less than 3×10^7 yr. Since the emission measure of this gas agrees with that inferred from the gas cooling at the higher temperatures which dominate the images and the SSS result, we must conclude that the flow is steady (Nulsen 1988). The shape of the continuum and line spectrum observed with the SSS is consistent with the same mass deposition rate at all X-ray temperatures (as expected) so we must again conclude that the flow is long-lived. It cannot be some intermittent or transient phenomenon only a billion years old.

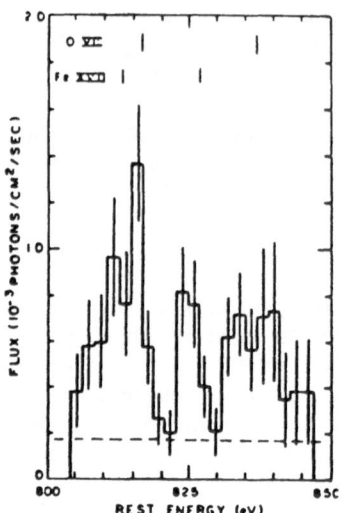

Figure 2. Part of the FPCS spectrum of the Perseus cluster (from Canizares, Markert & Donahue 1988). Note the prominent emission lines of OVIII and Fe XVII. The emission measure of the gas producing the Fe XVII lines is consistent with about 200 $\mathcal{M}_\odot$ yr^{-1} of gas cooling through $\sim 5.10^6 - 10^6$ K. This is the $\dot{M}$ found from imaging and SSS studies, which are most sensitive to higher temperature gas (typically $5.10^6 - 3.10^7$ K).

3.3 SUMMARY OF X-RAY DATA

The overwhelming evidence of the images and spectra shows that cooling does occur at a steady rate over long times (at least several billion years). Since gas is then cooling out of the hot phase at rates of hundreds of solar masses per year, an inflow must occur. We do not expect yet to have direct evidence of any inward flow since the velocity is highly subsonic at ~ 10 km s^{-1}.

Cooling flows are common and all of the nearest clusters (Virgo, Centaurus, Hydra, Fornax, Perseus) contain one apart from the Coma cluster. Clusters such as the Coma cluster and CA 0340-54, with 2 large central galaxies orbiting each other, are the main class of clusters that does not show strongly peaked emission. Even they could contain disrupted flows. The motion of the central galaxies means that there is no focus for the flow (Fabian, Nulsen & Canizares 1984). Many flows are observed out to a redshift of 0.1. A more distant one is in 3C295 at $z \sim 0.5$ (Henry & Henriksen 1987) and there is evidence from optical spectra for cooling flows being common around radio-loud quasars (Crawford & Fabian 1989). Most of the clusters detected by the Medium Sensitivity Survey of the *Einstein Observatory* must have the peaked, surface-brightness profile characteristic of cooling flows in order to have been detected (Pesce *et al.* 1989). Probably *all* clusters of galaxies will be found to have gas cooling out at rates exceeding a few solar masses per year when we have the improved spectral and spatial response of future missions such as AXAF.

The current values of $\dot{M}$ are probably good to a factor of 2 (Arnaud 1988) and could be higher if there are denser blobs beyond r_{cool} (Thomas *et al.* 1987). I am not aware of any alternative interpretation of the X-ray data which explains both the peaked images and the X-ray spectra.

4. The Fate of the Cooled Gas

The accumulated mass of cooled gas can be considerable;

$$\dot{M} t_a = 10^{12} \left(\frac{\dot{M}}{100 \, \mathcal{M}_\odot \, \mathrm{yr}^{-1}} \right) \left(\frac{t_a}{10^{10} \, \mathrm{yr}} \right) \mathcal{M}_\odot. \tag{4}$$

It is a significant fraction of the mass of the central galaxy, suggesting that we are witnessing the continued formation of that galaxy, typically one of the largest galaxies known. The accumulated mass is only a small fraction of the total mass of gas in the cluster.

4.1. DISTRIBUTED MASS DEPOSITION

The cooling gas cannot all flow into the very centre of the dominant galaxy and accumulate there or it would rapidly dominate the mass and velocity dispersion of the galaxy core. This would indeed be a way of ruling out long-lived homogeneous cooling flows. Fortunately, the data indicate that the cooled gas is deposited by the flow in a spatially-distributed manner. The evidence for distributed mass deposition is obtained principally from the X-ray surface brightness profiles, which are less peaked than expected if all the gas flowed into the very centre of the cluster. Further evidence is provided by the Fe XVII line, the total intensity of which is so high that the emitting gas must be widely distributed with a small filling factor or its surface brightness would exceed that observed in the images (Canizares *et al.* 1988).

These points mean that the hot gas is inhomogeneous and consists of denser blobs and clouds embedded in lower density gas. Since the hot gas at any radius must all be at the same pressure, the denser blobs must be cooler than the lower density gas. We envisage that there is a continuous spread of cloud densities above some low value corresponding to some upper temperature of order the mean intracluster gas temperature.

The mean conditions of an inhomogeneous flow are represented by the equations of a homogeneous flow. We have the equation of continuity;

$$\dot{M} = 4\pi r^2 \rho v, \tag{5}$$

the pressure equation (ignoring highly subsonic flow terms)

$$\frac{dP}{dr} = -\rho \frac{d\phi}{dr}, \tag{6}$$

and an energy equation,

$$\rho v \frac{d}{dr} \left(\frac{5}{2} \frac{kT}{\mu m} + \phi \right) = n^2 \Lambda, \tag{7}$$

where Λ is the cooling function. If the cooling region (where $t_{cool} < H_0^{-1}$) is at constant pressure $(d\phi/dr = 0)$, then $n \propto T^{-1}$ and

$$\rho v \frac{d}{dr} \left(\frac{5}{2} \frac{kT}{\mu m} \right) = n^2 \Lambda \tag{8}$$

and if

$$\Lambda \propto T^\alpha, \tag{9}$$

then

$$n \propto R^{-3/(3-\alpha)}. \tag{10}$$

This is proportional to $R^{-6/5}$ for bremsstrahlung. The density rises inward as the temperature falls. Constant pressure is a fair approximation to the core region of a cluster. Gravity is not particularly important, except perhaps for focussing the flow, until the gas has cooled to about the virial temperature of the central galaxy. Then the gas heats up as it flows in further and the pressure rises (Fabian & Nulsen 1977). The flow velocity $v \simeq r/t_{cool}$, which is highly subsonic.

When the flow is inhomogeneous we can estimate $\dot{M}(r)$ by assuming that the gas is composed of a number of phases, the densest of which cools out of the flow at the radius under consideration (Thomas et al. 1987), or by model fitting (White & Sarazin 1987). In the first approach, the cooling region is divided into a number of concentric shells of size compatible with the instrumental resolution. The luminosity, δL_i of the ith shell can then be considered to be the sum of the cooling luminosity of the gas cooling out at that radius from the mean temperature T_i at rate $\delta \dot{M}$ and the luminosity of gas flowing across the shell experiencing temperature and potential changes ΔT_i and $\Delta \phi_i$;

$$L_i = \delta \dot{M}_i \frac{5}{2} \frac{kT_i}{\mu m} + \left(\frac{5}{2} \frac{k\Delta T_i}{\mu m} + \Delta \phi_i \right). \tag{11}$$

In our most detailed approach (Thomas, Fabian & Nulsen 1987), we have allowed for as many phases at a radius as there are shells within that radius and have integrated the cooling function and spectrum carefully. A typical mass deposition profile is shown in Fig. 3. It agrees fairly well with that obtained by assuming that the gas is homogeneous, principally because most of the energy is lost on cooling from the average cluster temperature T_X at temperatures close to T_X. This approach does allow us to measure the spread of densities in the gas at any radius. It is this which determines the manner in which mass is deposited (Nulsen 1986). We infer that the intracluster gas must contain a density spread of at least a factor of two. This may not be surprising when it is recalled that it has been enriched in metals which must have mixed different gases together.

The result that $\dot{M}(r) \propto r$ means that the deposited matter has $\rho \propto r^{-2}$ which is essentially an isothermal halo such as inferred for the dark matter around galaxies. It is assumed that whatever condenses out of the cooled gas orbits about, or through, the central galaxy such that its mean radius is similar to that where it was formed (see §5.4).

If the gas forms stars, then cooling flows are some of the largest and strongest regions of star formation in our part of the Universe. Even a casual comparison of a central cluster galaxy and a spiral galaxy such as our own, which is thought to be forming stars at a rate of $3 - 10 \, \mathcal{M}_\odot \, \text{yr}^{-1}$, shows that cooling flows must form low-mass stars (Fabian, Nulsen

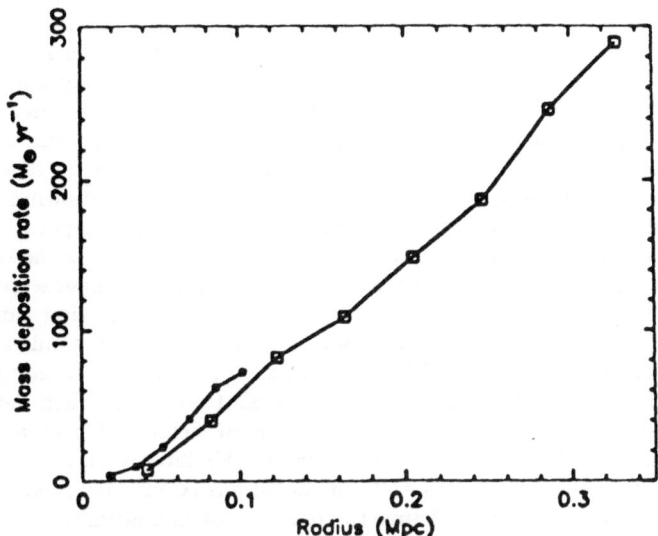

Figure 3. Mass deposited within radius r by the cooling flow in A2199, from Thomas *et al.* (1987).

& Canizares 1982; Sarazin & O'Connell 1983). Massive stars would make central cluster galaxies much bluer than they are. The absence of massive blue stars means that star formation with a spiral galaxy initial-mass-function (imf) is almost non-existent in the central galaxies of cooling flows. This is the main topic of this paper and is discussed in depth in §5.

It should be stressed that the cooling gas is not directly detected once it has cooled below about 3×10^6 K. If it recombines and forms low-mass stars ($\langle M_* \rangle < 0.5 \mathcal{M}_\odot$) in a distributed manner ($M(r) \propto r$) then there is no reason for it to have been seen.

There is, however, plenty of evidence for dark matter in clusters and low-mass stars are one plausible form of dark matter. The manner of the mass deposition with radius, $\dot{M}(r) \propto r$, leads to an isothermal halo which is consistent with the dark matter distribution around large galaxies (*eg* M87, Stewart *et al.* 1984a; Mould *et al.* 1987). Cooling flows are a source of baryonic dark matter.

4.2. HEATING

Since the implied star formation rates are so large and there is little sign of it optically, there have been a number of studies suggesting that the rates have been grossly overestimated. Some heat source that balances the cooling is the obvious solution. Cosmic rays (Tucker & Rosner 1982; Pedlar *et al.* 1988; Bohringer & Morfill 1988), conduction (Bertschinger & Meiksin 1986), supernovae (Silk *et al.* 1986) and galaxy motions (Miller 1986; see also Pringle 1989) have all been invoked as heat sources. Unfortunately for these models, the X-ray spectra indicate cooling without heating. None of the models proposed so far is able (or even attempts!) to account for the X-ray line emission. There are other problems with these heat sources as well (see Fabian 1988b; Bregman & David 1988).

To consider heat sources in more detail, we note that it is generally difficult to keep the gas stable whilst heating it (Stewart *et al.* 1984a). Most heat sources, such as cosmic rays, heat at a rate proportional to the gas density, whereas the cooling varies as the density squared. Heating may then cause the gas to become more unstable by tending to increase the temperature and pressure of the lowest density phases but allowing the denser gas to carry on cooling. If conduction occurs unimpeded then the temperature gradient set up by cooling is offset by a conductive heat flux. This situation is, however, restricted to only a small part of parameter space (initial density and temperature). For typical parameters, cooling dominates at the centre of a flow and conduction can be important further out (Nulsen *et al.* 1982). If conduction is dominant, then it tends to make the gas almost isothermal, in disagreement with observations. The X-ray spectroscopic observations (Canizares *et al.* 1979, 1982, 1988; Mushotzky *et al.* 1981, 1988) then demonstrate that conduction is inhibited, probably by tangled magnetic fields which greatly decrease the effective electron mean-free-path. We shall return to this later in §6.2.

Supernova heating from stars formed from cooled gas (Canizares *et al.* 1982; Lea *et al.* 1982; Silk *et al.* 1986) can at most change the mass deposition estimates by a factor of two. This is because the energy from a supernova can only heat a mass of gas sufficient to form another supernova progenitor (and an IMF's worth of lower-mass stars) to 8.10^6 K. There is no evidence for supernovae around the central galaxies in cooling flow clusters (see e.g. Caldwell & Oemler 1981).

Finally on the topic of heating, it is worth remembering that cooling flows occur in a wide variety of clusters, both with and without strong radio sources (e.g. Cyg A vs. AWM4) and in deep and shallow potential wells (e.g. Perseus vs. Hydra). Any heat source necessary to counteract the radiative cooling would represent a major heat flow, of $\gtrsim 10^{62}$ erg per cluster. In my view, the current lack of understanding of star formation in our own Galaxy means that it is not a simple business to extrapolate to other situations. Star formation can proceed at hundreds of solar masses per year without necessarily being evident optically, provided that only low mass stars are formed. There is no problem with the total mass deposited, which is distributed out to 100 - 300 kpc from the central galaxy and does not pile up within its centre. Clusters are full of dark matter and there can be no problem with some (or even all) of it being baryonic.

It has been suggested by Hu (1988) that cooling flows began only recently. This reduces the total accumulated mass but does not explain the lack of blue stars. The imf must be different from our local imf even if the flow is recent. Meiksin (1989) argues that the flows may be so time-dependent that the appearance of $\dot{M}(r) \propto r$ is an illusion (the spectroscopic data argue against this).

The total level of heating necessary to balance the cooling is very large, $\sim 10^{62}$ erg for a large flow over t_a and so if some heat source is found that can accommodate the X-ray spectral measurements successfully, it must be one of the major (unseen!) energy flows in the Universe! Whilst the luminosity of a cooling flow may be only 10 per cent of the total cluster X-ray luminosity and the mass lost through cooling a negligible drain on the enormous outer atmosphere, the cooling luminosity is a major loss of energy from the cluster core. Whatever is eventually decided about cooling flows, they cannot be an insignificant process.

5. Star Formation in Cooling Flows

5.1. CAN THE ACCUMULATED MATTER REMAIN AS DIFFUSE GAS?

Optical observations of the central galaxies in many cooling flows show patches of line-emitting nebulosity in the inner few kpc (Hu, Cowie & Wang 1985; Johnstone, Fabian & Nulsen 1987; Heckman *et al.* 1989). The spectra indicate that the gas has a relatively low ionization ([OII]>[OIII]). The Hβ luminosity of the gas ranges from 10^{39} to 10^{42} erg s^{-1} which means, at the pressure of intracluster gas, that the mass of warm, line-emitting gas ($10^3 < T < 10^4$ K) is in the range of $\sim 10^5 - 10^8 \mathcal{M}_\odot$. To be spread over the observed regions, the gas must be in the form of small clouds or filaments with a volume filling factor of $10^{-5} - 10^{-7}$. The emission lines show doppler widths indicating both random and systematic (rotation?) velocities of up to several hundred km s^{-1}.

The origin of this gas and its source of ionization is unclear. It probably originates from the cooling flow, but why only such a small fraction is observed is puzzling. When observed (there are flows in which no optical emission lines are observed, *eg* A2029), the emission is strongly peaked to the centre (Heckman *et al.* 1989) and is not consistent with, say, $L_{H\beta}(< r) \propto r$. It is far too luminous to be plain recombination of the cooling gas. Each proton must recombine 100 to 1000 times to produce the observed luminosity. Most simple sources of ionization, such as photoionization by massive stars or by an active nucleus, have been ruled out (Johnstone & Fabian 1988).

Whatever the origin or amount of the warm gas, it can be used to probe the properties of the hotter gas, such as its pressure and its motions. The gas pressure, for example, can be obtained from the [SII] lines, which are collisionally de-excited at densities above a few 100 cm^{-3}. In the Perseus cluster, these lines change their ratio indicating high density and thus high pressure within 5kpc of the nucleus (Johnstone & Fabian 1988). Since the pressure has risen above the X-ray inferred pressure (from the X-ray surface brightness) at 20 kpc, the mean gas temperature must be down to the virial temperature of the central galaxy, NGC1275, of about 10^7 K. (Hydrostatic equilibrium requires that the pressure increases inward once the gas has cooled to the local virial temperature.) This is further confirmation that the gas has cooled there below the outer temperature of $\sim 6 \times 10^7$ K. Since the gas pressure is so high, the magnetic pressure cannot be more than about twice the gas pressure (and is probably less than that).

Whether there is colder (recombined) gas also associated with the warm gas is unknown. Limits on the amount of HI from its emission are not very restrictive ($\sim 10^9 \mathcal{M}_\odot$; Burns, White & Haynes 1981; Valentijn & Giovanelli 1982), especially when it is realized the HI is probably optically thick (Loewenstein & Fabian 1989; see also the discussion of Bregman *et al.* 1989 on M86). HI has been seen in absorption against NGC1275 in the Perseus cluster by Crane *et al.* (1982) and by Jaffe *et al.* (1988). Lazareff *et al.* (1989) and Mirabel *et al.* (1989) have also observed CO in NGC1275 indicating that there is a very large quantity of gas ($> 5 \times 10^9 \mathcal{M}_\odot$) in the central 10 kpc or so, if a Galactic CO to H_2 conversion factor is used. (This factor is probably a gross overestimate, since it assumes that line widths are due to self-gravitation, which is unlikely in a turbulent cooling flow where the velocity widths just reflect the motions of the hotter gas which carries the cold gas around). Observations of the CO line in other cooling flows have led only to upper limits (Grabelsky & Ulmer 1989).

If the observed absorbing HI in NGC1275 is associated with the observed CO and is all contained within the inner 10 kpc, then, given that most of the gas cools out of the flow at much larger radii ($\dot{M}(< r) \propto r$), the gas must turn into some unobservable form relatively quickly. This unobservable form may be small, very optically-thick, HI clouds with a low covering fraction (so no absorption against the central nucleus) or low-mass stars. Since small HI clouds are likely to collapse into stars anyway, we shall proceed by assuming that we are dealing with stars. The term 'stars' will also be taken to include brown dwarfs and jupiter–mass objects.

5.2. DETECTABLE STAR FORMATION IN COOLING FLOWS ($M_\star > 1\,M_\odot$)

The general statement about the necessity for low mass stars applies to the bulk of the cooled gas. The optical emission line clouds or filaments often seen in the centre are atypical of most of the flow. These clouds may give rise to higher mass stars. There is some excess blue light observed at the centres of many cooling flows (see *eg* Hintzen & Romanishin 1988) and it does correlate in strength with the mass deposition rate (Johnstone, Fabian & Nulsen 1987; O'Connell & McNamara 1989). The spatial extent of the blue light (~ 5 kpc; Hintzen & Romanishin 1988) is similar to that of the warm cloud region (Heckman *et al.* 1989). Spectral fits of the blue light together with upper limits from IUE spectra show that the upper mass limit for stars must be around $1.5 - 2\,M_\odot$ there (Crawford 1988; Crawford *et al.* 1989), *ie* the excess blue light is attributable to F and early G stars only.

Small cooling flows occur continuously in most elliptical galaxies which are not in rich clusters. Ram-pressure stripping removes the gas from galaxies within clusters. This is observed in the Virgo cluster where M86 has a plume of X-ray emission to the NW (Forman *et al.* 1979). A faint diffuse patch of optical light is observed coincident with this plume and may be due to stars formed from the cooling gas (Nulsen & Carter 1986).

5.3. LIMITS AND DETECTION OF LOW-MASS STARS IN COOLING FLOWS ($M_\star \ll 1\,M_\odot$)

Simple considerations of the total colours and magnitudes of the central cluster galaxies in cooling flows show that the mean mass of a star formed in a cooling flow, $\langle M_\star \rangle$, must be less than $1\,M_\odot$ (Cowie & Binney 1977; Fabian, Nulsen & Canizares 1982; Sarazin & O'Connell 1983). Obtaining lower limits on $\langle M_\star \rangle$ is more difficult, since, in terms of colours and integrated magnitude, a few red giants can mimic a swarm of low mass stars. More detailed studies are necessary, in which dwarfs can be distinguished from giants. The CO absorption feature at 2.3μ is stronger in giants than in dwarfs and has been observed in the centres of several cooling flow galaxies by Arnaud & Gilmore (1987). They show that the low-mass dwarf population ($0.25 < M_\star < 0.5\,M_\odot$) must be similar to that in 'non-cooling-flow' galaxies. There is no such constraint at larger radii, where the bulk of the star formation must occur.

An exciting possibility has emerged from the Giant, Red Envelope Galaxy (GREG) found from the Einstein Medium Sensitivity Survey by Maccagni *et al.* (1988). This object could be showing low-mass stars formed in a cooling flow (Johnstone & Fabian 1989). This galaxy lies in the centre of an X-ray luminous poor group of galaxies and has a normal de Vaucouleurs profile in the V-band. A large r^{-1} envelope appears in the i-band (Fig.4), which can plausibly be explained as due to $0.5\,\mathcal{M}_\odot$ stars. If due to starlight, then the envelope requires an exceptional initial-mass-function of the kind required by cooling flows.

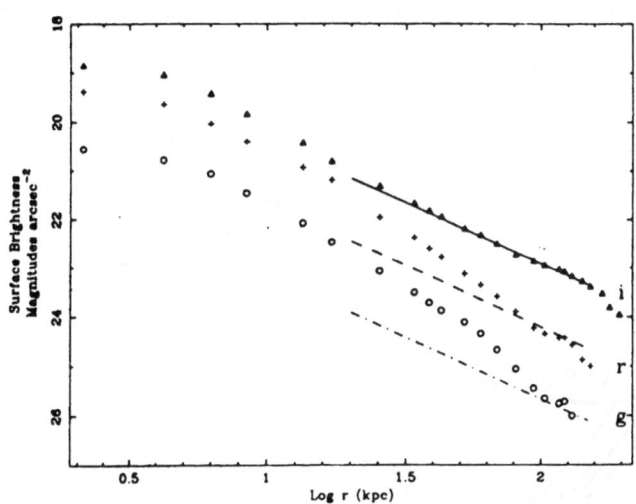

Figure 4. Surface brightness profiles of GREG in the Gunn g, r and i bands (Johnstone & Fabian 1989, with data from Maccagni *et al.* 1988). Note the large r^{-1} envelope in the i-band profile.

Its profile is consistent with mass deposition in a cooling flow, which must have been more massive in the past. No other similar objects have yet been reported, although it is possible that all cD envelopes are due to cooling flows.

The picture that is emerging so far is that most of the cooling flow deposits matter in some dark, 'isothermal' halo of low-mass stars (or even Jupiters) with $\rho_* \propto r^{-2}$. The innermost regions may produce a small fraction of higher-mass stars and to be consistent with the mass profiles of dark haloes there must be a core radius to the mass deposition distribution of $\sim 10 - 20\,\mathrm{kpc}$. The advection of angular momentum and the production of turbulence may be important in changing the behaviour within this core, as discussed in §6.2.

Limits on the mean mass of stars formed in the bulk of the cooling gas, $\langle M_* \rangle$, can be obtained from a comparison of the mass-to-light ratio for dwarf stars with the whole dark halo (Fig. 5). Assuming that the mass-to-light ratio in the B band exceeds 100 in a halo (it is not much larger in the whole cluster), then we see that the luminosity-weighted $\langle M_* \rangle$ cannot exceed $\sim 0.4\,\mathcal{M}_\odot$.

More detailed limits can be obtained by assuming that $\rho_* \propto r^{-2}$ so that the surface brightness $S(r) \propto r^{-1}$ and that the imf of the stars formed from the flow is a power-law of slope α in the mass range $M_l < M_* < M_u$. Johnstone & Fabian (in preparation) have fitted the maximum r^{-1} envelope to the observed surface brightness profiles of central cooling flow galaxies (*eg* from Schombert 1987) and obtain the constraints shown in Fig. 6. M_l is fixed at $0.08\,\mathcal{M}_\odot$ (the lower end of the main sequence). In most objects, M_u must be $\sim 1\,\mathcal{M}_\odot$ or less (unless the imf is steep) in order to match a total mass $\dot{M} \times 10^{10}$ yr. M_u must be less than $\sim 0.8\,\mathcal{M}_\odot$ to accommodate a $10^{13}\,\mathcal{M}_\odot$ dark halo in the case of NGC 6166 in A2199. A much stronger limit ($M_u < 0.2\,\mathcal{M}_\odot$) is obtained for the (very) dark halo of M87, which has little evidence for any r^{-1} component (Fig. 6). We conclude that, at their present mass deposition rates, cooling flows form stars of masses predominately below $1\,\mathcal{M}_\odot$. If they deposited all the inferred dark matter (if $\dot{M}$ were much greater in the past), then they must form very low-mass stars.

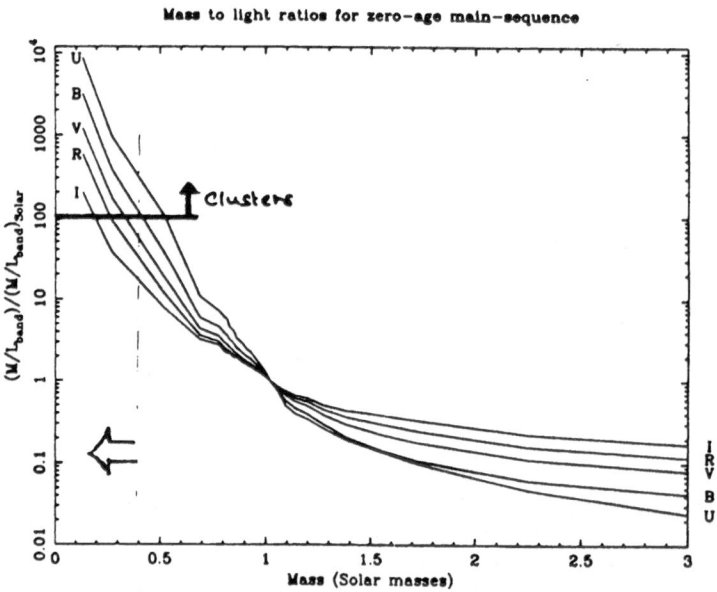

Figure 5. The mass-to-light ratio of main sequence stars.

5.4. DO STARS TRACE $\dot{M}(< r)$?

The above limits on the imf of stars formed in cooling flows assumes that the stars trace the mass deposition profile. This is reasonable if the $\dot{M}(< r) \propto r$ profile extends to very large radii without a distinct edge. White & Sarazin (1987) discuss the stellar profile produced if there is a distinct edge (say at r_{cool}). $S_*(r)$ can then be steeper than r^{-1} which would relax the constraints in the previous section (but not by very much).

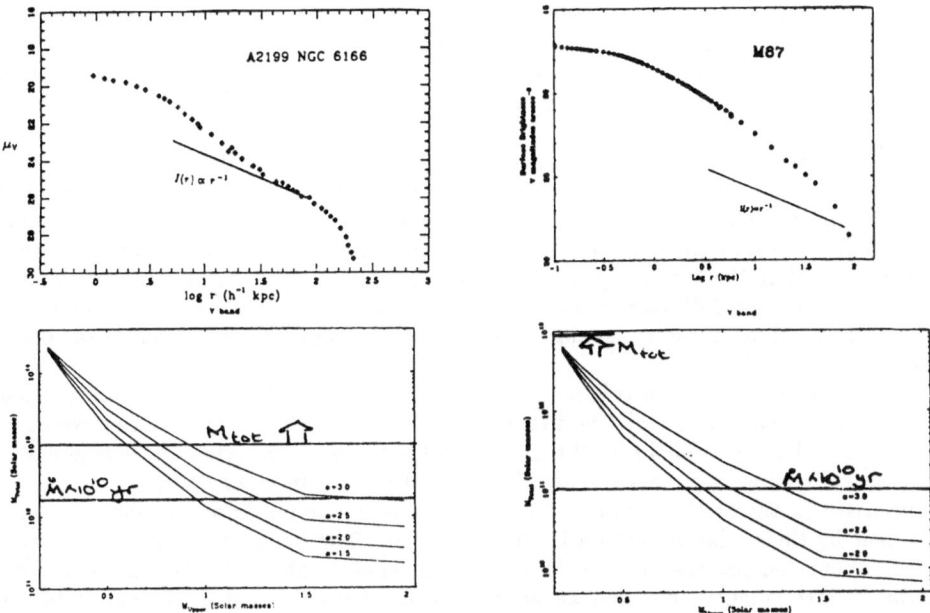

Figure 6. Upper limits to star formation in the envelopes of A2199 and M87.

The shape of the stellar profile depends upon the orbits of the stars formed in the flow. If the flow is very slow moving (an ideal flow) then the stars will be formed almost at rest and will then fall into the galaxy on radial orbits. This could produce a central cusp of light and mean that the best place to search for the stars is in the very centre. Soares & Sanders (1987) have suggested that this could produce a central cusp in M87 (see Fabian *et al.* 1985a for a discussion of the formation of a central star cluster). Another property of stars on radial orbits is that the situation is unstable to the formation of a bar. Bar-like structures are seen in some elliptical galaxies and an internal origin via a cooling flow is a possibility. Since no enormous bars are seen in most cooling flow galaxies, we can presumably expect that the orbits are not very radial. If the flow is turbulent (as discussed for at least the centre of many flows in §6.2), then the young low-mass stars may have substantial tangential velocities and the orbits be more nearly circular.

5.5. WHY *LOW-MASS* STAR FORMATION?

The conditions under which stars form at large radii in a cooling flow are quite different from those in our Galaxy. Perhaps we should not then be surprised that the imf is quite different. In the absence of any predictive theory of star formation, we can look for differences in the environments of cooling flows and regions of star formation in our own Galaxy. The most important difference is that there are *no molecular clouds* at large radii in cooling flows. In the Milky Way, massive stars ($M_\star > 1\,\mathcal{M}_\odot$) form only in large, warm molecular clouds. Low mass star formation is much more pervasive. The observations of 'naked' T Tauri stars shows that lower mass stars can form outside molecular clouds. It should also be noted that not all molecular clouds form massive stars. The nearest cloud to us, the ρ Oph cloud, has very few young stars above a few solar masses, although it has plenty of lower mass ones, ie its imf is deficient in intermediate-mass stars (Young, Lada & Wilking 1986).

There are several reasons for this; the most important of which is the inability of a large cloud to remain coherent at large radii. A second reason is that a lack of dust at large radii (whether or not there is any in the core[1], gas that has cooled from $T > 10^6$ K is dust-free; Draine & Salpeter 1979), presumably means that molecules have difficulty forming. Other reasons may be that the thermal pressure of the gas is 100 – 1000 times higher than in our interstellar medium (although similar to that in collapsing molecular clouds) and that the differential motions, masses, angular momentum and magnetic field strengths of clouds may be different.

A large cloud cannot remain coherent since it would fall relative to the hotter gas and break into pieces. Only if a cloud was large enough for it to be self-gravitating would it have sufficient coherence to hold itself together. (This could be the origin of some globular clusters around cooling flow galaxies; see Fabian, Nulsen & Canizares 1984).

As discussed in the next Section (6.1), denser blobs in cooling flows are shredded into smaller ones by their relative motion in the hotter gas. The limiting cloud size from this process is about ten parsecs ($T > 10^7$ K) when the mass is about $1\,\mathcal{M}_\odot$. More massive clouds break into smaller ones and an accumulation of clouds of mass around $1\,\mathcal{M}_\odot$ is expected. We must appeal to fragmentation in the cooled solar-mass clouds, or at least the formation of a core, in order to obtain very low mass stars since, at the pressure in a cooling flow, individual low-mass stars require that the gas cools below 3K to be above the Jeans limit. The pressure increases within a collapsing cloud due to the collapse and the Jeans limit drops so that low-mass stars can be formed. The clouds in cooling flows may then undergo opacity-limited fragmentation as outlined by Hoyle (1953) and shown by Low & Lynden-Bell (1976) and Rees (1976) to give stars of mean mass about 0.01 to $0.1\,\mathcal{M}_\odot$. If fragmentation does not occur due to the effects of angular momentum and magnetic fields (which slow the collapse process), then a core may form which gradually accretes more of the cloud. The final star then has a much larger mass then the core and perhaps much of the initial cloud mass. This situation is thought to occur in our Galaxy, where the cloud is supported against infall in the Galactic gravitational field by Galactic rotation. In the case of a cooling flow, however, the cloud is supported by thermal pressure. The dense core of the cloud cannot be supported in this way and will fall out of the cloud and accrete very little more gas. The final mass of the star will therefore be similar to the mass of the cloud core and much less than the mass of the initial cloud. Whether fragmentation occurs or not, the mean mass of a star formed at a large radius in a cooling flow is therefore much lower than the mean mass of stars formed in our Galaxy.

We return to the central regions of the flow in §6.2 and show that more massive stars can form there. Indeed, as discussed in §5.1, there do appear to be molecular clouds in the

---

[1] Hintzen & Romanishin (1988) find that the central parts of the cooling flow galaxy 3A0335+096 are red. Whether this is due to red stars or to dust is not clear. NGC 4696 in the Centaurus cluster, which has a cooling flow, also has a dust lane, so some dust is found in the middle of cooling flows. It may just be due to stellar mass loss from the galaxy itself.

inner 10 kpc of NGC1275 and some formation of stars above a solar mass. Most of the gas in a cooling flow, however, must form very low-mass stars.

6. Inhomogeneous Intracluster Gas

6.1. THE BEHAVIOUR OF GAS BLOBS

The distributed manner of the mass deposition shows that the cooling flow is inhomogeneous. This means that it contains blobs of gas that are denser than the surrounding gas. Density inhomogeneities will have been introduced early into the cluster gas by the production of metals, by the activity of any quasars in the cluster and from the stripping of gassy galaxies and winds. Later it will be mixed and inhomogenized by the infall of subclusters. A further spread of density will occur because of the motion of the central cluster galaxy and turbulence. Convection, with magnetic fields and viscosity, then creates a limited range of densities throughout the gas.

How the cooling gas blobs behave is ill-understood. Malagoli *et al.* (1987) and Balbus (1988) have shown that cooling flows are not expected to be thermally unstable and cannot generate sizable blobs from initially infinitesimal perturbations. A region that is slightly overdense with respect to its surroundings will fall ahead of the flow under gravity and join a region of similar properties to itself. Computations of the oscillations of overdense blobs are discussed by Loewenstein (1989); further computations have been made by Hattori *et al.* (1989) and by Brinkmann & Massaglia (1989).

In the limit of zero viscosity the gas in a cooling flow should be homogeneous and cool into a central singularity. Gravity causes any denser gas to fall ahead and join gas with similar properties. Only near the centre of the flow, where $v \sim v_{ff}$ the free-fall velocity, would the flow become inhomogeneous. However a real flow does have viscosity, is turbulent and contains magnetic fields. Consequently, the linear perturbation analyses are not particularly relevant to a real cooling flow. The non-linear behaviour of gas blobs in a flow has been explored by Nulsen (1986). A large gas blob of size r and overdensity $\delta\rho$ will try to move ahead of the mean flow and reach a terminal velocity

$$v_T \simeq v_{Kepler} \sqrt{\left(\frac{\delta\rho}{\rho_0}\frac{r}{R}\right)}. \tag{13}$$

The relative motion causes the blob to spread out and fragment (Nittman, Falle & Gaskell 1982). r/R is reduced as is the relative velocity of the overdense gas. Magnetic fields can help to pin the gas to the mean flow so that it comoves (see also Loewenstein 1989b). The net result is that large, slightly overdense blobs at large radii from the centre of the flow are turned into an emulsion of smaller and very overdense blobs at smaller radii. The densest gas will cool out of the flow (i.e. $T \to 0$ K) at intermediate radii. The typical mass of a small blob that can survive is about $1\,\mathcal{M}_\odot$, The density distribution of gas at a given radius evolves a 'cooling tail' (volume filling fraction $f \propto \rho^{-(4-\alpha)}$, where α is the exponent of the cooling function; Nulsen 1986; Thomas, Fabian & Nulsen 1987). This allows mass to be deposited in a distributed manner. If a spread of densities exists throughout the cluster, then gas may be deposited by cooling well beyond the radius where the mean cooling time is H_0^{-1}.

6.2. WHY DO SOME CLOUDS SURVIVE IN THE CENTRE?

As discussed in §5.1, warm clouds of optically-emitting gas are observed in the centres of many cooling flows. Their mere existence argues that thermal conductivity is suppressed by a large factor below the 'Spitzer value', κ_S. Tangled magnetic field lines are the obvious cause of this since the electrons that dominate the conductivity spiral around the field lines and can be trapped by magnetic mirrors[2]. The magnetic pressure in the intracluster gas is less than one per cent of the thermal energy so that the field is very tangled if the gas is turbulent. The cooled blobs may have their own separate magnetic field structure that is little related to their surroundings. A cloud would not have cooled out if it was thermally coupled to the hotter gas.

The motions of the clouds implied by the optical line-widths pose an enormous survival problem for the clouds if they are accelerated (by gravity) through a static hot medium since the clouds break-up into smaller pieces which prevents them attaining high velocities ($eg\ v_T > 100\,\mathrm{km\ s^{-1}}$, see equation 13). This problem is resolved by concluding that the velocity linewidths are the result of emission from small clouds formed in cluster gas with large-scale chaotic motions, ie the hotter gas is turbulent (Loewenstein & Fabian 1989). Provided that the motions exceed $\sim 100\,\mathrm{km\ s^{-1}}$, clouds formed by cooling from the hot gas will have the required velocity width.

The amount of turbulence throughout the cooling flow region is limited by the rate of energy dissipated by the smallest eddies, $\varepsilon_d \sim u^3/\ell$. If the turbulent velocity scale, u, and the length scale, l, are 100 km s^{-1} (the order of observed linewidths) and 10 kpc (the order of the observed maximum scale of magnetic field tangling from Faraday depolarization studies; Garrington et al. 1988, Dreher et al. 1987; Laing 1988), respectively, then ε_d exceeds the inferred cooling rate $\rho\Lambda$ at radii greater than 20 – 50 kpc. If such a strong constant heating source were present throughout the flow it would give a much flatter surface brightness profile than observed and reduce the strength of the FeXVII line.

This suggests that the turbulent velocity field within the flow is such that the velocity decreases outward. We envisage that turbulence is only an important physical process in the *central* regions (the inner few kpc) of cooling flows where the *observed* optical emission-line gas lies and where the X-ray emission is unresolved. The gravitational field in the dark halo of the central galaxy inhibits the propagation of turbulence outward. (Turbulence may, however, be important in the main body of the cluster on scales of $\sim 0.5\,\mathrm{Mpc}$, Fabian 1990)

Pringle (1989) has pointed out that the steep negative density gradient in a cooling flow can focus sound waves (noise) in the intracluster gas into the core of the cooling flow. This can help to heat the cooling gas there and may provide a source of energy for the warm clouds and could even 'shut-off' the flow if the noise energy can be dissipated in the hot gas at the necessary (large) rate. Balbus & Soker (1989) have considered the propagation of gravity waves in the core of a cluster. These can transport energy outward, meaning that the mass deposition rates can be higher than originally estimated.

---

[2] Tribble (1989) has produced an interesting model of thermal conductivity in a plasma with very tangled magnetic field. The electrons random walk rather than flow. The resulting conductivity coefficient involves the length scale of the temperature gradient, meaning that the conventional approach has totally broken down.

There are several sources of energy that could drive turbulent motions, and noise, in the inner 10-50 kpc in cooling flows. In addition to galaxy motions (including motion of the central galaxy), there is likely to be dissipation of magnetic energy from fields frozen in and advected inwards with the cooling flow as well as dissipation of rotational energy and angular momentum. As discussed by Nulsen, Stewart & Fabian (1984), turbulent viscosity is the most likely transport process of angular momentum in a cooling flow. Even small amounts of angular momentum carried in with the flow would become dynamically important at radii of a few kpc (Cowie, Fabian & Nulsen 1980). The tangential motions provoked by the angular momentum should pump turbulence in the hot gas.

The turbulent hot gas with entrained cool clouds in the core of a cooling flow can explain the observed velocity widths and the unaccounted for heating mechanism of these clouds if the bulk kinetic energy can be successfully converted into heat. This is thought to occur in our own ISM (see Cox 1979; Spitzer 1982). Heated clouds may also 'hang around' for longer, collide and coalesce into larger, molecular, clouds and lead to higher mass star formation (as for Galactic molecular clouds). In this picture, the optical emission-line filaments and the higher mass star formation are related and occur only in the centre of a turbulent region.

7. The Lifetime of Cooling Flows

The X-ray observations of cooling flows in most nearby clusters and the necessity for them in the MSS clusters (Pesce *et al.* 1989), which span a redshift range out to at least $z = 0.3$, means that cooling flows must last at least several billion years. There is also direct evidence for one in the cluster around 3C 295 (Henry & Henriksen 1986) at $z = 0.5$. Limits on the numbers of MSS clusters and on clusters in the Einstein Observatory Deep Survey show that cooling flows do not evolve faster than $(1 + z)^2$ (Pesce *et al.* 1989), unless they contain Active Galaxies which outshine them. (Then some of the identifications with active galaxies should be joint with an underlying cluster.)

It is likely that radio-loud quasars are surrounded by cooling flows since their host galaxies are ellipticals that are preferentially found in clusters. Yee & Green (1984; 1987) have directly shown that radio-loud quasars out to $z \sim 0.6$ are often in cluster central galaxies. The richness of the cluster appears to increase with redshift. We have used the ratio of extended emission lines around quasars to show that radio-loud ones are embedded in gas at high pressure (Fabian *et al.* 1987; Crawford *et al.* 1988; Crawford & Fabian 1989). Romanishin & Hintzen (1989) find that the colours of the underlying galaxies are blue in the same way that many central cluster galaxies are blue. They have also shown (1987) that 3C 275.1 at $z = 0.55$ is probably in a cooling flow.

We (Forbes *et al.* 1989) have recently found that the quasars 3C 254 and 309.1 are embedded in very massive cooling flows ('cooling floods') with mass deposition rates of the order of many hundreds to thousands of solar masses per year *within 30 kpc of the centre*. This means that the total mass deposition rate may be sufficient to make all the underlying galaxy *and all its dark halo* in a few billion years.

The central galaxies in many nearby flows contain active objects (usually active in the radio) such as Per A, Cyg A, Hydra A and Vir A. It is possible that the cooling matter helps to fuel the central engine. More distant and more powerful versions of this could easily be surrounded by more massive cooling flows. Distant radio-loud quasars and radio galaxies are probably embedded in cooling flows forming dark haloes (Fabian *et al.* 1985b; Fabian *et al.* 1986; Fabian 1988a). Some evidence for this view has been given just above and in the similarity of the emission-line spectra of nearby cooling flow filaments and the extended emission around distant radio galaxies (see *eg* Spinrad & Djorgovski 1984). The aligned, extended optical continuum around such radio galaxies can be due to electron scattering in the hot gas if it is part of a very massive cooling flow (Fabian 1989).

These ideas imply that cooling flows may have been still more common in the past and much more massive. They form most (all?) of the dark matter around the host galaxies, which must therefore be baryonic and formed into its present state relatively recently. If clusters form hierarchically from subclusters, and only one (or no) cooling flow survives a merger, then there could have been many more cooling flows in the past, many more than seems possible from the present number of central cluster galaxies.

8. How Much Baryonic Dark Matter Can Form In Cooling Flows?

Since I have made the case (or have attempted to do so) that much of the dark matter around central cluster galaxies is baryonic, and is low mass stars formed in cooling flows, then it seems sensible to ask just how much dark matter can be made in this way.

As already mentioned, there could have been many more clusters in the past, so we can envisage forming more than just the dark matter in the central couple of hundred kpc of a present rich cluster. Previous discussions of the formation of dark matter from cooling flows are Fabian, Arnaud & Thomas (1985); Fabian *et al.* (1986); Ashman & Carr (1988); Thomas (1988) and Thomas & Fabian (in preparation; much of the following discussion is from that work).

The role of cooling in galaxy formation was first discussed by Rees & Ostriker (1977) and by Silk (1977) (see also Hoyle & Gold 1959). They identify the protogalactic gas clouds for which $t_{cool} = t_{grav}$ as the boundary between those that form stars immediately and those that do not. The precise mechanisms by which a cloud falling outside this boundary does not form a galaxy, or forms a different kind of galaxy, are left unclear. Peter Thomas and I note that this boundary is really where the cooling flows begin and suggest that it marks a change in the M/L of the stars formed from the gas. The boundary $t_{cool} = t_{grav}$ corresponds to masses $\sim 10^{12} - 10^{13} \mathcal{M}_{\odot}$, where the exact mass depends on metallicity ($\sim 10^{12} \mathcal{M}_{\odot}$ for no metals, $\sim 10^{13} \mathcal{M}_{\odot}$ for solar metallicity). Low mass clouds have $t_{cool} < t_{grav}$ if heated to the virial temperature (indeed it is difficult to see how they ever heat up by gravitational collapse under this condition). More massive protogalaxies form cooling flows.

We postulate that lower-mass protogalaxies rapidly collapse into large pieces, without the formation of any substantial hotter medium, and form massive stars which produce a visible galaxy. On the other hand, most of a higher mass protogalaxy is heated to its virial temperature and cools from $T > 10^6$ K. In this case, density inhomogeneities are restricted to small sizes by relative motion in the hotter gas so that there only are low-mass blobs and consequently only low-mass stars (dark matter) formed. Of course, the core of a large cloud behaves somewhat like a small cloud so a visible galaxy is still produced in the centre, but is surrounded by a massive dark halo from the remaining cooled gas.

In order to explain the common observation of dark matter we must assume that the Universe is build up of subunits, each of which is massive enough to have formed a cooling flow and so $> 10^{12} - 10^{13} \mathcal{M}_\odot$. Dark matter in the local group would have to be due to a past massive cooling flow in the whole group (there could still be a substantial intragroup medium). In this picture, dark matter in a very isolated, low-mass galaxy (total mass of galaxy and any associated group $< 10^{12} \mathcal{M}_\odot$) would be very difficult to explain. We are mainly concerned, however, with the formation of dark matter around massive elliptical galaxies (see Fabian *et al.* 1986b for a discussion of the mass of early-type galaxies). Mergers of subunits lead to the mixing and spreading of the low mass star component throughout the core of a cluster and so can provide all the dark matter in a cluster. We just require that there was an epoch of massive cooling flows in subclusters.

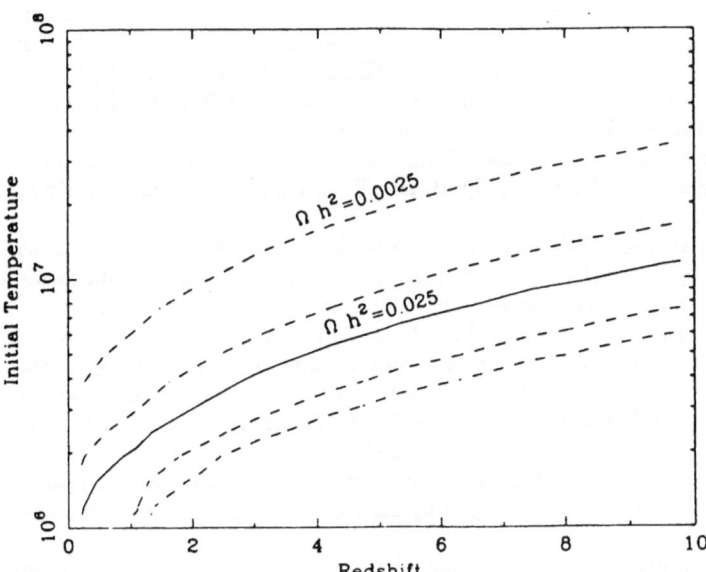

Figure 7. Upper limit to fraction of closure density that can have cooled from temperature T at redshift z.

Since the subclusters had virial temperatures of 10^7 K and below, the main constraint on how much cooling could have taken place is obtained from the soft X-ray background below 1 keV. A strong constraint on the extragalactic component to the soft X-ray background is found from the work of the Wisconsin group (McCammon *et al.* 1976). They searched for dips in the soft background flux due to a distant component being absorbed by known hydrogen clouds in our Galaxy. A limit is obtained because no such dips are seen (most of the Soft X-ray background is Galactic). We use the limit (for the value and more discussion see Fabian, Canizares & Barcons 1989) to constrain the total fraction of the closure density that can turn into dark matter by cooling from temperature $T > 10^6$ K at redshift z (Fig. 7). All the dark matter in clusters and groups of galaxies can be made from cooling flows provided that $T < 10^7$ K and $z > 5$ or so. If we need it to be formed more recently ($z < 2$, say) then $T < 3.10^6$ K.

9. Summary

The X-ray data show that cooling flows are common, continuous and long-lived. Some type of condensed dark matter is being formed around many central cluster galaxies at the present epoch. The widespread evidence for dark matter in clusters shows that there is no problem caused by the continuing accumulation of fresh dark matter from cooling flows.

Obtaining more direct evidence that it is baryonic is difficult, however. Deep studies of the light, particularly the infra-red light, of the outer envelopes of central cluster galaxies should be important here. Future X-ray missions such as ROSAT will find many more flows and map out their evolution. ASTRO-D, SPECTRUM-X and AXAF, with better X-ray spectroscopy, will define the mass cooling rates and the distribution of the mass deposition more precisely. ISO working in the infra-red should help to reveal the liberation of binding energy in the low-mass stars formed from the flows. The discovery of more Giant Red Envelope Galaxies would be very helpful, as will further study (particularly spectroscopy) of the one already known.

A scenario for star formation in cooling flows has been presented here, in which most of the flow (probably > 95%) forms low mass stars whilst a small fraction forms stars up to a few solar masses in the central few kpc. The main difference between star formation in our Galaxy and in a cooling flow is that clouds in our Galaxy are supported against infall by Galactic rotation, whereas clouds in a cooling flow are supported by thermal pressure. Only small clouds, of about a solar mass or less, can be supported in a cooling flow; larger clouds are broken into smaller ones by their motion in the hotter surrounding medium. Consequently, no molecular clouds are formed, and no massive stars. In the centre of a flow, angular momentum and cloud collisions are probably important so that larger clouds, and sometimes molecular clouds (helped by dust from the giant stars in the central galaxy), form.

Finally, this picture has been extended to include galaxy formation. It is proposed that the ratio of t_{cool}/t_{grav} determines the M/L of most of the condensed matter formed by a protogalactic cloud. Massive galaxies thereby have massive, dark baryonic haloes formed by an early cooling flow. The merger of these galaxies and their associated subgroups formed the present day clusters, complete with their baryonic dark matter.

I thank Carolin Crawford, Roderick Johnstone, Mike Loewenstein, Paul Nulsen and Peter Thomas for help and discussions and the Royal Society for supporting my work.

References

Arnaud, K.A., 1988. In *Cooling Flows in Clusters and Galaxies*, ed. A.C.Fabian, Reidel, 31.

Arnaud, K.A. & Gilmore, G., 1987. In *Structure and Dynamics of Elliptical Galaxies*, ed T. De Zeeuw, Reidel, p 445.

Ashman, K.M. & Carr, B.J, 1988. *Mon. Not. Roy. Astron. Soc.* **234**, 219.

Balbus, S., 1988. *Astrophys. J.* **328**, 395.

Balbus, S. & Soker, N., 1989. Preprint

Bertschinger, E. & Meiksin, A., 1986. *Astrophys. J. Lett.* **306**, L1.

Bohringer, H. & Fabian, A.C., 1989. *Mon. Not. Roy. Astron. Soc.* **247**, 1147.

Bohringer, H. & Morfill, G., 1989. *Mon. Not. Roy. Astron. Soc.* **330**, 609.

Burns, J. O., White R. A. & Haynes, M. P., 1981. *Astron. J.* **86**, 1.

Bregman, J.D. & David L.P., 1988. *Astrophys. J.* **326**, 639.

Bregman, J.N., Roberts, M.S. & Giovanelli, R., 1988. *Astrophys. J. Lett.* **330**, L93.

Brinkmann, W.P. & Massaglia, 1989. Preprint

Caldwell, C.N. & Oemler, A., 1981. *Astron. J.* **86**, 1424.

Canizares, C.R., Clark, G.W., Markert, T.H., Berg, C., Smedira, M., Bardas, D., Schnopper, H. & Kalata, K., 1979. *Astrophys. J. Lett.* **234**, L33.

Canizares, C.R., 1981. In *X-ray Astronomy with the Einstein Satellite* ed. R. Giacconi, Reidel, 215.

Canizares, C.R., Clark, G.W., Jernigan, J,G. & Markert, T.H., 1982. *Astrophys. J. Lett.* **262**, L33.

Canizares, C.R., Stewart, G.C. & Fabian A.C., 1983. *Astrophys. J.* **272**, 449.

Canizares, C.R., Markert, T.H. & Donahue, M.E., 1988. In *Cooling Flows in Clusters and Galaxies*, ed. A.C.Fabian, Reidel, 63.

Cowie, L.L. & Binney, J., 1977. *Astrophys. J.* **215**, 723.

Cowie, L.L., Fabian, A.C. & Nulsen, P.E.J., 1980. *Mon. Not. Roy. Astron. Soc.* **191**, 399.

Cox, D.P., 1979. *Astrophys. J.* **234**, 863.

Crane, P., van der Hulst, J. & Haschick, A., 1982. In: *Proc. IAU Symp. No. 97, Extragalactic Radio Sources*, eds. Heeschen, D.S. & Wade, C.M., Reidel, Dordrecht, Holland, p307.

Crawford, C.S., Fabian, A.C. & Johnstone, R.M., 1988. *Mon. Not. Roy. Astron. Soc.* **235**, 183.

Crawford, C.S., 1988. PhD Thesis, University of Cambridge.

Crawford, C.S., Arnaud, K.A., Fabian, A.C. & Johnstone, R.M., 1989. *Mon. Not. Roy. Astron. Soc.* **236**, 277.

Crawford, C.S. & Fabian, A.C., 1989. *Mon. Not. Roy. Astron. Soc.* **239**, 219.

Draine, B.T. & Salpeter, E.E., 1979. *Astrophys. J.* **231**, 77.

Dreher, J.W., Carilli, C.L. & Perley, R.A., 1987. *Astrophys. J.* **316**, 611.

Edge. A.C., 1989. Ph.D. Dissertation, Univ. of Leicester.

Edge, A.C., Stewart, G.C., Fabian, A.C. & Arnaud, K.A., 1989. Preprint

Fabian, A.C. *et al.* , 1974. *Astrophys. J. Lett.* **189**, L59.

Fabian, A.C. & Nulsen, P.E.J., 1977. *Mon. Not. Roy. Astron. Soc.* **180**, 479.

Fabian, A.C., Nulsen, P.E.J. & Canizares, C.R., 1982. *Mon. Not. Roy. Astron. Soc.* **201**, 933.

Fabian, A.C., Nulsen, P.E.J. & Canizares, C.R., 1984. *Nature* **311**, 733.

Fabian, A.C., 1985a. *Nature* **307**, 343.

Fabian. A.C. *et al.* 1985b. *Mon. Not. Roy. Astron. Soc.* **216**, 923.

Fabian, A.C., Arnaud, K.A. & Thomas, P.A., 1986. In *Dark Matter in the Universe*, eds. J. Kormendy & G.R. Knapp, Reidel, 201.

Fabian, A.C., Arnaud, K.A., Nulsen, P.E.J. & Mushotzky, R.F., 1986a. *Astrophys. J.* **305**, 9.

Fabian, A.C., Thomas, P.A., Fall, S.M. & White, R.A., 1986b. *Mon. Not. Roy. Astron. Soc.* **221**, 1049.

Fabian, A.C., Crawford, C.S., Johnstone, R.M. & Thomas, P.A., 1987. *Mon. Not. Roy. Astron. Soc.* **228**, 963.

Fabian, A.C., 1988a. In *Cooling Flows in Clusters and Galaxies*, ed. A.C.Fabian, Reidel, 315.

Fabian, A.C., 1988b. In *Hot Thin Plasmas in Astrophysics*, ed. R. Pallavicini, Reidel, 293.

Fabian, A.C., 1989 *Mon. Not. Roy. Astron. Soc.* **238**, 41P.

Fabian, A.C., 1990. In *Ultrahot Plasmas in Astrophysics*, ed Brinkmann, W.P., Kluwer in press.

Fabian, A.C., Canizares, C.R. & Barcons, X., 1989. *Mon. Not. Roy. Astron. Soc.* **239**, 15P.

Forbes, D.A., Fabian, A.C., Crawford, C.S. & Johnstone, R.M., 1989. *Mon. Not. Roy. Astr. Soc.* submitted.

Forman, W., Schwarz, J., Jones, C., Liller, W. & Fabian, A.C., 1979. *Astrophys. J. Lett.* **234**, L27.

Garrington, S. T., Leahy, J. P., Conway, R. G., Laing, R. A., 1988. *Nature* **331**, 147.

Gold, T. & Hoyle, F., 1958. In *Paris Symposium on Radio Astronomy*, ed. RN Bracewell, Stanford Univ. Press, 574.

Gorenstein, P., Fabricant, D., Topka, K., Tucker, W. & Harnden, F.R., 1977. *Astrophys. J. Lett.* **216**, L95.

Grabelsky, D. & Ulmer, M. 1989. Preprint

Hattori, M., & Habe, A., 1989. *Mon. Not. Roy. Astr. Soc.* in press.

Haynes, M. P., Brown, R. L. & Roberts, M. S., 1978. *Astrophys. J.* **221**, 414.

Heckman, T.M., Baum, S.A., van Breugel, W.J.M. & McCarthy, P.,1989. *Astrophys. J.* **338**, 48.

Helmken, H., Delvaille, J.P., Epstein, A., Geller, M.J., Schnopper, H.W. & Jernigan, J.G., 1978. *Astrophys. J. Lett.* **221**, L43.

Hintzen, P. & Romanishin, W., 1986. *Astrophys. J. Lett.* **311**, L1.

Hintzen, P. & Romanishin, W., 1988. *Astrophys. J. Lett.* **327**, L17.

Henry, J.P & Henriksen, M.J., 1986. *Astrophys. J.* **301**, 689.

Hoyle, F., 1953. *Astrophys. J.* **118**, 513.

Hu, E.M., 1988. In *Cooling Flows in Clusters and Galaxies*, ed. A.C.Fabian, Reidel, 73.

Hu, E.M., Cowie, L.L. & Wang, 1985. *Astrophys. J. Suppl.* **59**, 447.

Jaffe, W., de Bruyn, A.G. & Sijbreng, D., 1988. In: *Cooling Flows in Clusters and Galaxies*, ed. Fabian, A. C., Reidel, Dordrecht, Holland, p145.

Johnstone, R.M., Fabian, A.C. & Nulsen, P.E.J., 1987. *Mon. Not. Roy. Astron. Soc.* **224**, 75.

Johnstone, R.M. & Fabian, A.C., 1988. *Mon. Not. Roy. Astron. Soc.* **233**, 581.

Johnstone, R.M. & Fabian, A.C., 1989. *Mon. Not. Roy. Astron. Soc.* **237**, 27P.

Jones, C. & Forman, W., 1984. *Astrophys. J.* **276**, 38.

Lahav, O., Edge, A.C., Fabian, A.C. & Putney, A., 1989. *Mon. Not. Roy. Astron. Soc.* **238**, 881.

Laing, R.A., 1988. *Mon. Not. Roy. Astron. Soc.* **331**, 149.

Lazareff, B., Castets, A., Kim, D-W. & Jura, M., 1989. *Astrophys. J. Lett.* **336**, L13.

Lea, S.M., Silk, J., Kellogg, E. & Murray, S., 1973. *Astrophys. J. Lett.* **184**, L105.

Lea, S.M., Mushotzky, R.F. & Holt, S.S., 1982. *Astrophys. J.* **262**, 24.

Loewenstein, M., 1989a. *Mon. Not. Roy. Astron. Soc.* **238**, 15.

Loewenstein, M., 1989b. Preprint

Loewenstein, M. & Fabian, A.C., 1988. *Mon. Not. Roy. Astr. Soc.* in press.

Low, C. & Lynden-Bell, D., 1975. *Mon. Not. Roy. Astron. Soc.* **176**, 367.

Maccagni, D., Garilli, B., Gioia, I.M., Maccacaro, T., Vettolani, G. & Wolter, A., 1988. *Astrophys. J. Lett.* **334**, L1.

Malagoli, A., Rosner, R. & Bodo, G., 1987. *Astrophys. J.* **319**, 632.

Mathews, W., G. & Bregman, J.N., 1978. *Astrophys. J.* **244**, 308.

McCammon, D., Meyer, P., Sanders, W. & Williamson, 1976. *Astrophys. J.* **209**, 46.

Meiksin, A., 1989. *Astrophys. J.* in press.

Miller, L., 1986. *Mon. Not. Roy. Astron. Soc.* **220**, 713.

Mirabel, F. *et al.* 1989. Preprint

Mitchell, R.J., Charles, P.A., Culhane, J.L., Davison, P.J.N. & Fabian, A.C., 1975. *Astrophys. J. Lett.* **200**, L5.

Mould, J.R., Oke, J.B. & Nemec, J.M., 1987. *Astron. J.* **92**, 53.

Mushotzky, R.F., Holt, S.S, Smith, B.W., Boldt, E.A. & Serlemitsos, P.J., 1981. *Astrophys. J. Lett.* **244**, L47.

Mushotzky, R.F. & Szymkowiak, A.E. , 1987b. In *Cooling Flows in Clusters and Galaxies*, ed. A.C.Fabian, Reidel, 47.

Nittmann, J., Falle, S.A.E.G. & Gaskell, P.H., 1982. *Mon. Not. Roy. Astron. Soc.* **201**, 833.

Nulsen, P.E.J., Stewart, G.C., Fabian, A.C., Mushotzky, R.F., Holt, S.S., Ku, W.H.M. & Malin, D.F., 1982. *Mon. Not. Roy. Astron. Soc.* **199**, 1089.

Nulsen, P.E.J., Stewart, G.C. & Fabian, A.C., 1984. *Mon. Not. Roy. Astron. Soc.* **208**, 185.

Nulsen, P.E.J., 1986. *Mon. Not. Roy. Astron. Soc.* **221**, 377.

Nulsen, P.E.J. & Carter, D., 1987. *Mon. Not. Roy. Astron. Soc.* **225**, 935.

Nulsen, P.E.J., 1988. In *Cooling Flows in Clusters and Galaxies*, ed. A.C.Fabian, Reidel, 378.

O'Connell, R. & McNamara, B., 1989. Preprint

Pedlar, A. *et al.* , 1988. In *Cooling Flows in Clusters and Galaxies*, ed. A.C.Fabian, Reidel, 149.

Pesce, J.E., Edge, A.C., Fabian, A.C. & Johnstone, R.M., 1989. *Mon. Not. Roy. Astr. Soc.* in press.

Pringle, J.E., 1989. *Mon. Not. Roy. Astron. Soc.* **239**, 479.

Rees, M.J., 1976. *Mon. Not. Roy. Astron. Soc.* **176**, 483.

Rees, M.J. & Ostriker, J.P., 1977. *Mon. Not. Roy. Astron. Soc.* **179**, 541.

Romanishin, W. & Hintzen, P., 1989. *Astrophys. J.* **341**, 41.

Rothenflug, R. & Arnaud, M., 1986. *Astron. Astrophys.* **147**, 337.

Sarazin, C.L., 1986. *Reviews of Modern Physics* **58**, 1.

Sarazin, C.L., 1988. *X-ray Emission from Clusters of Galaxies*, C.U.P.

Sarazin, C.L. & O'Connell, R.W., 1983. *Astrophys. J.* **258**, 552.

Schombert, J.M., 1986. *Astrophys. J. Suppl.* **60**, 603.

Schwartz, D.A., Schwarz, J. & Tucker, W.H., 1980. *Astrophys. J. Lett.* **238**, L59.

Silk, J., 1976. *Astrophys. J.* **208**, 646.

Silk, J., 1977. *Astrophys. J.* **211**, 638.

Silk, J., Djorgovski, G., Wyse, R.F.G. & Bruzual, G.A., 1986. *Astrophys. J.* **307**, 415.

Singh, K.P., Westergaard, N.J. & Schnopper, H.W., 1986. *Astrophys. J. Lett.* **308**, L51.

Soares , D.S.L. & Sanders, R.H., 1987. *Mon. Not. Roy. Astron. Soc.* **229**, 119.

Spinrad, H. & Djorgovski, G., 1984. *Astrophys. J. Lett.* **280**, L9.

Spitzer, L. 1982. *Astrophys. J.* **262**, 315.

Stewart, G.C., Canizares, C.R., Fabian, A.C. & Nulsen, P.E.J., 1984a. *Astrophys. J.* **278**, 536.

Stewart, G.C., Fabian, A.C., Jones, C. & Forman, W., 1984b. *Astrophys. J.* **285**, 1.

Thomas, P.A., 1988. *Mon. Not. Roy. Astron. Soc.* **235**, 315.

Thomas, P.A., Fabian, A.C. & Nulsen, P.E.J., 1987. *Mon. Not. Roy. Astron. Soc.* **228**, 973.

Tribble, P. 1989. *Mon. Not. Roy. Astron. Soc.* **238**, 1247.

Tucker, W.H. & Rosner, R., 1982. *Astrophys. J.* **267**, 547.

Valentijn, E. A. & Giovanelli, R., 1982. *Astron. Astrophys.* **114**, 208.

White, R.E. & Sarazin, C.L., 1987. *Astrophys. J.* **318**, 612.

White, R.E. & Sarazin, C.L., 1987. *Astrophys. J.* **318**, 621.

White, R.E. & Sarazin, C.L., 1987. *Astrophys. J.* **318**, 629.

Yee, H.K.C. & Green, R.F., 1984. *Astrophys. J.* **280**, 79.

Yee, H.K.C. & Green, R.F., 1987. *Astrophys. J.* **319**, 28.

Young, E.T., Lada, C.J. & Wilking, B.A., 1986. *Astrophys. J. Lett.* **304**, L45.

QSO ABSORPTION LINES, EARLY EVOLUTION OF GALACTIC HALOS AND THE METAGALACTIC UV FLUX

Wallace L. W. Sargent and Charles C. Steidel[1]
Palomar Observatory
California Institute of Technology
Pasadena, CA 91125
USA

ABSTRACT. Recent observations of QSO absorption lines are described which indicate that a large fraction of the baryonic content of the Universe seen at redshifts $z \sim 3$ was in the form of tenuous highly ionized clouds in the outer halos of galaxies. The heavy element composition of the clouds covered the range now exhibited by the globular clusters in the halo of our own Galaxy. The observations also indicate a secular increase in the average heavy element content by a factor of 3 in the first 3 billion years. The QSO absorption lines indicate that the metagalactic flux of ionizing radiation has a hard spectrum over the whole redshift range $1.3 \leq z \leq 3.6$. The observed luminous QSOs at high redshifts cannot produce the inferred ionizing flux and the hot stars expected to be present in newly forming galaxies cannot account for the high ionization levels observed. It is therefore conjectured that the early Universe contained a large population of low–luminosity active galactic nuclei. These observations provide the first direct information about the evolution of the gaseous halos of galaxies at early times.

1. Outline

The problem of the formation and early evolution of galaxies is now one of the central issues of observational cosmology. The most direct method of attack is provided by observations of the colors and luminosities of the highest redshift galaxies; however, such observations are necessarily limited to the most luminous objects, which may be untypical in their behavior. It is expected that as we go back in time an increasing proportion of the mass of a galaxy is in the form of interstellar gas which can be studied via emission lines if excited by hot stars. Quasar absorption lines have the advantage that their sensitivity as probes of the interstellar gas in distant galaxies depends only on the brightness of the background QSO. Thus, in principle, it is possible to use the absorption lines to study the evolution of typical galaxies, rather than the brightest objects. Until recently it had not

---

[1] Now at the Department of Astronomy, University of California, Berkeley, CA. 94720.
The talk was delivered by the first author.

D. Lynden-Bell and G. Gilmore (eds.), Baryonic Dark Matter, 223–235.
© 1990 *Kluwer Academic Publishers.*

been possible to obtain the large body of statistical information on the systematic behavior of QSO absorption lines with increasing redshift which is necessary in order to obtain even the first crude insights into galaxy evolution by this method. However, recent large surveys conducted by Boksenberg and the present authors has at last led to significant insights into the composition and physical state of the gas in galactic halos at large redshifts. It turns out that such gas makes a significant contribution to the the cosmological density in baryons, Ω_B, at redshifts $z \sim 3$. This talk is devoted to a summary of these new developments. It turns out that a major ingredient in the discussion is the question of the nature of the metagalactic ionizing flux. Accordingly, it is necessary to go into the method of estimating the UV flux before embarking on an account of the behavior of the absorption lines due to galaxies. For later comparisons with the contributions due to gas and stars in galaxies to Ω_B we begin in the following Section by discussing the the contribution of generally distributed intergalactic gas to the cosmological density parameter.

It is a mixed compliment to be the last speaker at this Workshop. We should perhaps remind you that the "clean-up hitter" in baseball is expected to hit a home run – however, he also strikes out frequently.

2. The Gunn–Peterson Limit

Using spectra of the first QSO known with a redshift z larger than 2, (Schmidt 1965), Gunn and Peterson (1965) argued that limits on the density of neutral hydrogen in the intergalactic medium could be obtained from a measurement of the apparent decrement of the QSO continuum shortward of the Lyman α emission line. This decrement would be caused by essentially continuous, overlapping Lyman α absorption if the medium contained smoothly distributed HI gas. The effective optical depth of any such medium at redshift z is given approximately by

$$\tau_{GP} \approx \frac{4.14 \times 10^{10} h^{-1} n_{HI}(z)}{(1+z)(1+2q_0 z)^{1/2}}, \tag{1}$$

where h is the Hubble constant in units of 100 km s^{-1} Mpc^{-1} and n_{HI} is the number density of neutral H. The early observations allowed the limit $\tau_{GP} < 0.5$ at $z \approx 2$, or $n_{HI} \sim 3 \times 10^{-11}$ for $q_0 = 1/2$. Of course, it is now well-known that a considerable amount of flux is removed shortward of Lyman α emission by the "Lyman α forest" of discrete absorption lines, and it is necessary to account for the discrete absorption in order to place sensitive limits on the quantity of generally-distributed gas at high redshift. Steidel and Sargent (1987) used spectrophotometric observations of 8 QSOs with $z_{em} \approx 3$ in combination with statistics of the incidence of Lyman α absorption lines obtained from high dispersion observations to conclude that $\tau_{GP} \leq 0.05$ (1σ upper limit) at $z \sim 2.7$, or

$$n_{HI} \leq 4.4 h \times 10^{-12} \quad \text{cm}^{-3} \tag{2}$$

for $q_0 = 0$ (the limit is approximately twice this value for $q_0 = 1/2$). By assuming that the discrete Lyman α clouds are intergalactic, a value of the maximum pressure of the IGM can be obtained using inferred properties of the Lyman α clouds. Further, the assumptions that the IGM is in photoionization equilibrium in the UV radiation field from QSOs at $z \sim 3$, and that the IGM expands adiabatically, allow one to infer limits on the total density of hydrogen in the IGM at the present epoch because once the medium is ionized, it will remain highly ionized even if the radiation field diminishes significantly over time. The result is that, at $z = 0$,

$$n_H < 6.0h \times 10^{-7} \quad cm^{-3} \tag{3}$$

for $q_0 = 1/2$, or

$$\Omega_{IGM} < 0.05h^{-1}. \tag{4}$$

This upper limit is consistent with the range of Ω_B allowed by the standard model of Big Bang Nucleosynthesis (see Section 8).

3. The Proximity Effect

As the statistical information on the Lyman α forest increased, several workers noted that, while the number of Lyman α lines per unit redshift increases substantially with redshift, in the spectrum of a given QSO there is a countervailing trend in the vicinity of the QSO emission redshift; that is, the number density of Lyman α lines *decreases* for $z_{abs} \approx z_{em}$ (Murdoch *et al.* 1986; Tytler 1987; Hunstead 1988). One explanation of this trend is that the radiation field produced by the QSO itself has altered the ionization state of clouds in the QSOs immediate vicinity–thus, the "proximity effect". Under this hypothesis, it is possible in principle to gauge the relative strengths of the metagalactic and QSO ionizing radiation fields by measuring the distance to which a significant decrement in the number of lines is actually observed in a given QSO spectrum. Bajtlik, Duncan, and Ostriker (1988) have applied this method to a sample of 18 QSOs for which Lyman α forest statistics exist; their conclusions are that the data are consistent with the excess radiation field hypothesis for the origin of the proximity effect, and that their model is best–fit by a metagalactic UV flux $\log J_{\nu_0} = -21.0 \pm 0.5$ ergs s^{-1} cm^{-2} Hz^{-1} ster^{-1}. This value does not appear to change significantly over the redshift range $1.8 \leq z \leq 3.8$. It is interesting to note that the inferred value of J_{ν_0} is significantly larger than the integrated UV flux from observed luminous QSOs (Bechtold *et al.* 1987; Lin and Phinney 1989), and is comparable to the value required to explain the absence of a Gunn–Peterson discontinuity at $z \sim 3$.

4. Evolution in Redshift

Since the universe is expanding, as we look out towards higher redshifts galaxies or intergalactic clouds should appear closer together. Correspondingly, the beam of light from a distant QSO should intercept more absorbing objects at large redshifts than nearby provided the product of the their co–moving space density and their mean cross–section remains constant. Quantitatively, the number of interceptions per unit redshift range, $\Delta z = 1$, should vary with redshift according to the expression

$$dN/dz \equiv N(z) = N_0(1 + z)[1 + 2q_0 z]^{-1/2} \tag{5}$$

for $\Lambda = 0$. Here $N_0 = \pi r_0^2 \phi_0 c / H_0$ is the local value of N(z). If, as seems in general likely, the absorbing objects evolve in cosmic time, $N_0 = N_0(z)$. A large amount of effort has been devoted over the last few years to determining $N(z)$ for particular classes of QSO absorption systems. It has become customary to express $N(z)$ in terms of a power law form, $N(z) \times (1+z)^\gamma$, where γ is the power to be determined from the empirical results. From above, for non–evolving absorbers, one expects $\gamma = 0.5$ if $q_0 = 1/2$ and $\gamma = 1.0$ if $q_0 = 0$.

The "Lyman α forest" systems, which are by far the most abundant class of absorption system, and which are believed to be associated with intergalactic clouds (e.g., Sargent *et al.*), exhibit strong evolution with redshift. The most recent results show that $\gamma = 2.30\pm0.40$ over the redshift range $1.6 \leq z \leq 3.8$ (Murdoch *et al.* (1986); Hunstead *et al.* 1987). In contrast, the Lyman limit systems, which are those systems with HI column densities exceeding $\sim 10^{17}$ cm^{-2} and can be discovered over a very large redshift range when IUE and ground–based observations are combined ($0.2 \leq z \leq 4.7$), are consistent with no evolution ($\gamma = 0.68\pm0.54$; Sargent, Steidel, and Boksenberg 1989). The absorption systems discovered on the basis of the Mg II $\lambda\lambda2796$, 2803 doublet can be observed over the redshift range $0.3 \leq z \leq 2.2$; the currently available data give $\gamma = 1.55 \pm 0.52$ (Sargent, Steidel, and Boksenberg 1989), thus they are consistent with no evolution of the absorbers, although a flattening of the $N(z)$ curve at $z \sim 1$ is suggested. More data are needed to confirm this effect. The absorption systems discovered on the basis of the C IV $\lambda\lambda1548$, 1550 doublet, which can be observed over the redshift range $1.3 \leq z \leq 4$, are found to evolve significantly with redshift in the opposite sense to the Lyman α forest absorbers: $\gamma = -1.3 \pm 0.5$ over the redshift range $1.3 \leq z \leq 3.7$ (Sargent, Boksenberg, and Steidel 1988; Steidel 1989). This discovery is extremely surprising and our interpretation of this effect will be discussed in the following Section.

5. Explanation of the behavior of C IV

Taken alone, the power γ obtained for the C IV systems can be somewhat misleading. The value of γ turns out to be a rather strong function of the minimum equivalent width threshold for the sample, in the sense that raising the threshold (i.e., including only strong C IV doublets) sharply steepens the decline in $N(z)$ with z (Steidel 1989). There is some indication that a sample which is sensitive to doublets two or three times weaker than the bulk of the current samples would yield a value of γ which would actually be consistent with no evolution whatsoever. The explanation of these trends is, therefore, that a changing *distribution* of C IV equivalent widths is the most significant factor in the observed evolution, and probably *not* the actual number of absorbers. This explanation is strengthened by the observed $N(z)$ relation for the Lyman limit systems (a subset of the heavy element absorbers), which we have seen above to be consistent with no evolution.

Thus, the observed evolution of the C IV absorption is interpreted as a curve of growth effect, whereby the mean C IV column density is a decreasing function of z. This interpretation is also supported by the observed increase of the mean C IV doublet ratio with z, indicating evolution toward the optically thin limit (Sargent, Boksenberg and Steidel

1988; Steidel 1989). Systematically changing C IV column densities could be naturally explained by some combination of changing ionization state in the clouds or by a changing chemical abundance of C; however, as yet there is no evidence for a substantial change in the metagalactic flux over the relevant redshift range ($1.3 \leq z \leq 3.7$), on the basis of the proximity effect discussed in Section **3** and from the absence of a significant change in, e.g., the ratio of Si IV to C IV equivalent widths with redshift (Sargent, Boksenberg and Steidel 1988; Steidel 1989). Therefore, our preferred interpretation of the evolution of the C IV absorption systems is that it is caused by a systematically evolving abundance of C in the clouds. Specifically, the data appear to be consistent with an increase in $\langle C/H \rangle$ by about a factor of 3 over the redshift range $3.5 \geq z \geq 1.5$ (Steidel 1989).

Under the chemical evolution hypothesis, it is interesting to consider the implied timescales for heavy element enrichment of the gas. For $q_0 = 0.5$ and $H_0 = 50$ km s^{-1} Mpc^{-1}, the epoch over which the evolution is inferred to have occurred corresponds to times between 1.3 and 4 billion years after the big bang. A simple "closed box" model (Searle and Sargent 1972) for galactic evolution suggests that gas was processed into stars at a roughly constant rate over this span of time (Steidel, Sargent, and Boksenberg 1988). In addition, for the assumed values of q_0 and H_0, the time period of interest is also coincidental with the inferred age of the Galactic halo as measured from globular cluster main sequence turnoffs. The absorption line data indicate that there was considerable evolution of the halo gas in galaxies in general during that time.

6. Physical Properties of the Lyman Limit Absorbers

The Lyman limit absorbers (LLSs), those HI absorption systems with $N(HI) \geq 10^{17}$ cm^{-2}, are very important to the interpretation of the systems selected on the basis of heavy element absorption lines. First, we have seen that the LLSs do not appear to evolve significantly in terms of $N(z)$, in contrast to the observed relation for C IV. However, the LLSs are a subset of the heavy element absorption systems, rather than of the "Lyman α forest" systems (see, e.g., Sargent, Steidel, and Boksenberg 1989), and indeed most if not all have associated C IV absorption at some level. Thus, it may be that the LLSs, which are easily identified in low–resolution spectra of high redshift QSOs, are an ideal sample of absorbers to use in attempting to infer the properties (and explain the evolution) of the heavy element absorption systems in general. In particular, since the statistical studies of the C IV systems have suggested that chemical evolution may be the most important effect in the evolution of the heavy element absorbers, it would be very interesting to infer true chemical abundances and ionization states; by selecting systems purely on the basis of their HI content, one should avoid any selection bias having to do with abundances of heavy elements. Moreover, measurement of chemical abundances requires a knowledge of the H content of the absorbers, and the LLSs allow a quite accurate determination of the HI column density.

Thus, a sample of LLSs with $z \sim 3$ has been studied in detail in an effort to make preliminary determinations of such absorber properties as chemical abundance, particle density, ionization state, and physical size and mass (Steidel 1990; Sargent, Steidel, and

Boksenberg 1990). Briefly, the method involves constraining photoionization models of the absorbing clouds by measuring the relative column densities of ions with differing ionization potentials, together with the HI column density, corresponding to each LLS. The grids of models have been calculated using Ferland's CLOUDY program, and are characterized in terms of the ionization parameter $\Gamma \equiv n_\gamma/n_H$, where n_γ is the number density of H-ionizing photons incident on the cloud and n_H is the number density of H atoms, ionized and neutral. For a given input HI column density, in principle a unique value of Γ can be found which produces the observed ion ratios. If the intensity of the metagalactic flux (assumed to be the dominant source of ionizing photons) were known, then n_H would emerge directly from the models; however, if the flux is expressed in terms of $J_0 \equiv J_{\nu_0}/10^{-21}$ ergs s-1 cm^{-2} Hz^{-1} ster^{-1} (see Section **3**), then

$$n_H = 6.3 \times 10^{-5} J_0/\Gamma \quad \text{cm}^{-3}. \tag{6}$$

For the LLSs studied, which had $17.0 \leq \log N(HI) \leq 19.0$, it turns out that

$$-3.5 \leq \log \Gamma \leq -2.0, \tag{7}$$

so that for $J_0 \approx 1$ (as suggested by Bajtlik, Duncan, and Ostriker 1988),

$$0.006 \leq n_H \leq 0.2 \quad \text{cm}^{-3}. \tag{8}$$

Accordingly, the clouds are extremely tenuous. The method we have used also provides fractional ionization values for all of the observed ions, from which ionization corrections can be made and the *total* H column density N_H and the chemical abundances follow. An estimate of a cloud's linear size also follows, since $D_c = N_H/n_H$, or

$$D_c = 5.2 N_{20} \Gamma_{-2}/J_0 \quad \text{kpc}, \tag{9}$$

where N_{20} is the total H column in units of 10^{20} cm^{-2} and Γ_{-2} is the ionization parameter in units of 10^{-2}. For spherical clouds with $R_c = D_c/2$,

$$M_c = 4.4 \times 10^7 N_{20}^3 \Gamma_{-2}^2 J_0^{-2} M_\odot. \tag{10}$$

The results for the 8 LLSs with $z \sim 3$ that were studied can be summarized as follows: The chemical abundances (C, Si, O) were found to be in the range

$$-3.0 \leq [M/H] \leq -1.5; \tag{11}$$

this is similar to the range observed for Population II stars in the halo of our Galaxy. The arguments presented in Section **5** suggest that by the epoch corresponding to $z \sim 1.5$, these abundances will probably have increased by $\Delta[M/H] \sim 0.5$. From the $N(z)$ data, one infers very large effective radii for the absorbing galaxies, $R \sim 100 kpc$, and thus we are probably looking at *early, extreme halo material*. For $J_0 \approx 1$, the cloud diameters are inferred to be in the range $1 < D_c < 15$ kpc, and the masses are in the range $10^6 < M_c < 10^9 \ M_\odot$. The sizes are consistent with limits obtained from observations of heavy element absorption in the spectra of gravitationally lensed QSOs.

The photoionized clouds will have $T \sim 3 \times 10^4$ K (consistent with the observed absorption line widths), and the Jeans length for the clouds will be

$$D_J \approx 15 \left(\frac{T_4}{n(H)_{-2}} \right)^{1/2} \quad \text{kpc}, \tag{12}$$

or $D_J \sim 15$ kpc and $M_J \sim 10^9 \, M_\odot$ for the inferred particle densities. If we consider a fragment in the halo of a massive galaxy, the tidal radius as estimated by the Jacobi limit will be

$$r_t \sim \left(\frac{M_c}{3 M_G} \right)^{1/3} R_G \quad , \tag{13}$$

where M_G is the galaxy mass, and R_G is the galactocentric radius of the cloud orbit. For $R_G \sim 100$ kpc and $M_G \sim 10^9 \, M_\odot$, $r_t \sim 10$ kpc ($r_t \sim 1$ kpc for $M_c \sim 10^6 \, M_\odot$). Thus, the deduced sizes and masses of the Lyman limit absorbing clouds are reasonable for early halo clouds. In fact, such clouds are *predicted* on theoretical grounds to have been present in the early evolution of galactic halos (e.g., Saslaw 1968; Larson 1969; Tinsley 1978; Fall and Rees 1985; Rees 1988; Lake 1988). It is suggested (Steidel 1990) that the clouds we observe as heavy element absorbers in the spectra of high redshift QSOs are the sub–galactic fragments which would eventually form the structure on the scale of globular clusters (cf. Chaffee *et al.* 1986) up to that of dwarf satellite galaxies (c.f. York *et al.* 1986).

7. Contributions to Ω_B

We now arrive at perhaps the most relevant (to the subject of this Workshop) section of the talk: Just how much baryonic material are we talking about? It is interesting that, despite a range of roughly a factor of 100 in HI column density for the LLS clouds discussed in Section 6, the inferred *total* H column densities are all within a factor of ~ 5 of one another, and indeed the typical value of $\log N_H$ for the LLSs is ~ 20.5 cm^{-2}, within a factor of a few of the typical value for the "damped Lyman α" systems, which are mostly neutral and which are thought to be produced by galactic disks (for a summary, see Wolfe 1988). If we assume that the typical total H column density for a heavy element absorption system is $\sim 2 \times 10^{20}$ (this includes systems with HI columns above and below 10^{17} cm^{-2}), then in a unit redshift range $2.5 \leq z \leq 3.5$, there will be approximately 6 such systems, giving a total hydrogen column of $\sim 1.2 \times 10^{21}$ cm^{-2}. For $q_0 = 1/2$, the unit redshift range corresponds to $\sim 210 h^{-1}$ Mpc measured at the epoch corresponding to $z \sim 3$. Thus, the contribution to Ω_B from the heavy element absorbing clouds is

$$\Omega_{HE} \approx \frac{1.2 \times 10^{21}}{6 h^{-1} \times 10^{26}} \frac{1.3 m_H}{1.9 h^2 \times 10^{-29}} \left(\frac{1}{1 + \langle z \rangle} \right)^3 \quad , \tag{14}$$

where $h = H_0 / 100$ km s^{-1} Mpc^{-1} and m_H is the mass of a hydrogen atom. Evaluated at $\langle z \rangle \approx 3$, $\Omega_{HE} \approx 4 h^{-1} \times 10^{-3}$. Wolfe (1988) has found that the contribution from the damped Lyman α systems (proposed to be indicative of primitive disks of galaxies) is $\Omega_{DLA} \sim 2.5 \times 10^{-3}$. These numbers are to be compared with the limits on Ω_B from standard nucleosynthesis given by Pagel earlier in this *Workshop*, namely,

$$9 \times 10^{-3} \leq \Omega_B h^2 \leq 30 \times 10^{-3} \tag{15}$$

and the baryonic material associated with visible galaxies (Yang *et al.* 1984),

$$0.01 < \Omega_{gal} < 0.02. \tag{16}$$

The conclusion we reach is that at early epochs, a very substantial fraction of the baryonic material associated with galaxies was in the form of diffuse gas clouds in extended halos.

8. Relation between Heavy Element and Lyman α Absorbers

It is interesting at this point to compare the inferred properties of the heavy element clouds with those which have been suggested for the Lyman α forest clouds. Sargent *et al.* (1980) first suggested that the Lyα clouds are pressure–confined by a hot IGM, and that they have masses in the range $10^7 - 10^8 \; M_\odot$, very close to the masses inferred for the heavy element clouds discussed in Section **6**. The characteristic sizes of the forest clouds have been constrained by observations of close pairs of QSOs and gravitationally lensed QSOs: Foltz *et al.* (1984) obtained a lower limit on the cloud sizes of $\sim$ 8 kpc based on observations of the probable gravitational lens Q2345+007A, B, and Shaver and Robertson (1983) have set upper limits of $\sim$ 400 kpc based on the absence of correlated absorption in the spectra of the UM 680/UM 681 pair. Using photoionization models comparable to those discussed in Section **6**, and again assuming $J_0 \approx 1$, a typical Lyα forest cloud with $N(HI) = 10^{14}$ cm^{-2} would have a mass of $\sim 10^8 \; M_\odot$, a particle density of $n_H \approx 10^{-4}$ cm^{-3}, and a size of $\sim$ 50 kpc. The total hydrogen column for the cloud would be $\sim 10^{19}$ cm^{-2}. The point is that one can build a consistent picture in which the only difference between the heavy element clouds and the forest clouds is in particle density, and hence ionization state (Steidel 1990).

However, a very strong distinction between the heavy element systems and the Lyman α forest lines is in their clustering properties. The heavy element systems are strongly clustered on scales up to 1000 km s^{-1} (as would be expected if they are indeed associated with galaxies; Sargent, Boksenberg, and Steidel 1989), whereas the latest results (Bechtold and Shectman 1989) show that there is no significant clustering of the Lyman α–only systems down to velocity splittings of 50 km s^{-1}. These observational results could be naturally explained if the more diffuse forest clouds were *anti–correlated* with galaxies. This anti–correlation could come about under a variety of scenarios; for example, diffuse clouds with sizes on the order of 50 kpc are likely to be tidally disrupted in the vicinity of a galaxy–sized potential (see Section **6**), or the clouds may be confined by dark "mini–halos" (Rees 1986), or the ambient pressure in the vicinity of galaxies may be too high to allow the low–density clouds' existence if they are pressure confined.

The chemical abundances of the forest clouds have been the subject of a great deal of scrutiny. Most recently, Chaffee *et al* (1986) have set upper limits of $[M/H] \leq -3.5$ for a particular cloud having an HI column density of $\sim 10^{16.5}$ cm^{-2}. We observe that this is remarkably close to the lower limit of the inferred abundances of the LLSs discussed in Section **6**. Chaffee *et al.* (1986) also pointed out that if the only difference between the forest clouds and heavy element clouds is in density, then a wide range of heavy element abundances must exist.

All of the above arguments suggest that there exists a *continuum of primordial objects*, ranging from the most diffuse forest cloud which may have true primordial abundances, to the most metal–rich heavy element absorbing cloud with abundances near the solar values. All of these properties may be dictated by the proximity to galaxy–sized gravitational potentials; the continuum in abundances may represent a function of galactocentric distance both because of increasing particle densities (and thus HI columns) within the clouds and because of the increased incidence of cloud–cloud collisions which may induce star formation. The different redshift evolution of the forest clouds and the heavy element clouds (see Section 4) could also be explained by a similar mechanism–the diffuse clouds, which exist in isolated regions, may evaporate as the universe expands, whereas the denser clouds, influenced by the gravitational field of a galaxy, presumably are falling in and becoming progressively more dense and more enriched in heavy elements. At lower redshift, one would then expect that most of the observed Lyman α lines *are* associated with galaxies, and thus should begin to exhibit clustering. This effect at low z may have been observed (Webb 1987; Bechtold and Shectman 1989).

9. Fate of the Halo Gas

There is little evidence that typical galaxies (including the Milky Way) have extended gaseous halos at the present epoch (see, for example, Savage 1988). However, Bergeron (1988) has shown that the Mg II absorbers at moderate redshifts ($0.2 < z_{abs} < 0.7$) are frequently found to be associated with *intrinsically bright galaxies which show signs of current star formation*. Taken together, the facts suggest that galaxies may be active (in terms of star formation) *because* they have gaseous halos, continuously supplying new gas to the disk (Steidel 1990). One may therefore preferentially select young systems (i.e., ones which still have gaseous halos) in selecting by the presence of absorption. Further, the epoch of the "disappearance" of gaseous halos may signal the most important epoch for star formation in young galaxies, and the results of the surveys for heavy element absorption described above suggest that this may have been relatively recently, i.e., $z < 1.5$.

10. The Origin of the Metagalactic UV Flux

We have seen in Section 3 that the intensity of the metagalactic UV radiation field (evaluated at the H Lyman limit) at high redshift is likely to be $\log J_{\nu_0} \approx -21$ ergs s^{-1} cm^{-2} Hz^{-1} ster^{-1}, but that it is not possible to account for all of these ionizing photons with the observed high redshift QSOs. In fact, by the latest estimate, when the opacity of the Universe due to intervening absorbing gas is taken into account, the QSOs fall short by a factor of ~ 20 (Lin and Phinney 1989). This raises the important question: What produced all of the ionizing photons? Possible additional sources of ionizing photons include young, star forming galaxies (e.g., Bechtold *et al.* 1987), or perhaps a population of AGNs which are either of low luminosity or which have escaped detection for some other reason (e.g., because of dust obscuration; Heisler and Ostriker 1988). The various possibilities, along with suggested tests for distinguishing amongst the possibilities, have been discussed recently by Miralda–Escude and Ostriker (1989).

We have seen in Sections 6 and 8 that the heavy element absorption systems detected in the spectra of high redshift QSOs are most likely photoionized by the same metagalactic flux which ionizes the "forest" clouds. Because lines of several ionization stages are often observed, such systems provide useful tools for constraining the shape and intensity of the ionizing radiation field. Steidel and Sargent (1989) have used observations of the heavy element absorption systems together with photoionization models to argue that the metagalactic flux cannot be dominated by hot stars in young galaxies, simply because such sources do not produce enough high energy photons relative to H–ionizing photons to produce the ionization levels observed in the absorbing clouds. Specifically, at most about 50 percent of the photons at the Lyman limit can be contributed by ordinary stellar sources; the remainder must be produced by sources which have spectral energy distributions similar to those of AGNs. Thus, it is suggested that one must find sources of ionizing photons with AGN–like spectra to produce the factor of ~ 10 more ionizing photons than can be accounted for by the observed QSOs. Moreover, if the number of AGNs is falling off rapidly for redshifts beyond ~ 3.5 as indicated by the QSO number counts, then one ought to begin to observe significant changes in the ionization level of the absorbing clouds. At present, no such changes have been observed out to $z \sim 4$. The above arguments point to the possible existence of a population of low–luminosity AGNs which does not evolve as quickly as the high–luminosity QSOs.

11. Conclusions

Our conclusions may be summarized as follows:–

a) Based on the "proximity effect", the metagalactic ionizing flux is $J_{\nu_0} \approx 10^{-21}$ ergs s^{-1} cm^{-2} Hz^{-1} ster^{-1}, and is roughly constant over the redshift range $1.8 < z < 3.8$ currently observable.

b) This UV flux cannot be accounted for by radiation from observed luminous QSOs.

c) There is no detectable Gunn–Peterson trough in the spectra of QSOs with $z > 2$.

d) Arguments based on the ionization of pressure confined Lyman α clouds lead to a limit on the density parameter for generally distributed intergalactic hydrogen gas of $\Omega_H(IGM) < 0.05h^{-1}$.

e) The Lyman α clouds contribute $\Omega_c < 0.001$.

f) The QSO absorption lines in the spectra of high redshift QSOs indicate that the mean C/H abundance ratio increases by a factor of ≈ 3 in the first 1 to 4 billion years (assuming h = 1/2, $q_0 = 1/2$) after the big bang. On a simple "closed box" model of galactic evolution the rate of consumption of interstellar gas into stars is roughly constant over this period.

g) A study of a sample of "Lyman limit" absorption systems shows that the interstellar gas in the outer parts of galaxies at the epoch corresponding to $z \approx 3$ was deficient in the heavy elements relative to solar abundances by factors of 30 to 1000. This range is interestingly similar to that exhibited by globular clusters in the outer parts of the Galaxy today.

h) Early galactic halos consisted of diffuse clouds with masses in the range 10^6 to 10^9 $M_{\odot}$ and radii in the range 1 to 15 kpc. The mass range extends from globular clusters up to dwarf galaxies. A sketchy picture emerges in which galaxies were formed from the aggregation of diffuse clouds in enormous halos. The rate of star formation at early times was presumably controlled by the supply of this halo gas to the primitive disk.

i) The halo gas associated with the heavy element absorbers in QSO spectra contributes significantly to Ω_B at $z \approx 3$. A comparable contribution comes from the "damped Lyman α" absorbers inferred to be galactic disks.

j) The relative abundances of heavy ions in the QSO heavy element absorption systems indicates that hot stars in young galaxies do not contribute significantly to the metagalactic ionizing flux up to the highest observed redshifts– $z \approx 3.6$. Since the observed high luminosity QSOs are not responsible, this points to the existence of a substantial population of low–luminosity active nuclei as the most likely source.

These results show the importance of the QSO absorption lines for detailed understanding of the early evolution of galaxies and the intergalactic medium. Although some of the details in our interpretations may be over simplified or in error, we are now confident that the absorption lines can be used to explore the evolution of the interstellar gas in galaxies from the earliest epochs down to the present day.

12. Acknowledgements

Many of the results described in this paper came from the long collaboration between W. L. W. S. and A. Boksenberg and the "Flying Circus" which provided science and entertainment in equal measure for 15 years starting in 1973. Many of the reduction and analysis techniques were developed, and the long term astronomical goals established, during their association with the incomparable Peter Young from 1976 until his tragic death in 1981. The beginning (but not the end) of this work was supported by the National Science Foundation under grants AST82–16544 and AST84–16704 to W. L. W. S..

References

Bajtlik, S., Duncan, R. C., and Ostriker, J. P. 1988. *Astrophys. J.* **327**, 570.

Bechtold, J., Green, R., and York, D. G. 1987. *Astrophys. J.* **312**, 50.

Bechtold, J., and Shectman, S. 1989 in Proc. IAU Symposium 134, *Active Galactic Nuclei* (in press)

Bergeron, J. 1988 in *QSO Absorption Lines: Probing the Universe*, proc. Space Telescope Science Institute Symposium No. 2, ed. J. C. Blades, D. Turnshek, and C. A. Norman, p. 127

Chaffee, F. H., Foltz, C. B., Bechtold, J., and Weymann, R. J. 1986. *Astrophys. J.* **301**, 116.

Fall, S. M., and Rees, M. J. 1985. *Astrophys. J.* **298**, 18.

Foltz, C. B., Weymann, R. J., Roser, H. J., and Chaffee, F. H., Jr. 1984. *Astrophys. J. Lett.* **281**, L1.

Gunn, J. E.. and Peterson, B. A. 1965. *Astrophys. J.* **142**, 1633.

Heisler, J., and Ostriker, J. P. 1988. *Astrophys. J.* **305**, 103.

Hunstead, R. W., Pettini, M., Blades, J. C., and Murdoch, H. S. 1987 in *Observational Cosmology*, eds. A. Hewitt *et al.* p. 799

Lake, G. 1988. *Astrophys. J.* **327**, 99.

Larson, R. B. 1969. *Mon. Not. Roy. Astron. Soc.* **145**, 405.

Lin, Z., and Phinney, E. S. 1989, *Mon. Not. Roy. Astr. Soc.* (in press)

Miralda–Escude, J., and Ostriker, J. P. 1989, *Astrophys. J.* submitted

Murdoch, H. S., Hunstead, R. W., Pettini, M., and Blades, J. C. 1986. *Astrophys. J.* **309**, 19.

Rees, M. J. 1986. *Mon. Not. Roy. Astron. Soc.* **218**, 25P.

Rees, M. J. 1988 in *QSO Absorption Lines: Probing the Universe*, proc. Space Telescope Science Institute Symp. No.2, eds. J.C. Blades, D. Turnshek, C. A. Norman, p. 107

Sargent, W. L. W., Young, P. J., Boksenberg, A., and Tytler, D. 1980. *Astrophys. J. Suppl.* **42**, 41.

Sargent, W. L. W., Boksenberg, A., and Steidel, C. C. 1988. *Astrophys. J. Suppl.* **68**, 539.

Sargent, W. L. W., Steidel, C. C., and Boksenberg, A. 1989. *Astrophys. J.* **69**, 703.

Sargent, W. L. W., Steidel, C. C., and Boksenberg, A. 1990, *Astrophys. J.* in press

Saslaw, W. C. 1968. *Mon. Not. Roy. Astron. Soc.* **141**, 1.

Savage, B. D. 1988 in *QSO Absorption Lines: Probing the Universe*, proc. Space Telescope Science Institute Symposium No. 2, ed. J. C. Blades, D. Turnshek, and C. A. Norman, p. 195

Schmidt, M. 1965. *Astrophys. J.* **141**, 1295.

Searle, L., and Sargent, W. L. W. 1972. *Astrophys. J.* **173**, 25.

Shaver, P. A., and Robertson, J. G. 1983 in *Proc. 24th Liege Symposium, Quasars and Gravitational Lenses* (Liege: Institut d'Astrophysique), p. 598

Steidel, C. C. 1989, *Astrophys. J. Suppl* (in press)

Steidel, C. C. 1990, *Astrophys. J. Suppl* (in press)

Steidel, C. C., and Sargent, W. L. W. 1987. *Astrophys. J. Lett.* **318**, L11.

Steidel, C. C., Sargent, W. L. W., and Boksenberg, A. 1988. *Astrophys. J. Lett.* **333**, L5.

Steidel, C. C., and Sargent, W. L. W. 1989. *Astrophys. J. Lett.* **343**, L33.

Tinsley, B. M. 1978 in Proc. IAU Symposium 84 *Large Scale Characteristics of the Galaxy*, ed. W. B. Burton (Boston: Reidel), p. 431

Tytler, D. 1987. *Astrophys. J.* **321**, 49.

Webb, J. K. 1987 in Proc. IAU Symposium 124 *Observational Cosmology* eds. A Hewitt, *et al.* (Dordrecht:Reidel), p. 803

Wolfe, A. M. 1988 in *QSO Absorption Lines: Probing the Universe*, proc. Space Telescope Science Institute Symposium No. 2, ed. J. C. Blades, D. Turnshek, and C. A. Norman, p. 297

Yang, J., Turner, M. S., Steigman, G., Schramm, D. N., and Olive, K. A. 1984. *Astrophys. J.* **281**, 493.

York, D. G., Dopita, M., Green, R., and Bechtold, J. 1986. *Astrophys. J.* **311**, 610.

BARYONIC DARK MATTER AND THE CHEMICAL EVOLUTION OF GALAXIES

B. E. J. Pagel*
Royal Greenwich Observatory
Herstmonceux Castle
Hailsham, E Sussex BN27 1RP
UK

> *"Oh dear, what can the matter be?"*
>
> *(Traditional English song)*

ABSTRACT. Reasonable inhomogeneous Big Bang nucleosynthesis models suggest (in usual notation) $.009 \leq \Omega_B h_0^2 \leq .04$ corresponding to typical mass:light ratios between 13 h^{-1} and 60 h^{-1} if there is no large undetected luminosity density due to obscured or low surface brightness galaxies. The lower limits are large in relation to what is seen in the solar neighbourhood, which latter is quite in accordance with a reasonable model of galactic chemical evolution. Larger mass:light ratios occur (and are expected) in elliptical galaxies, but these make only a minor contribution to the overall average.

The conclusion is that baryonic dark matter exists either in some form that is radiatively and nuclearly inactive (*eg* 'Jupiters') and separate spatially from visible stellar populations – *eg* in galactic dark halos – or in the form of galaxies that are still undetected (or whose luminosity has been underestimated) in optical surveys owing to low surface brightness and/or obscuration. Chemical evolution considerations offer only a few negative clues as to its nature and distribution.

1. Introduction

Big Bang nucleosynthesis theory permits and possibly even demands the existence of substantial baryonic dark matter (BDM). On the other hand, substantial BDM is more of an embarrassment than anything else from the point of view of the theory of chemical evolution (*ie* abundance evolution) of galaxies (GCE), in so far as we have one. In this talk I consider what seem to me to be the constraints on BDM imposed by both standard Big Bang nucleosynthesis theory (SBBN) and non-standard (*ie* inhomogeneous) BBNS and present two GCE models, one for elliptical galaxies (Arimoto & Yoshii 1987; Yoshii & Arimoto 1987) with substantial BDM arising naturally in the form of white dwarf and neutron star remnants and the other for our own Galaxy in the solar neighbourhood (Pagel 1989a) which leaves virtually no room for contributions from any BDM beyond the usual

* Present address: NORDITA, Blegdamsvej 17, DK-2100 Copenhagen 0, Denmark.

D. Lynden-Bell and G. Gilmore (eds.), Baryonic Dark Matter, 237–256.

neutron star and white dwarf remnants already counted in existing surveys that lead to a mass:light ratio of the order of 2.

The conclusion will be that dark baryonic halos of galaxies may well exist, but that the material must then be in some form that is neither directly observable nor capable of influencing the composition of visible stars in any way. Jupiter-sized objects (Carr 1989) are clearly good candidates; alternatively, it seems quite an attractive idea that galactic halos may be non-baryonic and that BDM exists in the form of low surface brightness galaxies like Malin 1 (Bothun et al. 1987; Impey & Bothun 1989) and/or obscured galaxies (Disney, Davies & Phillips 1989).

2. Constraints from Big Bang Nucleosynthesis Theory (BBNS)

Three sorts of BBNS theory are currently in debate, namely:
1. SBBN (Yang et al. 1984; Boesgaard & Steigman 1985) which assumes a homogeneous Friedman universe and small lepton numbers.
2. BBN with density fluctuations from the quark-hadron phase transition.
3. BBN with decaying particles that substantially modify results from earlier synthesis according to SBBN.

I reject the third type of theory out of hand, partly because one specific form that has attracted attention recently (Dimopoulos et al 1988) can perhaps be ruled out on the basis of predicting too much $^6Li/^7Li$ (Audouze & Silk 1989), but mainly because it perversely throws away what is to my mind one of the major attractions of SBBN, namely its ability to make very specific predictions, notably the restriction on $\mathcal{N}_\nu$, the number of light neutrino flavours. I shall also ignore other avant-garde ideas like that of superconducting strings (Malaney & Butler 1989), again on grounds of imposing too few constraints in the present state of knowledge.

Predictions of SBBN theory are shown in Figure 1. The primordial abundances of the trace elements D, ^3He and ^7Li depend primarily on the baryon: photo ratio η (unchanged since $e^\pm$ annihilation a few seconds after the Big Bang), which in turn is related through the known temperature of the microwave background to the average mass density of baryons in the universe and through the known (?) luminosity density of galaxies to the average baryonic mass:light ratio:

$$\Omega_B h_0^2 = \bar{\rho}_{bo}\frac{8\pi G}{3}\left(3.09 \times 10^{17}s\right)^2 = 5.32 \times 10^{28}\bar{\rho}_{bo}$$
$$= 3.73 \times 10^{-3}(T_0/2.75)^3\eta_{10} \tag{1}$$
$$= 3.6 \times 10^{-12}l\langle M_b/L\rangle$$

where Ω_B is the baryon density $\bar{\rho}_{bo}$ (gm cm^{-3}) in units of the Einstein-de Sitter closure density, h_0 is the Hubble constant in units of 100 km s^{-1} Mpc^{-1}, T_o the present temperature of the microwave background, l the luminosity density in solar luminosities per cubic megaparsec and M_b/L the baryonic mass: light ratio in solar units.

The primordial helium mass fraction, Y_p, increases only slowly with η and it also depends significantly on N_ν (which I assume to be 3; Pagel & Simonson 1989) and on the half-life τ

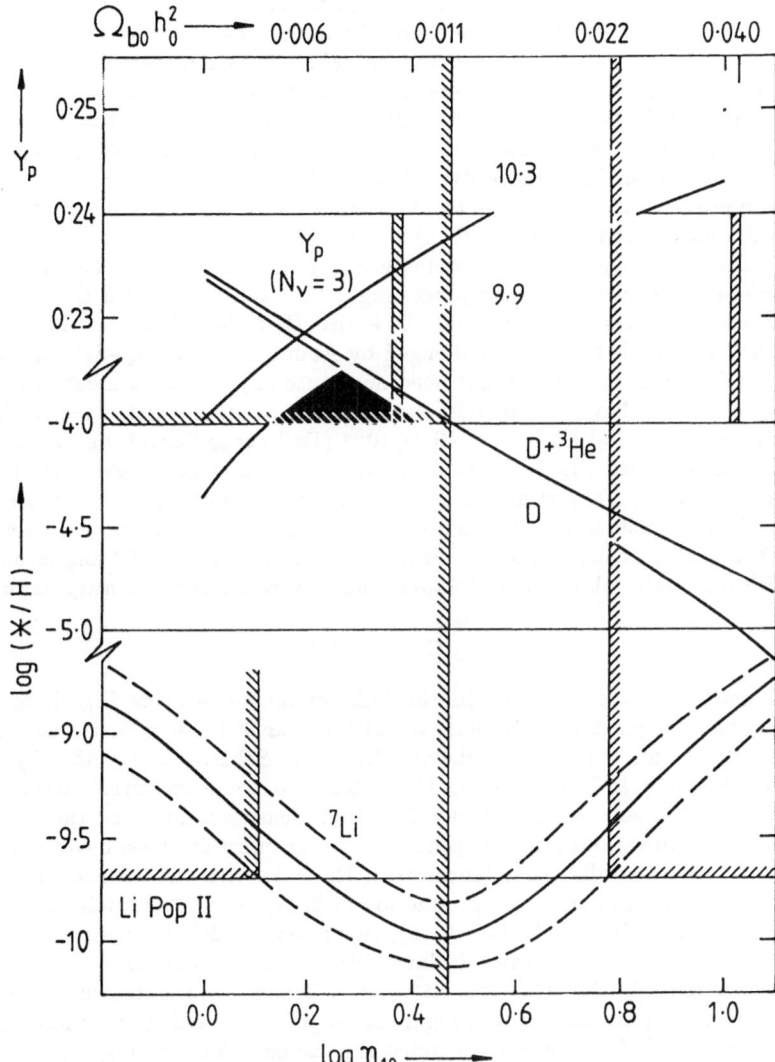

Figure 1. Primordial abundances predicted by SBBN theory after Yang *et al.* (1984), Steigman (1989) and Deliyannis *et al.* (1989), ^{7}Li from the latter reference being shown with $\pm 2\sigma$ error limits, as functions of η and $\Omega_B h_0^2$. Helium abundances by mass are given for reasonable upper and lower limits for the neutron half-life τ. Horizontal lines show upper limits based on observation and reasonable GCE considerations for ^{4}He, D + ^{3}He and ^{7}Li respectively. Tall vertical lines show the corresponding limits on η or $\Omega_B h_0^2$ from SBBN, while the shorter double vertical lines show the limits from inhomogeneous BBNS models currently considered plausible.

of the neutron, for which a recent measurement gives 10.1 ± 0.2 minutes (Last *et al.* 1988), substantially less than some older estimates. From figure 1 we can read off the following restrictions on η based on estimates of primordial abundances. The oldest one comes from $D/H \gg 10^{-5} \Rightarrow \Omega_{b_o} h_0^2 < 0.04$, but this limit can be tightened by considerations of the lithium abundance in subdwarfs, for which there is strong circumstantial evidence supporting the original suggestion of Spite & Spite (1982) that this is very close to primordial (Rebolo, Molaro & Beckman 1988; Deliyannis *et al.* 1989), implying $\Omega_B h_0^2 < 0.022$. A very similar limit comes from primordial helium below 0.24 (Pagel & Simonson 1989).

Lithium also provides a lower limit to the baryonic density, $\Omega_{b_o} h_0^2 > 0.004$, but Yang et al (1984) derived a more restrictive limit, $\Omega_B h_0^2 > 0.011$, from considerations of the astration of $D + {}^3He$, the point being that, when deuterium is destroyed, it is changed into ${}^3He$, some of which survives further stellar processing. The well-determined $D + {}^3He$ in the Solar System then restricts the primordial $(D + {}^3He)/H$ to less than 10^{-4} if at least 1/4 of ${}^3He$ survives. This limit has been challenged by Audouze, Delbourgo-Salvador & Salati (1988), but not for good reasons, since it need not cause any embarrassment to cosmology and in any case neither they nor anybody else have ever come up with a reasonable GCE model permitting primordial $(D + {}^3He)/H > 10^{-4}$ (Delbourgo-Salvador et al 1985, 1987; Vangioni-Flam & Audouze 1988). In models with inflow of unprocessed material (Pagel 1989a), not considered by Yang et al, the specific argument about ${}^3He$ loses its force, but at the same time it is virtually impossible to reach astration factors for deuterium itself greater than about 3 in such models, so the conclusion is the same as that of Yang *et al.* SBBN, combined with reasonable observational limits, therefore restricts the density parameter to the range

$$.011 \leq \Omega_B h_0^2 \leq .022. \tag{2}$$

These limits need to be re-evaluated in the light of inhomogeneous Big Bang theories which assume that the quark-hadron phase transition near 200 Mev is first order and leads to significant density fluctuations (Applegate, Hogan & Scherrer 1987; Alcock, Fuller & Mathews 1987; Malaney & Fowler 1988). The outcome depends on three parameters: the density contrast R, the volume filling factor f_v of high-density regions and the length scale d; and on the details of various diffusion processes. Much effort has been devoted to finding models that can fit primordial abundances with $\Omega_b = 1$, with the additional possibility that they may also lead to a non-zero production of heavy elements (Applegate, Hogan & Scherrer 1987) and Be (Boyd & Kajino 1989), two recent models of this sort being those of Mathews et al (1988) and Terasawa & Sato (1989). Three problems with these models suggest that it is unlikely that they can fit what is known about primordial abundances with $V_b = 1$: the density contrast is very high, the length scale needs to be finely tuned and even then the smallest ${}^7Li$ abundance predicted is a factor of 3 or so above the Population I abundance observed in young stars, the interstellar medium and the (solid) Solar System; even if one rejects the strong circumstantial evidence that Population II Li is primordial, and assumes no stellar ${}^7Li$ production, one is still faced with the need to envisage destruction of Li through astration by at least as large a factor as is expected for deuterium. Fowler (1990) points out that there are many nuclear reactions relevant to the low-density, neutron-rich regions, that have not yet been incorporated in nuclear reaction networks, and the proper inclusion of which can change things considerably; but pending settlement of this question I shall ignore the possibility that $\Omega_b = 1$ and just consider more conservative inhomogeneous Big Bang models (IBBN) with a moderate density contrast (Kurki-Suonio et al 1989; Reeves

1989). These give somewhat wider limits on η or Ω_b than does SBBN, but not dramatically so, the new limits from D + ^{3}He and from ^{7}Li, 4He being approximately

$$.009 \leq \Omega_B h_0^2 \leq 0.04. \tag{3}$$

The lower limit raises the question: Do we need BDM, over and above expectations from conventional stellar evolution and initial mass functions? The upper limit raises the question: Can BDM account for what we know about Ω from direct measurements, or do we need non-baryonic dark matter? Clearly if h_0 is small enough, BDM can account for direct measurements of Ω_0 up to 0.15 or so and the second question hinges on getting better determinations of both Ω_0 and h_0. The first question will be examined further in what follows.

3. Comparison of BBNS constraints with observed mass:light ratios

Dark matter in the universe has been recently reviewed in Kormendy & Knapp (1987), Trimble (1987, 1988) and Audouze & Van (1988), where mass:light ratios and related dark matters are extensively discussed. A recent discussion by Efstathiou, Ellis & Peterson (1988) of various galaxy surveys leads (in agreement with earlier studies referenced therein) to a blue luminosity density from observed galaxies (to ±30 per cent)

$$l_B = 1.9 \times 108 h_0 L_{B_\odot} \text{Mpc}^{-3} \tag{4}$$

which when combined with the value of the critical closure density

$$\rho_{cr} = 2.85 \times 10^{11} h_0^2 \mathcal{M}_\odot \text{Mpc}^{-3} \tag{5}$$

relates Ω to a typical mass:light ratio (in solar units)

$$\langle M_b/L_B \rangle = 1500 \Omega_B h_0 \tag{6}$$

whence the BBNS limits on $\Omega_B h_0^2$ derived in the previous section lead to

$$13 < 13 h_0^{-1} \leq \langle M_b/L_B \rangle \leq 60 h_0^{-1} \leq 120, \tag{7}$$

but the lower limit will be weakened if there is substantial undetected luminosity in low surface-brightness or obscured galaxies (see Introduction).

Table 1 shows some estimates of mass:light ratios based on dynamics in different objects and the corresponding values of Ω_0 if those are typical. Entries in the middle section of the table, between the two horizontal lines, are completely compatible with equations (3) and (7) derived from BBNS, if any of these are typical of the matter in the universe. Those below the lower line could have dark matter in non-baryonic form, but if non-typical they could also have it as BDM in the form of low-mass, non-luminous stars or "Jupiters", the latter being favoured by Fabian, Nulsen & Canizares (1982) and by Sarazin & O'Connell (1983) as a sink for the hot gas in cooling flows (Fabian, Nulsen & Canizares 1984). Such objects

TABLE 1

Some dynamical mass:light ratios, and values of Ω_O

	M/L_B	Ω_O^a	References
Solar cylinder to ± 2 kpc	$3.0 \pm .7$	$.002\ h_0^{-1}$	Bahcall 1984
	$2.0 \pm .5$	$.001\ h_0^{-1}$	Kuijken & Gilmore 1989
Irr. within optical radius	2.3 ± 1.7	$.002\ h_0^{-1}$	Faber & Gallagher 1979
Cores of E's and bulges	$(12 \pm 6)\ h_0$	.008	Pickles 1985
Spir. within optical radius	$(9 \pm 5)\ h_0$	.006	Rubin 1987
Spir. to 21cm limit	$(20 \pm 7)\ h_0$	.014	Sancisi & van Albada 1987
Milky Way to 100 kpc	20 ± 10	$.014\ h_0^{-1}$	Ostriker 1987
Binary gal. and small groups	$(120 \pm 60)\ h_0$	.08	Trimble 1987
M87 (incl. halo)	$300\ h_0$	.2	Fabricant & Gorenstein 1983
Coma cluster	$(330 \pm 50)\ h_0$	.2	Hughes 1989

[a] If the corresponding M/L is typical of matter in the universe.

would contribute negatively to GCE by locking up heavy elements previously present in the gas. Existing flow rates are generally too low to provide more than a small fraction of the mass inferred dynamically in, *eg*, M87; but Carr & Ashman (1988) suggest that higher flow rates on differing scales in the past could account for all the mass in dark galactic halos.

Above the upper line in Table 1 we have mass:light ratios associated with ordinary luminous matter, the inclusion of cores of ellipticals and bulges being justified both on the basis of their well-defined mass: luminosity relation (Faber *et al.* 1987) and on the basis of GCE arguments (see below). These in themselves are above the BBNS lower limit, but with ellipticals contributing only 40 per cent as much light as spirals and irregulars (Sandage, Binggeli & Tammann 1985) they are not significant enough overall to pull the average over that limit. It follows that either some (maybe all) of the dark matter below the upper line is baryonic, with at least as much dark matter disassociated from visible stars as is associated with them, or there is substantial luminosity hidden away in obscured and low surface brightness galaxies and not counted in equation (4): dim rather than dark matter.

Baryonic or not, the dark halos contributing to large mass:light ratios below the upper line have quite a different density distribution from the luminous matter. In the case of disks this is virtually self-evident, but it is also true of visible spheroids such as the bulge and Population II halo of our own Galaxy in which - in accordance with the Hubble or de Vaucouleurs surface brightness distribution - the volume density of RR Lyrae stars (taken as tracers of the old metal-deficient population) decays approximately as r^{-3} outside a core radius that is well below 1 kpc (Oort & Plaut 1975) and out to 10 kpc or so, whereafter it appears to steepen yet more (Saha 1985). A similar law can be fitted to the height distribution of halo globular clusters (Pagel 1989b). Dark halos, on the other hand, have been fitted with different laws of the form $(1 + r^2/a^2)^{-\gamma/2}$ with $0 \leq \gamma \leq 4\gamma = 2$ for an asymptotically flat rotation curve), but all models with $\gamma \geq 3$ lead to core radii a of the order of 10 kpc or more (Kent 1986, 1987; Lake & Feinswog 1989), indicating that the dark matter has had a different dynamical history from that of the visible halo. The same thing is suggested by the low mass:light ratios observed in globular clusters (Pryor et al. 1989 a,b). In fact, since Kuijken & Gilmore (1989) have successfully cast doubt on the existence of unseen gravitating matter in the solar neighbourhood, it could be the case that all dark matter shown in the table is non-dissipative, which would impose some constraints on its origin. Alternatively it might have been puffed up by some dynamical process such as loss of a large fraction of the mass of the system in galactic winds, or through mergers.

It follows, in any case, that any BDM in galactic halos is most unlikely to have, or to have had, a continuing role in GCE because low-mass stars, Jupiters or compact remnants formed in the evolution of visible stellar material are expected to have the same spatial distribution as the stars that we see. BDM could have affected the initial conditions, eg by having a generation of massive stars that produced hot, metal-enriched gas which mostly escaped causing a dynamical distension of the BDM halo and enriching the intergalactic medium, while at the same time depositing an initial abundance of heavy elements in the material that later formed the visible spheroid of an elliptical galaxy or the visible spheroid and disk of a galaxy like our own. Hegyi & Olive (1986, 1989) and Ryu, Olive & Silk (1989) have described severe constraints on this type of effect and the GCE models to be described in the next section leave no clear role for any such process. However, there is no convincing argument that rules out small, inactive baryonic objects like Jupiters ($cf.$ Carr 1989), which affect neither radiation nor production (or consumption) of heavy elements.

4. Some Galactic Chemical Evolution Models

I shall describe mainly two GCE models, one by Arimoto & Yoshii for ellipticals and one that I have developed for the Galactic halo and disk in the solar neighbourhood. Both give mass:light ratios in essential agreement with those quoted for the relevant stellar populations in the upper part of Table 1, the first involving substantial BDM in the form of compact stellar remnants and the second rather little in accordance with local observations.

The problem with GCE models is that they try to combine ideas on stellar evolution and nucleosynthesis, on which there is some sort of theory albeit with substantial grey areas, with assumptions about galactic evolution for which there is very little hard physical basis, particularly on questions like the law of star formation, the initial mass function (IMF) and its constancy or variability. So, as Cameron (1989) has recently emphasised, there is not really a theory.

Under these circumstances I deem it best to use the simplest possible assumptions consistent with various observational constraints. In our own neighbourhood there are quite a lot of these, but much fewer in external galaxies, notably ellipticals. Considerable insight can be derived from the 'Simple' or 'Closed Box' model combined with the Instantaneous Recycling approximation which assumes that stars can be divided into two classes: massive stars which return a substantial fraction of their mass to the interstellar medium (ISM) with nucleosynthesis products added, leaving neutron star and white dwarf remnants, in a negligible time after their birth; and low-mass stars which live for ever, keep all their mass and merely serve to lock up diffuse material. In this approximation, the relevant consequences of stellar evolution, folded into the IMF, can be summarised in a few parameters of which the most important is the yield p for some element, defined as the mass of that element freshly synthesised and ejected by a generation of stars divided by the mass locked up in low-mass stars and compact remnants. For the IMF in the solar neighbourhood (Scalo 1986), assuming that all stars having initially between 10 and 100 times the mass of the Sun undergo supernova explosions from core collapse, the yield for a primary element like oxygen produced by such stars is of the order of its solar abundance, but the uncertainties are quite large. (The word 'primary' refers to elements for which the yield is assumed to be insensitive to the chemical composition of the progenitor stars.)

The Simple model assumes that we start with primordial gas which forms stars with unvarying IMF and constant yields, that the system is well mixed at all times and that nothing flows in or out. The Instantaneous Simple Model leads to the well-known equation.

$$z \equiv Z/p = ln(m/g) = ln(1/\mu) \tag{8}$$

(Searle & Sargent 1972) where Z is the mass fraction of a primary element in the ISM and in newly formed stars, z is the same thing in units of the yield, m is the total mass of the system, g is the mass remaining in the form of gas and dust and μ is the gas fraction. This model has some success when compared with abundances and gas fractions in irregular and blue compact galaxies (Lequeux et al 1979; Pagel 1981, 1986; Axon et al 1988), though with the rather small yield of 0.2× solar oxygen abundance. Computation of the gas fraction requires an estimate of the total mass which is affected by dark matter when this is done on dynamical grounds, particularly when virial or rotational velocities of extended HI 21 cm emission (Lequeux & Viallefond 1980; Bergvall & Jörsöter 1987; Brinks & Klein 1988; Lake & Skillman 1989) are used and one should really use an estimate of just the stellar mass for this purpose, either from the dynamics of the visible portion or from the luminosity with a predicted mass:light ratio. There is in fact a good relation between chemical composition and absolute blue magnitude for elliptical, dwarf spheroidal and irregular galaxies over their full range of luminosities (Pagel & Edmunds 1981; Mould 1984; Skillman, Kennicut & Hodge 1989). The small effective yield (in comparison to that expected from the solar neighbourhood IMF) found at the lower galaxian luminosities can arise from: (i) outflow of processed hot gas due to shallowness of the potential wells (Larson 1974); (ii) a BDM component of low-mass stars associated with the stellar population because of low metallicity (Peimbert & Serrano 1982) or for some other reason; or (iii) some effect of metallicity on the yields from massive stars. In spiral galaxies, McCall (1982), Edmunds & Pagel (1984) and Garnett & Shields (1987) find a good correlation between oxygen abundance in HII regions and surface mass density based on the maximal disk hypothesis.

The upshot is that dark matter in halos is probably not directly relevant to the abundances, but it could be relevant indirectly through its effect on the depth of the potential well.

The abundance-luminosity relation in elliptical galaxies is explained in the models of Larson (1974), Arimoto & Yoshii (AY 1987) and Yoshii & Arimoto (YA 1987) by appealing

TABLE 2

Theoretical properties of elliptical galaxies, after AY87 & YA87

Initial mass $\mathcal{M}_\odot$	ν^{-1} Myr	t_{GW} Myr	μ_{GW}	$\mathcal{M}_*$ $\mathcal{M}_\odot$	$\mathcal{M}_{rem}$ $\mathcal{M}_\odot$	M_v	B–V	$\mathcal{M}_*/L_B$	$\langle[Fe/H]\rangle_l$
2×10^{12}	130	845	.06	1.9×10^{12}	8×10^{11}	-23.6	1.04	23	.33
10^{11}	93	350	.35	9×10^{10}	4×10^{10}	-20.1	.99	25	.12
10^9	55	32	.65	3.5×10^8	2×10^8	-14.3	.83	19	-0.75
10^7	14	5	.74	2.6×10^6	1.7×10^6	-9.2	.66	13	-2.41

to supernova-driven galactic winds. Basically they apply the Simple model up to a time when the hot interstellar gas has accumulated enough energy to escape, whereafter chemical evolution stops, and the details depend on how the rate of star formation and the time when gas escapes each depend on the structural properties of the parent galaxy.

An important feature of such models is the metallicity distribution function among the component stars. If instantaneous recycling applies, the differential distribution is readily shown from equation (8) to follow the law

$$\frac{ds}{dz} \propto e^{-z}; \ z \leq ln(1/\mu) \tag{9}$$

or, in the more usual representation against logarithmic abundance,

$$\frac{ds}{d\log z} \propto z e^{-z}; \ z \leq ln(1/\mu). \tag{10}$$

Yoshii & Arimoto have carried out detailed numerical calculations (not assuming instantaneous recycling) which predict colour evolution as well as abundance evolution. Theirs is the first evolutionary synthesis, to my knowledge, in which chemical and colour evolution are modelled self-consistently. Physical assumptions are that the total binding energy increases as mass $^{1.45}$ and that the star formation rate (SFR) is proportional to the total mass of gas with coefficients ν proportional to the cooling time ($\nu \propto$ mass$^{-.1}$) for large galaxies ($\geq 10^9 M_\odot$) and to the free fall time ($\nu \propto$ mass$^{-.325}$) for smaller galaxies, with constants adjusted to give observed absolute magnitudes and colours at extremes of the mass range after 15 Gyr.

A basic requirement of this type of model is that the true yield (based on stellar evolution and assumed IMF) has to be high, about 3 times solar abundance, in order to reproduce the high metallicities found in giant galaxies and in the X-ray emitting intra-cluster gas (Mushotzky 1989). To get this, they assume a quite flat power-law IMF

$$\frac{dN}{d\log m} \propto m^{-.95}; \ .05 \leq m \leq 60 \tag{11}$$

which accordingly leads (in combination with rapid star formation) to quite large mass:light ratios of about 20, half or so of which is supplied by compact remnants.

These models lead to broad, skewed metallicity distributions in qualitative agreement with equation (10) and some other results are summarised in Table 2.

Here t_{GW} is the time when the galactic wind occurs, μ_{GW} the corresponding gas fraction lost, M. the final mass in stars + remnants, Mrem the final mass in remnants and $\langle[\text{Fe/H}]\rangle_l$ the luminosity-weighted mean metallicity. Metallicities and colours, as a function of absolute magnitude, are fairly well reproduced. M/L ratios are rather high, but can be reduced by increasing the lower mass limit of the IMF and allowing for secondary post-wind star formation (Arimoto 1989), bringing them into line with average observational values for the cores of ellipticals. The dynamical effects of terminal mass loss are shown to be able to explain some features of the relation between diameter and surface brightness, but one cannot expect a single-parameter theory to account for all the generic relationships of elliptical galaxies (Djorgovski & Davis 1987; Faber et al 1987).

Turning now to the solar neighbourhood in our own Galaxy, I give here a brief description of a new analytical model for the evolution of primary elements (Pagel 1989a) which is the only one I know of to satisfy all of the following constraints in one model:-
1. Provision of a consistent scenario for (visible, metal-deficient) halo and disk.
2. Relative numbers of long-lived halo and disk stars in a cylinder through the Sun perpendicular to the Galactic plane (halo $\leq$ 5 per cent of disk).
3. A gas fraction in the solar cylinder of at least 10 per cent.
4. A SFR history giving a reasonably large (> 0.5) ratio of present to average past SFR in the disk.
5. An age-metallicity relation giving a reasonable fit to the widely scattered relevant observational data (not a very strong constraint).
6. Metallicity distributions in the halo and disk (G dwarf problem).
7. The behaviour of relative abundances of oxygen, α-particle elements, europium and barium as functions of metallicity.
8. Radial abundance gradients in oxygen and iron (beyond the scope of this article).

A ninth constraint, based on nuclear cosmochronology data, is not considered because there are too many ambiguities (Clayton 1988; Pagel 1989b; Malaney, Mathews & Dearborn 1989; Pagel 1990).

Figure 2 shows the metallicity distributions of Galactic globular clusters (Zinn 1985; Armandroff & Zinn 1988) and of field stars in the halo with [Fe/H] < -1.4 (Beers, Preston & Shectman 1986; Beers 1987; Laird et al 1988). The distribution for globular clusters is bimodal, with distinct halo and thick disk components characterised by different spatial distributions (Zinn 1985), but the halo clusters and field stars approximate the $z e^{-z}$ distribution of equation 10 (Hartwick 1976; Searle & Zinn 1978) and the interpretation I adopt is essentially that of Hartwick (1976) who noted that the effective yield (given by the peak in the $z\,e^{-z}$ curve at $z = 1$) is very low, corresponding to [Fe/H] = -1.6 in Fig. 2, and interpreted this in terms of a modified Instantaneous Simple model in which gas is continually lost from the system at a rate of Λ times the net rate of star formation ($\Lambda = $ const.), resulting in a $z\,e^{-z}$ distribution with an effective yield $p/(1 + \Lambda)$. However, one also has to take into account the iron:oxygen anomaly in metal-deficient stars, where [O/Fe]$\approx$ 0.5, which is usually interpreted on the basis that only about 1/3 of the iron in the Sun comes from rapidly evolving core-collapse supernovae and the rest from Type 1a supernovae having an evolution time equal to or exceeding the duration of star formation in the halo (Tinsley 1977; Matteucci & Greggio 1986; Matteucci 1988). Furthermore, I assume that the gas continuously lost from the halo settles into the disk where it provides an initial abundance [Fe/H] $= -1.6$ equal to the average halo abundance and considerably less than the metal-

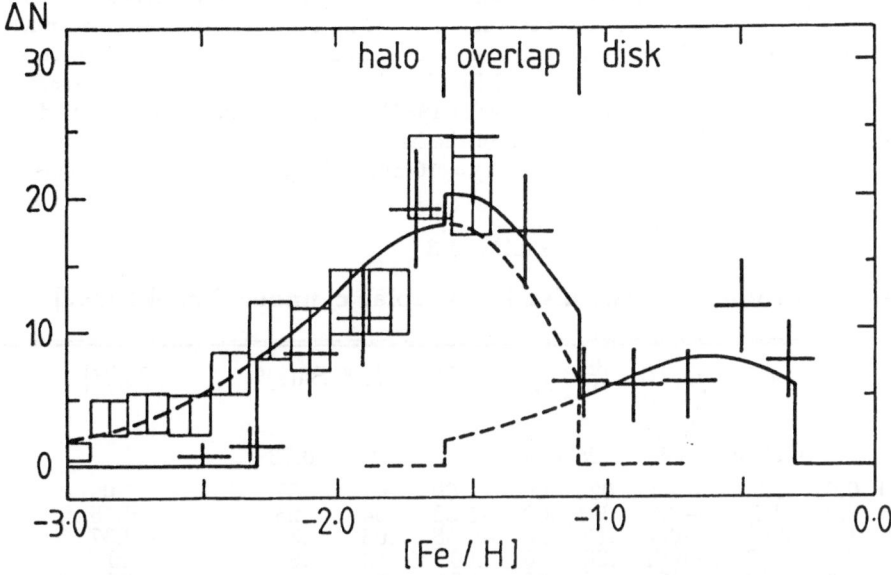

Figure 2. Metallicity distribution for 120 Galactic globular clusters (Zinn 1985; Armandroff & Zinn 1988), shown by crosses, and for field stars of the halo (Laird et al 1988) with [Fe/H]≤ −1.3, normalised to the same scale, shown by rectangular boxes. The solid curve shows a theoretical distribution for globular clusters composed of two simple models with different effective yields, truncated at the low metallicity ends at [Fe/H] = −2.3 and −1.6 respectively. Broken lines and curves show the separate distributions for the two components and (in the left part of the diagram) the continuation of the Simple model for the halo to indefinitely low metallicities. The heights of the crosses and boxes are ±1σ Poissonian errors for the numbers ΔN in each metallicity bin; their widths are those of the bins.

licity [Fe/H] = −1 or so of the latest and most metal-rich stars to be formed in the halo. This leads to an overlap in abundance between disk stars and halo stars for [Fe/H]≥ −1.6 and hence to a jump in the overall metallicity distribution shown by the solid curve. The overlap, which arises naturally in this model (as opposed to models like that of Matteucci & Francois (1989) which assumes that the disk starts at a metallicity where the halo leaves off, in accordance with the sudden wind scenario), provides a plausible explanation within the collapse picture (Eggen, Lynden-Bell & Sandage 1982) for the relatively small number of metal-deficient stars with disk-like kinematics and −1.6 ≤ [Fe/H] < −1.0 discovered by Norris, Bessell & Pickles (1985) and by Morrison, Flynn & Freeman (1989).

Figure 2 illustrates the fact that, while the globular clusters show a marginally significant deficit of numbers (relative to the z e⁻ᶻ curve) at the lowest metallicities ([Fe/H]≤ −2.3), there is no such deficit among the field stars. The data, therefore, while not excluding a very minute heavy-element enrichment due to inhomogeneous BBNS, leave little or no room for any enrichment by a hypothetical pregalactic generation of massive stars. Another point related to this question is the relationship between barium and europium. 88 percent of barium in the Solar System comes from the s-process, the remaining Ba and virtually all europium from the r-process (Cameron 1982), but in the most metal-deficient stars of the halo with [Fe/H]$\leq$ −2.5 or so, all heavy elements observed come from the r-process (Truran 1981; Gilroy et al 1988; Lambert 1989) so that there is no need for s-process seed nuclei or neutron sources over and above what is expected from normal stellar evolution. In my model, the observed relation between Ba and Eu abundances is very nicely fitted assuming that the s-process is primary (Malaney & Fowler 1989), but sets in only after a time delay equal to 0.15 of the period of star formation in the halo (whatever that may be), assuming suitable yields for the r-and s-process, $(1 + \Lambda) = 10$ and SFR proportional to the mass of gas.

TABLE 3

Chemical evolution of the disk in the solar cylinder [Pagel 1989a]

ωt	$g(t)$	$s(t)$	$\mu(t)$	[O/H] [α/H] [Eu/H]	f_1^a	f_2^a	[Fe/H]	Ba(r)^a	Ba(s)^a	[Ba/H]
			Yields^a	0.8	0.25	0.55		.1	0.73	
0.0	1.0	0.0	1.0	−1.1	.025	.00	−1.60	.01	.07	−1.09
0.5	1.5	0.6	0.7	−0.5	.095	.00	−1.02	.04	.25	−0.55
1.0	1.9	1.5	0.56	−0.4	.13	.20	−0.55	.05	.37	−0.37
4.0	1.5	7.3	0.17	−0.02	.30	.70	.00	.12	.88	.00
5.0	1.0	8.6	0.10	.05	.36	.87	.09	.14	1.04	.07
$\omega t = 5.5$			Instantaneous recycling no longer a good approximation							

^{a}Expressed linearly in units of total solar abundance of that element

Evolution in the disk is more complicated owing to the need to account for two features: the 'G dwarf problem' (Pagel & Patchett 1975; Pagel 1989b) and the relationship between iron abundance on one hand and abundances of oxygen and α−particle elements on the other (see Wheeler, Sneden & Truran 1989; Lambert 1989). For G dwarfs it is convenient to consider the distribution of oxygen abundances rather than iron abundances (i.e. metallicities as deduced from ultra-violet excess) because the instantaneous recycling approximation works better for oxygen (a product of short-lived massive stars) than for iron (which apparently has a significant contribution from Type Ia supernovae with substantial evolution times). I deal with the dG problem by assuming time-decaying inflow of unprocessed material, which in the form proposed by Lynden-Bell (1975) gives a superb fit to the oxygen abundance distribution function (Pagel 1979b), but for convenience in an analytical treatment taking time delays into account I use the related though slightly less well fitting linear formalism of Clayton (1985, 1988) with his k parameter equal to 4, and assume that oxygen, α-elements, r-process elements and one component of iron are produced in instantaneous recycling, but that there is another component of iron produced after a time delay Δ such that $\omega\Delta = 0.5$, where $\omega(\approx (3Gyr)^{-1})$ is the coefficient relating the net SFR (after stellar mass loss) to the mass of gas. True yields have been assumed constant throughout.

Some steps in the progress of the model are shown in Table 3, in which ωt is the time variable, assumed to reach 4.0 at the time of formation of the Solar System and 5.5 at present, but evolution is assumed to stop at $\omega t = 5.0$ because afterwards the gas fraction

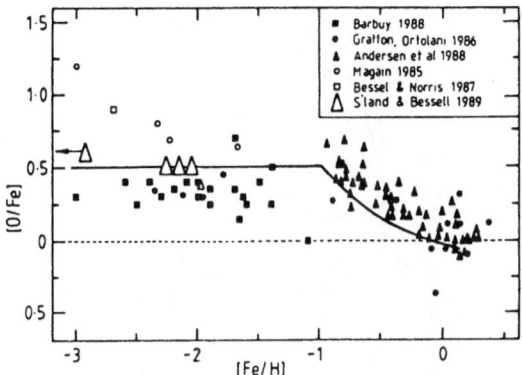

Figure 3. Relation between iron and oxygen abundances after Wheeler, Sneden & Truran (1989), references therein and Sutherland & Bessell (1989). The thick horizontal line and descending curve show predictions of the GCE model (Pagel 1989a).

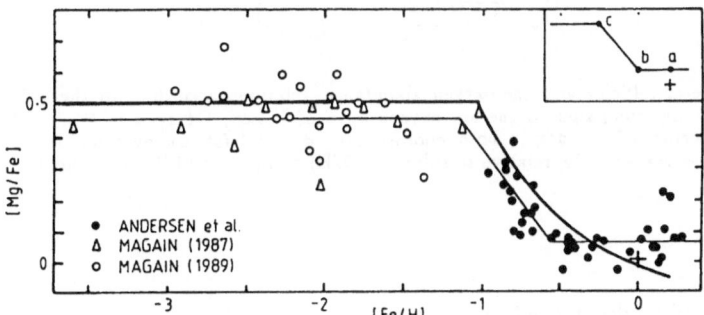

Figure 4. Relation between iron and magnesium abundances after Lambert (1989) and references therein. The thick horizontal line and descending curve (identical to that in Fig. 3) show predictions of the GCE model (Pagel 1989a). The thin straight lines and inset are the relation suggested by Lambert on an empirical basis.

becomes so small that outflow from old stars begins to be significant and so instantaneous recycling cannot be assumed any more. $g(t)$ and $s(t)$ are the masses of gas and stars + remnants, respectively, in units of the initial mass of the system, f_1 and f_2 are abundances of prompt and delayed iron, so that $[Fe/H] = \log(f1 + f2)$, and $Ba(r)$ and $Ba(s)$ are barium abundances contributed by r and s-process respectively.

Figure 3 shows the resulting relationship between iron and oxygen abundances (recent results of Abia & Rebolo (1989) suggesting a continuous increase in [O/Fe] as [Fe/H] goes down are not shown) and Figure 4 shows the same theoretical relationship (since r-process, O and α-elements are all assumed to vary in lockstep) applied to observational data for magnesium, a typical α-element. The fits are fairly crude, but discrepancies in the various

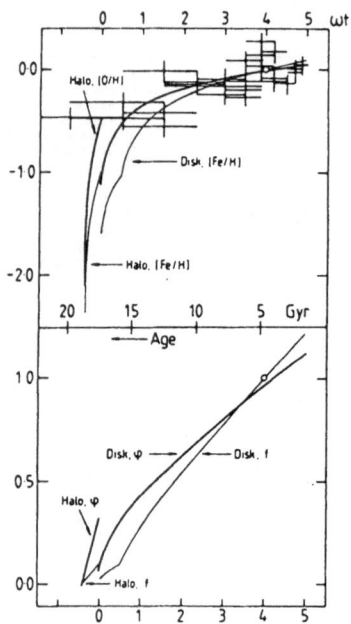

Figure 5. Age-metallicity and age-oxygen abundance relations according to the GCE model (Pagel 1989a), compared to the age-oxygen data of Nissen, Edvardsson & Gustafsson (1985) assuming errors of ±.1 dex in both coordinates. ϕ and f (shown by thick and thin curves, abundances respectively, relative to solar, *ie* [O/H] ≡ log ϕ and [Fe/H] ≡ log f.

data sets make it hard to do any better.

Figure 5 shows the oxygen and iron abundances as a function of stellar age and Figure 6 shows the distribution function of oxygen abundances with the corresponding [Fe/H] scale assuming

$$[Fe/H] = 2[O/H]; [Fe/H] \geq -1 \tag{12}$$

$$[Fe/H] = [O/H] - 0.5; [Fe/H] < -1. \tag{13}$$

The fit is fairly satisfactory, in contrast to that provided by a closed model with the same initial enrichment from the halo. The region $-1.0 \leq [Fe/H] -0.5$ roughly corresponds to the 'thick disk' (Gilmore & Wyse 1986) while that in the range $-1.6 \leq [Fe/H] < -1.0$ corresponds to what has been called the 'metal-weak thick disk' (Morrison, Flynn & Freeman 1989), but in my view (cf. Norris 1987) both of these components should be looked upon as parts of a continuous distribution covering the entire disk as modelled here and not as separate 'populations'.

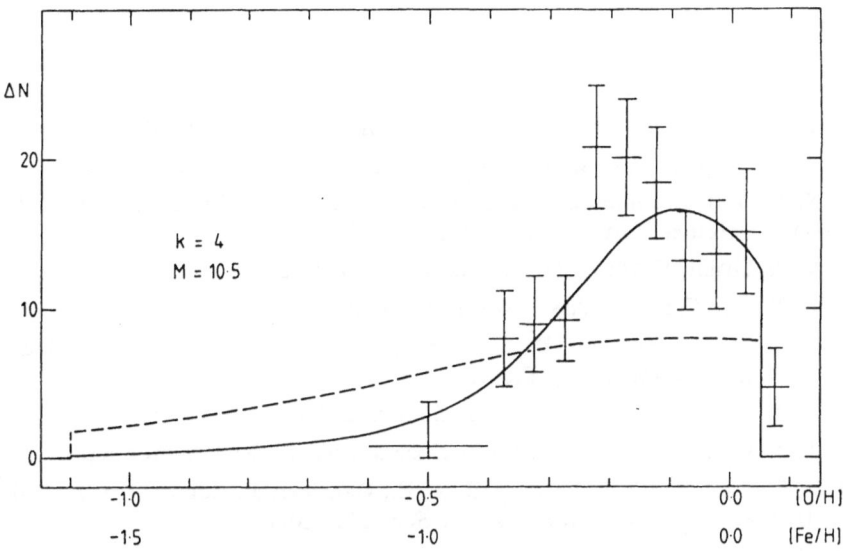

Figure 6. The oxygen abundance distribution of 132 G dwarfs in the Solar cylinder. The solid curve shows the prediction of the GCE model for the disk (Pagel 1989a) based on Clayton's formalism (M = 10.5 is the factor by which the initial mass is ultimately multiplied by inflow) and the broken curve shows the prediction of a closed model with the same initial abundance, yield, final gas fraction and total number of stars. Crosses show the data, after Pagel (1989b), with Poissonian ±1σ error bars for the number ΔN in a bin of 0.1 in [Fe/H] or 0.05 in [O/H].

5. Conclusion

The successes of the two GCE models described here suggest that - while the uncertainties are still great - there are no abundance data that defy explanation on the basis of conventional BBNS and stellar evolution with reasonable IMF's corresponding to mass:light ratios actually observed in luminous regions which - as we have seen - are quite small in relation to indications from BBNS which are exceeded in turn by some estimates based on dynamics. The physical significance of the inflowing unprocessed material postulated for the disk is not completely clear. Originally it was interpreted as a consequence of disk formation going outwards from the central regions by gradual accretion (Larson 1976), but now an alternative is available in the form of contraction of larger proto-disks manifested by absorption line systems at high red-shift with large HI column densities (Wolfe 1989), possibly aided by viscous processes (Sommer-Larsen & Yoshii 1989; Clarke 1989). But, however this may be, the composition of visible matter shows no signs of direct influence from dark matter, although some of this (possibly even all if H_0 is small and one does not insist on believing that we live in an Einstein - de Sitter universe) is very likely baryonic.

252

Chemical evolution offers only a few negative clues as to where and what such matter might be.

References

Abia, C., & Rebolo, R. 1989, preprint

Alcock, C., Fuller, G.M., & Mathews, G. J. 1987. *Astrophys. J.* **320**, 439..

Applegate, J. H., Hogan, C.J., & Scherrer, R. J. 1987, *Phys. Rev.* **D35**, 1151.

Arimoto, N. 1989, in *Evolutionary Phenomena in Galaxies*, J. Beckman & B. E. J. Pagel (eds), Cambridge Univ. Press, p. 341.

Arimoto, N., & Yoshii, Y. 1987. *Astron. Astrophys.* **173**, 23.

Armandroff, T., and Zinn, R. 1988. *Astron. J.* **96**, 92.

Audouze, J., Delbourgo-Salvador, P., & Salati, P. 1988, in *Dark Matter*, J. Audouze & J. T. T. Van (eds), Paris: Ed. Frontieres, p. 277.

Audouze, J., & Silk, J. 1989. *Astrophys. J. Lett.* **342**, L5.

Audouze, J., & Van. J. T. T. (eds) 1988 in *Dark Matter*, Paris: Ed. Frontieres.

Axon, D. J., Staveley-Smith, L.,Fosbury, R. A. E., Danziger, J., Boksenberg, A., & Davies, R. D. 1988. *Mon. Not. Roy. Astron. Soc.* **231**, 1077..

Bahcall, J. N. 1984. *Astrophys. J.* **287**, 926..

Beers, T. C. 1987, in *Nearly Normal Galaxies, From the Planck Time to the Present*, S. M. Faber (ed.)., New York: Springer, p. 41.

Beers, T. C., Preston, G. W., & Shectman, S. A. 1986. *Astron. J.* **90**, 2089.

Bergvall, N., & Jörsäter, S. 1988. *Nature* **331**, 589.

Boesgaard, A. M., & Steigman, G. 1985. *Ann. Rev. Astron. Astrophys.* **23**, 319.

Bothun, G. D., Impey, C. D., Malin, D. F., & Mould, J. R. 1987. *Astron. J.* **94**, 23.

Boyd, R. N., & Kajino, T. 1989. *Astrophys. J. Lett.* **336**, L55.

Brinks, E., and Klein, U. 1988. *Mon. Not. Roy. Astron. Soc.* **231**, 63P.

Cameron, A. G. W. 1982, *Astrophys. Space Sci.* **82** 123.

Cameron, A. G. W. 1989, in *Cosmic Abundances of Matter*, C. J. Waddington (ed.), New York: Amer. Inst. Phys., p. 349.

Carr, B. J. 1989, *Comments on Astrophysics* in press

Carr, B. J., & Ashman, K. M. 1988, in *Cooling Flows in Clusters and Galaxies*, A. C. Fabian (ed)., Kluwer, p.353.

Clarke, C. J. 1989. *Mon. Not. Roy. Astron. Soc.* **238**, 283.

Clayton, D. D. 1985, in *Nucleosynthesis: Challenges and New Developments*, Arnett, W. D., and Truran, J. W. (eds), Univ. of Chicago Press, p.65.

Clayton, D. D. 1988. *Mon. Not. Roy. Astron. Soc.* **234**, 1.

Delbourgo-Salvador, P., Gry, C., Malinie, G., and Audouze, J. 1985. *Astron. Astrophys.* **150**, 53.

Delbourgo-Salvador, P., Audouze, J., & Vidal-Madjar, A. 1987. *Astron. Astrophys.* **174**, 365..

Deliyannis, C. P.. Demarque, P., Kawaler, S. D., Krauss, L. M., & Romanelli, P. 1989, Phys. Rev. Lett., 62, 1583. Dimopoulos, S. Ezmailzadeh, R., Hall, L. J., & Starkman, G. J. 1988. *Astrophys. J.* **330**, 545.

Disney, M., Davies, J., & Phillips, S. 1989. *Mon. Not. Roy. Astron. Soc.* **239**, 939.

Djorgovski, S., & Davis, M. 1987. *Astrophys. J.* **313**, 59.

Edmunds, M. G., & Pagel, B. E. J., 1984. *Mon. Not. Roy. Astron. Soc.* **211**, 507.

Efstathiou, G., Ellis, R. S., & Peterson, B. A. 1988. *Mon. Not. Roy. Astron. Soc.* **232**, 431.

Eggen, O. J., Lynden-Bell, D., & Sandage, A. R. 1982. *Astrophys. J.* **136**, 748.

Faber, S. M., Dressler, A., Davies, R. L., Burstein, D., Lynden-Bell, D., Terlevich, R., & Wegner, G. 1987, in *Nearly Normal Galaxies*, 'From the Planck Time to the Present', S. M. Faber (ed.)., New York: Springer, p. 175.

Faber, S., & Gallagher, J. 1979. *Ann. Rev. Astron. Astrophys.* **17**, 135.

Fabian, A. C., Nulsen, P. E. J., & Canizares, C. R. 1982. *Mon. Not. Roy. Astron. Soc.* **201**, 933.

Fabian, A. C., Nulsen, P. E. J., & Canizares, C. R. 1984. *Nature* **310**, 733.

Fabricant, D., & Gorenstein, P. 1983. *Astrophys. J.* **267**, 535.

Fowler, W. A. 1990, in *Baryonic Dark Matter*, D. Lynden-Bell & G. Gilmore (eds), p 257

Garnett, D., & Shields, G. 1987. *Astrophys. J.* **317**, 82.

Gilmore, G., & Wyse, R. 1986. *Astron. J.* **90**, 2015.

Gilroy, K. K., Sneden, C., Pilachowski, C., & Cowan, J. J. 1988. *Astrophys. J.* **327**, 298.

Hartwick, F. D. A. 1976. *Astrophys. J.* **209**, 418.

Hegyi, D. J., & Olive, K. A. 1986. *Astrophys. J.* **303**, 56.

Hegyi, D. J., & Olive, K. A. 1989, preprint

Hughes, J. P. 1989. *Astrophys. J.* **337**, 21.

Impey, C., & Bothun, G. 1989. *Astrophys. J.* **341**, 89.

Kent, S. M., 1986. *Astron. J.* **91**, 1301.

Kent, S.M. 1987. *Astron. J.* **93**, 816.

Kormendy, J., & Knapp, G. R. (eds)., 1987, IAU Symp. No. 117: *Dark Matter in the Universe*, Reidel.

Kuijken, K., & Gilmore, G. 1989. *Mon. Not. Roy. Astron. Soc.* **239**, 605.

Kurki-Suonio, H., Matzner, R. A., Olive, K. A., & Schramm, D. N. 1989, preprint

Laird, J. B., Rupen, M. P., Carney, B. W., and Latham, D. W. 1988. *Astron. J.* **96**, 1908.

Lake, G., & Feinswog, L. 1989. *Astron. J.* **98**, 166.

Lake, G., & Skillman, E. 1989, preprint

Lambert, D. L. 1989, in *Cosmic Abundances of Matter*, C. J. Waddington (ed.)., New York: Amer. Inst. Phys., p. 168.

Larson, R. B. 1974. *Mon. Not. Roy. Astron. Soc.* **169**, 229.

Larson, R. B. 1976. *Mon. Not. Roy. Astron. Soc.* **176**, 31.

Last, J., Arnold, M., & Döhner, J. 1988, *Phys. Rev. Lett.* **60** 995.

Lequeux, J., Peimbert, M., Rayo, J. F., Serrano, A., & Torres-Peimbert, S. 1979. *Astron. Astrophys.* **80**, 155.

Lequeux, J., & Viallefond, F. 1980. *Astron. Astrophys.* **91**, 269.

Lynden-Bell, D. 1975, *Vistas in Astron.* **19** 299.

Malaney, R. A., and Butler, M. N. 1989, preprint

Malaney, R. A., & Fowler, W. A. 1988. *Astrophys. J.* **333**, 14.

Malaney, R., & Fowler, W. A. 1989. *Mon. Not. Roy. Astron. Soc.* **237**, 67.

Malaney, R. A., Mathews, G. J., & Dearborn, D. 1989, preprint

Mathews, G. J., Fuller, G. M., Alcock, C. R., & Kajino, T. 1988, in *Dark Matter*, J. Audouze & J. T. T. van (eds), Paris: Ed. Frontieres, p. 319.

Matteucci, F. 1988, in *Origin and Distribution of the Elements*, G. J. Mathews (ed.), Singapore: World Scientific, p. 186.

Matteucci, F., & Francois, P. 1989. *Mon. Not. Roy. Astron. Soc.* **239**, 885.

Matteucci, F., & Greggio, L. 1986. *Astron. Astrophys.* **154**, 279.

McCall, M. L. 1982, Thesis, University of Texas, Austin.

Morrison, H. L., Flynn, C., & Freeman, K. C. 1989, preprint

Mould, J. R. 1984. *Publ. Astron. Soc. Pacific* **96**, 773.

Mushotzky, R. 1989, in *Cosmic Abundances of Matter*, C. J. Waddington (eds), New York: Amer. Inst. Phys., p. 325.

Nissen, P. E., Edvardsson, B. & Gustafsson, B. 1985, in *Production and Distribution of the CNO Elements*, I. J. Danziger, F. Matteucci and K. Kjar (eds), ESO, Garching, p. 131.

Norris, J. 1987. *Astrophys. J. Lett.* **314**, L39.

Norris, J., Bessell, M. S., & Pickles, A. 1985. *Astrophys. J. Suppl.* **58**, 463.

Oort, J. H., & Plaut, L. 1975. *Astron. Astrophys.* **41**, 71.

Ostriker, J. B. 1987, in IAU Symp. No. 117: *Dark Matter in the Universe*, J. Kormendy & G. R. Knapp (eds), Reidel, p.85.

Pagel, B. E. J. 1981, in *The Structure and Evolution of Normal Galaxies*, S. M. Fall & D. Lynden-Bell (eds), Cambridge University Press, p. 211.

Pagel, B. E. J. 1986, in *Cosmogonical Processes*, W. D. Arnett, C. Hansen, J. Truran and S. Tsuruta (eds), Utrecht: VNU Science Press, p.66.

Pagel, B. E. J. 1989a, preprint

Pagel, B. E. J. 1989b in *Evolutionary Phenomena in Galaxies*, J. E. Beckman and B. E. J. Pagel (eds), Cambridge University Press, p.201.

Pagel, B. E. J., 1990, in *Astrophysical Ages and Dating Methods*, J. Audouze, M. Cass & E. Vangioni-Flam (eds), Paris: Ed. Frontieres, in press.

Pagel, B. E. J., & Edmunds, M. G. 1981. *Ann. Rev. Astron. Astrophys.* **19**, 77.

Pagel, B. E. J., & Simonson, E. A. 1989, preprint

Peimbert, M., and Serrano, A. 1982. *Mon. Not. Roy. Astron. Soc.* **198**, 563.

Pickles, A. J. 1985. *Astrophys. J.* **296**, 340.

Pryor, C., McClure, R. D., Fletcher, J. M., & Hesser, J. E. 1989a, preprint

Pryor, C., McClure, R. D., Fletcher, J. M. & Hesser, J. E. 1989b, preprint

Rebolo, R., Molaro, P., & Beckman, J. E. 1988. *Astron. Astrophys.* **192**, 192.

Reeves, H. 1989, *Physics Reports* in press.

Rubin, V. C. 1987, in IAU Symp. No. 117: *Dark Matter in the Universe*, J. Kormendy & G. R. Knapp (eds), Reidel, p. 51.

Ryu, D., Olive, K. A., & Silk, J. 1989, preprint

Saha, A. 1985. *Astrophys. J.* **289**, 310.

Sancisi, R., & van Albada, T. S. 1987, in IAU Symp. No. 117: *Dark Matter in the Universe*, J. Kormendy & G. R. Knapp (eds), Reidel, p. 67.

Sandage, A. R., Binggeli, B., & Tammann, G. 1985. *Astron. J.* **90**, 1759.

Sarazin, C. L., & O'Connell, R. W. 1983. *Astrophys. J.* **268**, 552.

Scalo, J. M. 1986. *Fundamentals of Cosmic Physics* **11**, 1.

Searle, L., & Sargent, W. L. W. 1972. *Astrophys. J.* **173**, 25.

Searle, L., & Zinn, R. 1978. *Astrophys. J.* **225**, 357.

Skillman, E., Kennicutt, R. C., & Hodge, P.W. 1989, preprint

Sommer-Larsen, J., & Yoshii, Y. 1989. *Mon. Not. Roy. Astron. Soc.* **238**, 133.

Spite, F. & Spite, M. 1982. *Nature* **297**, 483.

Steigman, G. 1989, Proc. II International Symposium for the 4th *Family of Quarks and Leptons.* in press

Sutherland, R. S., & Bessell, M. S. 1989, in press.

Terasawa, N., & Sato, K. 1989, *Phys. Rev. D.***39** 2893.

Tinsley, B. M. 1979. *Astrophys. J.* **229**, 1046.

Trimble, V. 1987. *Ann. Rev. Astron. Astrophys.* **25**, 425.

Trimble, V. 1988, *Contemp. Phys.* **29** 373.

Truran, J. W. 1981. *Astron. Astrophys.* **97**, 391.

Vangioni-Flam, E., & Audouze, J. 1988. *Astron. Astrophys.* **193**, 81.

Wheeler, J.C., Sneden, C., & Truran, J. W. 1989. *Ann. Rev. Astron. Astrophys.* **27**, 279.

Wolfe, A. M. 1989, in *The Epoch of Galaxy Formation*, C. S. Frenk *et al.* (eds), Kluwer, p. 101.

Yang, J., Turner, M. S., Steigman, G., & Olive, K. A. 1984. *Astrophys. J.* **281**, 493.

Yoshii, Y., & Arimoto, N., 1987. *Astron. Astrophys.* **188**, 13.

Zinn, R. 1985. *Astrophys. J.* **293**, 424.

NUCLEAR REACTIONS IN INHOMOGENEOUS COSMOLOGIES

WILLIAM A. FOWLER
W. K. Kellogg Radiation Laboratory
California Institute of Technology
Pasadena, CA   91106 U. S. A.

ABSTRACT. This paper presents additions to the standard nuclear reaction network which are necessary for investigations of primordial nucleosynthesis in inhomogeneous cosmological models. Also presented are the reaction rates associated with these new reactions. the use of these rates may help remove the ambiguity of whether results reported by different groups are, to some extent, an artifact of the different reaction networks employed. It is stressed that many of these reactions play important roles in the nucleosynthesis, but their rates are poorly determined and thus deserve further experimental study. Using the reaction network of Malaney and Fowler (1987, 1988) supplemented by the reactions presented here, it is shown that the production of ^{9}Be in inhomogeneous models is strongly dependent on the density contrast R, and on the high-density volume fraction f_v. Consequently, observations of ^{9}Be in stars could lead to vital information on the value of these crucial, but poorly determined, parameters. It is concluded that ^{9}Be may be used as a possible indicator of inhomogeneity at the time of nucleosynthesis only for a certain range of $f_v R$. For an $\Omega_b = 1$ universe the appropriate range would be $f_v R \gtrsim 10$. Ω_b is the ratio of the universal baryon (b) density to the "critical" density between an open and closed universe which can be calculated from Hubble's constant (H_0). For $H_0 = 58$ km s^{-1} Mpc^{-1} the critical density now is 6.3×10^{-30} g cm^{-3} (Fowler 1987).

It is currently believed that our observable universe originated about 10^{10} years ago as an expanding bubble in an otherwise steady state universe consisting originally of a false vacuum with $\rho = -p/c^2 \approx 5 \times 10^{93}$ g cm^{-3} at the Planck time. Since $\rho c^2 + P = 0$ the total energy of the false vacuum is zero. The false vacuum occurs in Einstein's equations as his cosmological constant. The bubble originated from a phase transition from the false vacuum to radiation and matter at high excitation during the early stage of inflation when $\rho \approx 10^{75}$ g cm^{-3}. The first matter to appear was 10^{14} GeV/c^2 particles termed $x\bar{x}$ which eventually decayed to gluons plus an excess of quarks over antiquarks (QCD-phase). Near the end of inflation the QCD-phase transformed into hadrons (pions plus nucleons) and leptons (electrons plus neutrinos). Throughout all of this the total energy of the universe remained zero because of gravitational attraction with negative potential energy.

257

D. Lynden-Bell and G. Gilmore (eds.), Baryonic Dark Matter, 257–264.
© 1990 *Kluwer Academic Publishers.*

Following the suggestion that density perturbations (Witten 1984) and chemical separation (Applegate and Hogan 1985) could occur after the universe underwent the phase transition from the QCD phase to the hadron phase, a series of calculations were carried out (Applegate, Hogan, and Scherrer 1987, 1988; Alcock, Fuller, and Mathews 1987; Malaney and Fowler 1987; Fuller, Mathews, and Alcock 1988a) which investigated the possibility of reconciling the observed abundances of the primordial isotopes with an $\Omega_b = 1$ universe. In these calculations the universe was separated into two decoupled regions, one of which was neutron-poor and the other neutron-rich. The results from the different groups were in good agreement in that it was found that the observed abundances of ^{2}H, ^{3}He, and ^{4}He could be reconciled with an $\Omega_b = 1$ universe, but that the production of ^{7}Li was apparently at least a factor ~ 5 larger than the observed value.

Following these early calculations, Malaney and Fowler (1988) showed that the diffusion of neutrons from one region to the other *during* the nucleosynthesis epoch played a much more important role than hitherto believed. Using a simplified diffusion model they showed that if neutrons diffused back into the proton-rich regions at early times, then the nucleosynthesis could be drastically altered (principally the n/p ratio is significantly increased in the proton-rich region thereby leading to too much ^{4}He production). However, they also pointed out that if the diffusion of neutrons back into the proton-rich region occurred only at late times, then the principal effect on the nucleosynthesis becomes the destruction of the ^{7}Be (which later decays to ^{7}Li via e^- capture) in the proton-rich regions via the ^{7}Be(n, p)^{7}Li(p, α)^{4}He sequence of reactions. The reason why the timescale for the onset of the back diffusion plays such an important role is simply due to the fact that the production of ^{4}He occurs at much earlier times relative to the production of ^{7}Be (see Figure 1 of Malaney and Fowler 1988). Although it must be stressed that the effect described by Malaney and Fowler (1988) is only found in a very limited region of a large parameter space, it did leave open the possibility that an $\Omega_b = 1$ universe could be reconciled with *all* of the observed primordial abundances.

There have subsequently appeared a large number of investigations which utilize more sophisticated treatments of neutron diffusion during the nucleosynthesis epoch. Now, however, the conclusions drawn by the different groups appear to be contradictory. While some groups confirm that in some regions of the parameter space a consistent fit to the observed primordial abundances with $\Omega_b = 1$ can indeed by found (Mathews *et al.* 1988, 1989; Teresawa and Sato 1989a,b), other groups conclude that the observed abundances lead to the constraint $\Omega_b \lesssim 0.3$ (Kurki-Sounio *et al.* 1988; Kurki-Sounio and Matzner 1989; Kurki-Sounio *et al.* 1989). We believe that the reason for these apparently contradictory results is largely due to the more limited parameter space allowed for by the latter investigators (the degree of sophistication utilized in modeling the neutron diffusion should play a more minor role). Teresawa and Sato (1989a,b) have investigated a large parameter space arising not only from uncertainties in the physics of the QCD phase transition, but also from uncertainties in the primordial ^{4}He value, the Hubble constant, and the neutron half-life.

However, since the nuclear reaction networks required for investigation of the primordial nucleosynthesis arising in inhomogeneous cosmologies require the addition to the standard networks of several new reactions (particularly at the crucial ^{6}Li–^{9}Be region), there has arisen some uncertainty as to exactly what reactions should be included and what are their reaction rates. This has led to some degree of confusion, and to the suspicion that the results reported by the different groups may, to some extent, be an artifact of the different reaction networks and rates employed. This problem is compounded by the fact that the reaction rates for many of these new reactions are either poorly determined, or have not been given in any previous compilation. In this paper an attempt is made to remedy this situation here by listing all the new nuclear reactions that should be included in non-standard big bang calculations, and by estimating their reaction rates.

Table 1 lists the new reactions that should be included in the inhomogeneous reaction networks below ^{16}O along with estimates of their rates. Those reactions marked with an asterisk are standard network reactions, but whose rates have been updated since their last publication. For a complete list of reactions included in the standard big bang network see Wagoner (1973). The importance of the $^7Li(^3H, n)^9Be$ reaction has recently been pointed out by Boyd and Kajino (1989).

Many of the reaction rates listed in Table 1 are poorly determined, and as such deserve experimental investigation. This is particularly so since many of them play crucial roles in determining key isotopic abundances. For example, it is believed that inhomogeneous cosmologies lead to the production of significant $A \geq 12$ isotopic production (Applegate, Hogan, and Scherrer 1987), and possibly even significant r-process production (Applegate, Hogan, and Scherrer 1987). Such r-process production may account for some s process observations in metal-poor stars (Malaney and Fowler 1988). The reaction flow leading to the production of heavy elements is mainly through

$$^4He(^3H, \gamma)^7Li(n, \gamma)^8Li(\alpha, n)^{11}B(n, \gamma)^{12}B(\beta)^{12}C(n, \gamma)^{13}C(n, \gamma)^{14}C$$

with a smaller flow through $^7Li(\alpha, n)^{11}B$ (Malaney and Fowler, 1987; Applegate, Hogan, and Scherrer 1988). Clearly, improvement on the determination of these reaction rates would be very valuable. The rate of the $^7Li(^3H, n)^9Be$ reaction is also poorly determined. As will now be shown, the ^9Be produced by this reaction should allow for a very important constraint on the QCD phase transition.

Using the above reaction network, an investigation has been made of the recent suggestion by Boyd and Kajino (1989) that the observed abundance of ^9Be could be used as a discriminatory test between homogeneous (standard) and inhomogeneous models. Boyd and Kajino predict a primordial $^9Be/^1H$ number density ratio in an inhomogeneous universe which is three orders of magnitude larger than the same ratio produced in a homogeneous universe. As such, they argue that the observations of the Be/H ratio in Pop II stars should provide a test of primordial nucleosynthesis theories which is independent of the questions associated with the ^7Li abundance. Here it is found that this is not *always* the case since the abundance of ^9Be, predicted by the inhomogeneous models, is very much dependent on the adopted density contrast parameter R, and on the high-density volume fraction f_v. It is shown that for a large range of these parameters the $^9B/^1H$ ratio predicted by the inhomogeneous models is in fact below that predicted by the standard model.

The parameter f_v refers to the volume fraction of the universe which is in the high-density phase at the time the quark-gluon plasma decouples from the universal expansion. Prior to the onset of nucleosynthesis, it is assumed that neutrons within the high-density regions can diffuse out and uniformly fill all space. This results in the formation of neutron-poor high-density regions surrounded by neutron-rich low-density regions. The density contrast parameter R, refers not to the density ratio of these regions following the neutron diffusion, but rather to their density ratio prior to any neutron diffusion. As most of the ^9Be production will take place in the low-density neutron-rich regions, consideration will not be given here to the process of *late-time* neutron diffusion back into the high-density neutron-poor regions (see Malaney and Fowler 1988).

Table 1

$${}^2\text{H}({}^2\text{H}, \gamma){}^4\text{He} = 4.84 \times 10\, T_9^{-2/3}\, \exp(-4.258/T_9^{1/3})$$

$${}^2\text{H}({}^2\text{H}, e^+e^-){}^4\text{He} = 7.09 \times 10^{-1}\, T_9^{-2/3}\, \exp(-4.258/T_9^{1/3})$$

$${}^{\dagger 6}\text{Li}({}^2\text{H}, n){}^7\text{Be} = 1.48 \times 10^{12}\, T_9^{-2/3}\, \exp(-10.135/T_9^{1/3})$$

$${}^{\dagger 6}\text{Li}({}^2\text{H}, p){}^7\text{Li} = 1.48 \times 10^{12}\, T_9^{-2/3}\, \exp(-10.135/T_9^{1/3})$$

$${}^6\text{Li}(n, \gamma){}^7\text{Li}^* = 5.10 \times 10^3$$

$${}^7\text{Li}({}^2\text{H}, p){}^8\text{Li} = 8.31 \times 10^8\, T_9^{-3/2}\, \exp(-6.998/T_9) \quad (Q < 0)$$

$${}^7\text{Li}(n, \gamma){}^8\text{Li}^* = 6.015 \times 10^3 + 1.141 \times 10^4\, T_9^{-3/2}\, \exp(-2.576/T_9)$$

$${}^7\text{Li}({}^3\text{H}, n){}^9\text{Be} = 1.46 \times 10^{11}\, T_9^{-2/3}\, \exp(-11.333/T_9^{1/3})$$

$${}^7\text{Li}({}^3\text{H}, 2n)2\,{}^4\text{He} = 8.81 \times 10^{11}\, T_9^{-2/3}\, \exp(-11.333/T_9^{1/3})$$

$${}^7\text{Li}({}^3\text{He}, np)2\,{}^4\text{He} = 1.11 \times 10^{13}\, T_9^{-2/3}\, \exp(-17.989/T_9^{1/3})$$

$${}^{\dagger 8}\text{Li}({}^2\text{H}, n){}^9\text{Be} = 2.89 \times 10^{11}\, T_9^{-2/3}\, \exp(-10.357/T_9^{1/3})$$

$${}^8\text{Li}(n, \gamma){}^9\text{Li} = 4.294 \times 10^4 + 6.047 \times 10^4\, T_9^{-3/2}\, \exp(-2.866/T_9)$$

$${}^8\text{Li}(\alpha, n){}^{11}\text{B}^* = 8.62 \times 10^{13}\, T_{9A}^{5/6}\, T_9^{-3/2}\, \exp(-19.461/T_{9A}^{1/3}), \text{ where } T_{9A} = \frac{T_9}{1 + T_9/15.1}$$

$${}^7\text{Be}({}^3\text{H}, np)2\,{}^4\text{He} = 2.91 \times 10^{12}\, T_9^{-2/3}\, \exp(-13.729/T_9^{1/3})$$

$${}^7\text{Be}({}^3\text{He}, 2p)2\,{}^4\text{He} = 6.11 \times 10^{13}\, T_9^{-2/3}\, \exp(-21.793/T_9^{1/3})$$

$${}^{11}\text{B}(n, \gamma){}^{12}\text{B}^* = 7.29 \times 10^2 + 2.40 \times 10^3\, T_9^{-3/2}\, \exp(-0.223/T_9)$$

$${}^{14}\text{C}(\alpha, \gamma){}^{18}\text{O} = 9.29 \times 10^{-8}\, T_9^{-3/2}\, \exp(-2.048/T_9) + 2.77 \times 10^3\, T_9^{-4/5}\, \exp(-9.876/T_9)$$

* standard reaction with modified rate

$\dagger$ S_0 determined from measurements on similar reactions. CFNASV-factors & τ-coefficients used for reaction indicated.

Table 2 contains the ratio of the ^{9}Be mass fraction produced in the inhomogeneous models to the mass fraction produced in the standard model, as a function of R for three different values of f_v. For the inhomogeneous calculations we have adopted $\Omega_b = 1$. For the standard model, however, we adopted $\Omega_b = 1/40$, which results in a ^{9}Be mass fraction of 10^{-16} (the ^{9}Be mass fraction predicted by standard model decreases rapidly with increasing Ω_b). A Hubble constant of 50 km s^{-1} Mpc^{-1} was assumed.

Consider first the calculation corresponding to $f_v = 0.1$ in Table 2. The amount of ^{9}Be produced in the inhomogeneous models for $f_v = 0.1$ becomes lower than the predicted standard model abundance for $R \lesssim 100$. The reason for the abrupt change in the ^{9}Be abundance at $R \sim 100$ lies in the dependence on R of the proton abundance in the neutron-rich regions of the universe. For example, if $R = 50$, then the n/p ratio in the neutron-rich region at the time of ^{4}He formation, is less than one. As such, not all available protons are consumed by the neutrons, and the substantial number left over can efficiently destroy the ^{9}Be via the ^{9}Be(p,d) 2^4He and ^{9}Be(p, α)^{6}Li reactions. However, if $R \gtrsim 100$ then the n/p ratio at ^{4}He formation is greater than 1, and almost all of the available protons are consumed by the neutrons. As a consequence of this lower proton abundance, the proton-induced destruction of ^{9}Be is greatly reduced in this case. This sensitivity of the proton abundance to initial conditions can readily be seen from Figure 1 of Malaney and Fowler (1987).

Also included in the Table 2 is the ^{9}Be production as a function of R for $f_v = 0.5$ and $f_v = 0.02$. It can be seen that these results are qualitatively similar to the $f_v = 0.1$ results, except that significant ^{9}Be production commences for different values of R in the three cases. This is largely a consequence of the dependence of X_n, the initial neutron-to-proton ratio at the onset of nucleosynthesis in the low density neutron-poor region, on the values of f_v and R. As well now be shown, smaller values of f_v require larger values of R in order for the critical value of X_n (i.e., the value of X_n at which significant ^{9}Be production commences) to be attained.

Consider the equations, in an $\Omega_b = 1$ universe, relating X_n with f_v and R (see Malaney and Fowler 1988), viz.,

$$1 = f_v R \Omega_n + (1 - f_v)\Omega_n \,, \qquad X_n = \frac{X_n^0}{X_n^0 + (1 - X_n^0)\Omega_n} \,. \tag{1}$$

Here, $X_n^0(\sim 0.15)$ and Ω_n are the n/p ratio and the nucleon density (in terms of the critical density), respectively, in the low-density neutron-rich region *prior* to the onset of the neutron diffusion. In the limit of $f_v R \gg 1 - f_v$ equations (1) can be reduced to

$$f_v R \approx \frac{X_n(1 - X_n^0)}{X_n^0(1 - X_n)} \,. \tag{2}$$

The critical value of X_n at which the proton destruction of ^{9}Be is significantly reduced is found to occur at $X_n \sim 0.65$. According to equation (2) then, the commencement of significant ^{9}Be production should occur at $f_v R \sim 10$. This is in good agreement with the detailed results shown in Table 2. For $\Omega_b \neq 1$, the corresponding relation would be $f_v R \sim 10 \, \Omega_b$.

We caution here that as we approach large values of R in Table 2, the simple neutron diffusion model we have adopted begins to break down. This is because for $R \gtrsim 10^4$ the

Table 2

$f_v = 0.02$		$f_v = 0.1$		$f_v = 0.5$	
R	^{9}Be $\dfrac{\text{INHOM}}{\text{HOMOG}}$	R	^{9}Be $\dfrac{\text{INHOM}}{\text{HOMOG}}$	R	^{9}Be $\dfrac{\text{INHOM}}{\text{HOMOG}}$
200	6.8E − 3	40	5.5E − 3	10	7.2E − 3
250	2.8E − 2	50	2.1E − 2	13	4.8E − 2
300	8.9E − 2	70	3.2E − 1	15	2.1E − 1
400	4.2	80	2.7	18	1.3E + 1
450	1.7E + 2	90	3.5E + 2	20	6.6E + 2
500	4.9E + 3	100	3.8E + 3	22	5.2E + 3
550	1.2E + 4	105	8.3E + 3	25	6.5E + 3
600	1.3E + 4	120	1.2E + 4	30	5.9E + 3
700	1.2E + 4	150	1.1E + 4	50	5.0E + 3
1000	1.0E + 4	200	9.5E + 3	80	4.5E + 3
		300	8.5E + 3	100	4.3E + 3
		1000	7.0E + 3	200	4.1E + 3
		5E + 4	6.8E + 3	500	3.9E + 3
				1000	3.8E + 3

density in the high-density region becomes large enough that neutron diffusion from this region is seriously inhibited. In this case the neutron-proton ratio in both the low- and high-density regions is not significantly altered from that determined by the neutron-proton weak interaction rates. The initial neutron-to-proton ratio in each region is then essentially the same in both regions (~ 0.15) at the onset of nucleosynthesis, and in such circumstances no significant ^{9}Be production takes place.

The main conclusion to be drawn from Table 2 then is the following. If observations of ^{9}Be do not show a high abundance relative to the standard model prediction ($\lesssim 10^{-16}$, by mass), this does not rule out inhomogeneity at the epoch of nucleosynthesis, but rather rules out only a certain region of the available parameter space. As can be seen from Table 2 the observation of ^{9}Be only becomes a discriminatory test between the different cosmological models if $f_v R \gtrsim 10$ (for $\Omega_b = 1$).

Clearly, observations of ^{9}Be could be used to provide constraints on the values of f_v and R. The value of R is one of the most important, and controversial, aspects of the inhomogeneous cosmologies. Some groups have argued that $R \lesssim 100$ (Kurki-Sounio et al. 1989), while others have argued that R could be as high as $\sim 10^6$ (Fuller, Mathews, and Alcock 1989b; Mathews et al. 1989). Similarly, f_v is poorly determined and is usually taken as a free parameter in the calculations. Observational insight into the value of R and f_v would therefore be of great value.

There is a general consensus from the detailed diffusion models, that the density contrast parameter R must be relatively high $R \gtrsim 300$; e.g., Teresawa and Sato 1989a) if an $\Omega_b = 1$ universe is to be made consistent with all of the observed primordial abundances. If it can be independently determined, from constraints on other primordial isotopes such as ^{7}Li (e.g., Malaney and Fowler 1988) that $f_v \gtrsim 0.1$, and if observations of metal-poor stars show the number-density ratio ^{9}Be/^{1}H is significantly below 10^{-13}, this would imply $R \lesssim 100$ and therefore rule out an $\Omega_b = 1$ universe. Alternatively, if in fact the ^{9}Be/^{1}H ratio is found to be at the 10^{-13} level in very metal-poor stars, then this would be dramatic confirmation of inhomogeneity at the epoch of nucleosynthesis, and would also leave open the possibility of an $\Omega_b = 1$ universe. It is exciting to note that present observations of beryllium in metal-poor stars are very close to the required sensitivity (Rebolo et al. 1988).

In summary, this paper lists the new nuclear reactions, and their associated rates, which should be included in inhomogeneous nucleosynthesis calculations. Using the full reaction network which has been presented, the significance of ^{9}Be production in inhomogeneous cosmologies has been investigated and it has been concluded that observations of this isotope in stars can only be used as a discriminant between inhomogeneous and homogeneous cosmologies if $f_v R \gtrsim 10$ (assuming $\Omega_b = 1$). Inhomogeneous cosmological models cover a very large parameter space, and no investigation has been made of the dependence of ^{9}Be production on Ω_b, the separation distance between high-density sites, or the geometry of the density perturbations. However, it is already clear that a low ^{9}Be abundance can in fact be very much consistent with a large range of inhomogeneous models.

The material presented in this paper was prepared jointly by R. A. Malaney and William A. Fowler. The nuclear reaction rates in Table 1 are mainly the responsibility of William A. Fowler. Fowler is grateful to Malaney for permission to use the somewhat modified textual material. Fowler's research is supported in part by NSF Grant PHY88-17296.

264

References

Alcock, C. R., Fuller, G. M., and Mathews, G. J. (1987), Ap. J. **320**, 439.

Applegate, J. H. and Hogan, C. J. (1985), Phys. Rev. D **31**, 3037.

Applegate, J. H., Hogan, C. F., and Scherrer, R. J. (1987), Phys. Rev. D **35**, 1151.

Applegate, J. H., Hogan, C. F., and Scherrer, R. J. (1988), Ap. J. **329**, 572.

Boyd, R. N. and Kajino, T. (1989), Ap. J. Lett. **336**, L55.

Caughlan, G. R. and Fowler, W. A. (1988), At. Data Nucl. Data Tables, **40**, 283.

Fowler, W. A. (1987), Q. Jl R. astr. Soc. **28**, 87.

Fuller, G. M., Mathews, G. J., and Alcock, C. R. (1988a), Phys. Rev. D **37**, 1380.

Fuller, G. M., Mathews, G. J., and Alcock, C. R. (1988b), in J. Audouze and J. Tran Thanh Van (eds.) *Dark Matter, Proc. of the XXIIIrd Recontre de Moriond*, Editions Frontiers, France, p. 303.

Kurki-Sounio, H., Matzner, R. A., Centrella, J. M., Rothman, T., and Wilson, J. R. (1988), Phys. Rev. D, **38**, 1091.

Kurki-Sounio, H. and Matzner, R.A. (1989), Phys. Rev. D **39**, 1046.

Kurki-Sounio, H. and Matzner, R.A., Olive, K. A., and Schramm, D. N. (1989), preprint UMN-TH-713/88).

Malaney, R. A. and Fowler, W. A. (1987), in G. J. Mathews (ed.) *The Origin and Distribution of the Elements*, World Scientific, Singapore, p. 76.

Malaney, R. A. and Fowler, W. A. (1988), Ap. J. **333**, 14.

Mathews, G. J., Fuller, G. M., Alcock, C. R., and Kajino, T. (1988), in J. Audouze and J. Tran Thanh Van (eds.) *Dark Matter, Proc. of the XXIIIrd Recontre de Moriond*, Editions Frontiers, France, p. 319.

Mathews, G. J., Meyer, B., Alcock, C. R., and Fuller, G. M. (1989), in preparation.

Rebolo, R., Molaro, P., Abia, C., and Beckman, J. E. (1988), Astr. Ap. **193**, 193.

Teresawa, N. and Sato, K. (1989a), Univ. of Tokyo preprint UTAP79/1988.

Teresawa, N. and Sato, K. (1989b), Univ. of Tokyo preprint UTAP83/1988.

Wagoner, R. V. (1973), Ap. J. **179**, 343.

Walker, T. P., Mathews, G. J., and Viola, V. E. (1985), Ap. J. **229**, 745.

Witten, E. (1984), Phys. Rev. D **30**, 272.

CONSTRAINTS ON BARYON–DOMINATED COSMOLOGICAL MODELS FROM LIGHT ELEMENT ABUNDANCES AND CMB FLUCTUATIONS

JOHN E. BECKMAN
Instituto de Astrofísica de Canarias
38200 La Laguna
Tenerife
Spain

ABSTRACT. This article is in two distinct sections. In the first, the evidence leading towards a value for the primordial abundance of ^{7}Li is discussed. The problem of interpreting the observed abundances in Population II subdwarfs of extremely low metallicity, [Fe/H]$\leq$ -1.4, is analyzed. The dependence of [^{7}Li/H] on [Fe/H] lends support to a primordial value for logN(^{7}Li) close to 2.2, on the conventional scale where logN(H)$\equiv$12. In the second section recent results of the Tenerife cosmic microwave background fluctuation experiment are presented. Detection of sky fluctuations at levels of order $\Delta T/T=4 \times 10^{-5}$ on angular scales of a few degrees are interpreted in terms of baryon–dominated cosmologies. The arcminute scale upper limit to $\Delta T/T$ of Readhead *et al.* (1989) permits two areas in the parameter space Ω_b–n, *i.e.* baryon density against slope of the fluctuation power spectrum. One corresponds to Ω_b $\leq$0.08, the other to Ω_b >0.8. The former would represent standard Big Bang cosmologies, the latter models with inhomogeneities at the quark–hadron phase transition. If the Tenerife fluctuations are primordial, sections of a well–defined locus in Ω_b–n space define the permitted cosmologies; if they are local, the locus restricts them, but as yet neither high Ω_b nor low Ω_b models can be excluded.

1. Introduction

As an observationalist, I want to give an opportunity for theorists to understand the complexities of two very different types of measurements which are freely, even blithely, quoted and used to constrain cosmological models. The first is the determination of the abundance of the light element ^{7}Li, the second is the measurement of cosmic background temperature fluctuations, each selected because my degree of personal involvement has been sufficient to make me painfully aware of the agonies and the pitfalls which lie behind the bare published numbers. There will be no need in the present context, to give a complete account of the underlying theoretical framework which these observations can be used to test. I will instead abstract those results of the appropriate chains of theoretical inference needed to confront the observations. In so doing, I hope that the theorists will forgive me where I have ignored their agonies and pitfalls. Since this paper is given in a school on Baryonic Dark Matter, I have confined my attention to baryonic cosmological models, with no more than passing reference to models dominated by non–baryonic matter. This is not to say that ideologically I have anything against non–baryonic matter, or, a priori, against any particular value of Ω for the universe. It does seem to me, however, that observers and experimentalists are steadily closing down the options on ranges of cosmologies and that, theoretical ingenuity notwithstanding, it makes realistic sense to set our caps at obtaining a reliable value for Ω.

D. Lynden-Bell and G. Gilmore (eds.), Baryonic Dark Matter, 265–278.
© *1990 Kluwer Academic Publishers.*

2. The ^{7}Li abundance

2.1. ^{7}LI AS A PRIMORDIAL DENSITY INDICATOR

Since the measurements by Spite and Spite (1982) of the abundances of ^{7}Li in Population II stars, with Iron metallicities [Fe/H] $\leq$-1, it has been possible to take ^{7}Li seriously as a constraint on cosmologies. They found three key results:

(i) ^{7}Li is present in the atmospheres of Population II subdwarfs even though their ages, (they are at least as old as the galactic disk) might well have led one to predict that this fragile nuclide would have been depleted to unmeasurable levels.

(ii) The abundance of ^{7}Li varies little with surface temperature in these objects, for 5800 K $\leq$ Teff $\leq$ 6300 K.

(iii) The value for the abundance, [^{7}Li/H] = 2.1, on the scale of logH = 12, i.e. ^{7}Li/H = 1.25x10^{-10}, lies in the range predicted by standard Big Bang nucleosynthesis, originally in Wagoner, Fowler and Hoyle (1967) and subsequently refined by these and others. This abundance lies close to the minimum in the production curve of ^{7}Li against the photon to baryon ratio η, and not only marked a new triumph for the standard Big Bang – after all the ^{4}He to ^{7}Li ratio is $\sim$2x10^9 and a theory giving even ball–park predictions of this range must be taken seriously – it also restricted η rather strongly, and broadly confirmed the ^{4}He–based range of values for the baryonic density Ω_b of the universe, i.e. 0.01 $\leq \Omega_b \leq$ 0.1.

The problem with ^{7}Li is the same generic problem as that with the other Big Bang nuclides, D, ^{3}He and ^{4}He. Can we account for the post–fireball changes in its abundance to pin down a reliable and acceptable primordial value? ^{4}He has the advantages that its abundance is large, and that it is highly stable, so not fragile in stars. All the accounting is on the black side of the ledger, and the problem is to extrapolate correctly the increase in abundance back to the lowest metallicity in order to find its primordial value. D is more difficult; it is generally believed that no D has been produced since the Big Bang; the accounting for D is all in the red. It is a question of trying to estimate the astration, and also chemical fractionation which contribute to determine its measurable abundances in different contemporary astrophysical contexts. It is considered more reliable to use the sum ^{3}He + D as a possible parameter fit to Big Bang nucleosynthesis, since at least some D is astrated to ^{3}He.

In the case of ^{7}Li there has been both production and destruction during the lifetime of the galaxy. Extrapolation to low metallicity to obtain [^{7}Li/H]$_p$, the primordial abundance, must somehow take both into account. This clearly daunting task led many to discount ^{7}Li as a reliable η and Ω_b measurer, in spite of the attractive minimum in the theoretical primordial production curve.

2.2. THE GALACTIC EVOLUTION OF ^{7}LI

If we measure ^{7}Li in young late F and early G main sequence stars of the local disc population, we find that [^{7}Li/H] is typically $\geq$ 3, i.e. ^{7}Li/H is a few times 10^{-9}. If we go back in age, using open clusters with well–determined ages, (Garcia Lopez et al. (1989)) we find values which differ little from this, with some tendency to decrease, especially at low surface temperatures. This apparent constancy has lead a school of observers, notably led by Boesgaard (1985), and including Hobbs and Pilachowski (1988), to postulate that the history of ^{7}Li in the galaxy could have started with a primordial value a little larger than 10^{-9}. According to this scenario, the value of 10^{-10} seen in the population II subdwarfs is the result of more than ten billion years on the main sequence, during which time considerable ^{7}Li depletion has occurred. Clearly 10^{-9} or 10^{-10} for ^{7}Li/H makes a decisive difference for any model cosmology. In Fig 1 we show a version of the abundance vs. η or Ω_b diagram for the standard model (Wagoner et al., 1967; Yang et al. , 1984). In this, we see that 10^{-9} for ^{7}Li is at variance with the range determined by ^{4}He, and supported by D + ^{3}He. If 10^{-9}

were indeed the primordial value, SBBN would be excluded. Some non–standard model, possibly with inhomogeneities, would be required.

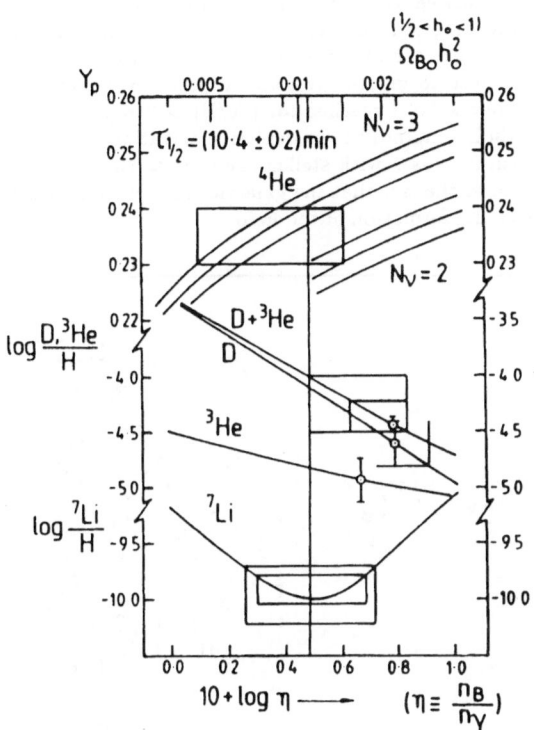

Figure 1. After Yang *et al.* (1984), plot of light element abundances against baryon/photon ratio, η, or baryon density Ω_b in units of the closure density. h_o is the Hubble parameter expressed as $H_o/100$ km s^{-1}Mpc^{-1}; $\tau_{\frac{1}{2}}$ is the neutron half–life, and N_ν the number of neutrino flavours. Current observational limits are within the error boxes shown.

Are observationalists in a position to distinguish between the two ^{7}Li cases? The frank response is no. However we are duty bound to weigh the detailed evidence and offer the best possible current answer. My answer will be that values in the 10^{-10} range are more probable than values in the 10^{-9} range. I will first offer the evidence in favour of this assertion, then bring up the counter evidence, and finally show why I find the counter–evidence less than convincing. In Fig 2 taken from Rebolo, Molaro and Beckman (1988), we see a thorough attempt to tackle the extrapolation of ^{7}Li measurements to low metallicities. You can see that for [Fe/H] > -1.4 the ^{7}Li abundances are spread over a wide range which increases with increasing metallicity, through the solar value ([Fe/H] = 0) to higher values. On the contrary below [Fe/H] = -1.4, the ^{7}Li abundances show remarkably little spread, over two decades of [Fe/H]; the lowest metallicity star measured so far, G64-12, with Fe/H -3.5, (Rebolo, Beckman and Molaro, 1987) shares this common ^{7}Li abundance of $\sim 1.5 \times 10^{-10}$. Although the range of surface temperatures for the stars with Fe/H ≤ -1.4 is restricted because stars with Teff > 6400 K have evolved away from the main sequence and stars with Teff $\leq$ 5400 K have depleted their ^{7}Li due to their deep convective zones (an effect seen in all MS stars with Teff in this range) the constancy of the ^{7}Li/H is still remarkable. It is hard to believe that depletion processes can have operated to give such a clean uniformity and there is theoretical backing for this (Delyannis *et al.*, 1989). A highly plausible scenario for Fig 2 is that 1.5×10^{-10} represents essentially undepleted primordial ^{7}Li, and the diagram for [Fe/H] > -1.4 shows the counterplay of galactic production and astration. In this picture the upper envelope would show the production curve, and the remaining stars have been subject to greater or lesser degrees of depletion. The alternative scenario says that the value of a few x 10^{-9} reached for [Fe/H] > 0 represents a true primordial, undepleted, value and that the whole diagram shows depletive astration. The upper envelope is a locus of minimum depletion, which increases with stellar age (i.e. as the metallicity falls) and the plateau below [Fe/H] $= -1.4$ is due to the fact that the galactic chronometer earlier than this covers a very short period indeed, that of the formation of the halo.

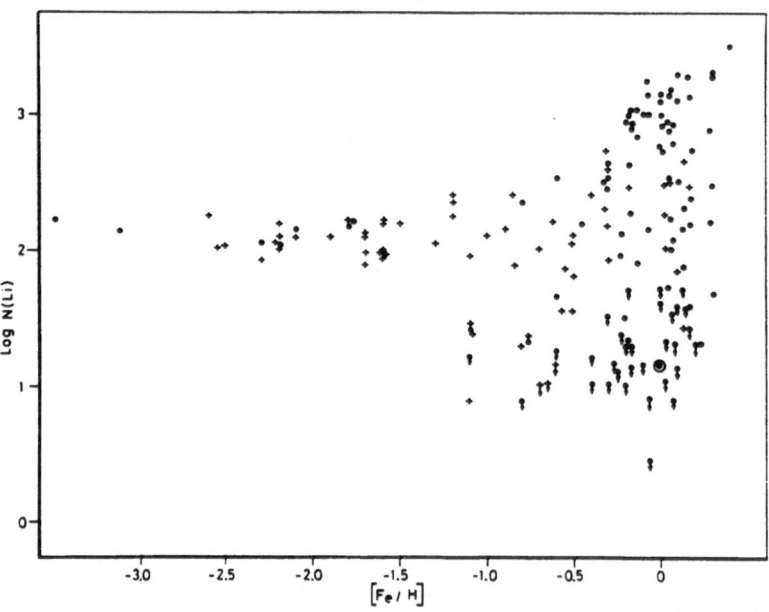

Figure 2. ^{7}Li abundance, logN(Li) *vs.* iron metallicity [Fe/H] for field stars in the solar neighbourhood. Note: (a) The low scatter for [Fe/H]≤ -1.4. (b) The rise in the upper envelope for [Fe/H]> -1.4. (c) The wide scatter at high metallicities. All stars here have $T_{eff} \geq 5500$K ($\bullet$: $T_{eff} > 6000$K; +: 5500K$< T_{eff} < 6000$K).

The scenario in which the Population II ^{7}Li abundance, i.e. 1.5×10^{-10}, is primordial implies a galactic source of ^{7}Li producing a subsequent increase of a factor greater than ten to yield the Population I value. This has been seen as a weakness, although there is no lack of candidate processes to give the ^{7}Li. These have been listed in the literature, and include spallation of C, N and O by galactic cosmic rays, as well as production in giants and novae. A recent paper by Dearborn *et al.* (1989) suggests that type II supernovae might be a source for ^{7}Li. The weakness is supposed to be that it is not clear just how the galactic production of ^{7}Li has occurred, but Fig 2 shows that the need to explain a rise in ^{7}Li is in fact another useful constraint on galactic evolution models, and one which so far from posing difficulties, fits other data rather well. Taking the upper envelope,

^{7}Li "takes off" at [Fe/H] = -1.4, and if we assume that this shows where the rising galactic ^{7}Li production curve crosses the primordial plateau, and that ^{7}Li is rising linearly with Fe, then the value at [Fe/H] = 0 would be ^{7}Li $\sim$3.5, in ^{7}Li/H = 3x10^{-9}. This is very close to the values in fact observed in objects of near solar or higher metallicity. Thus, whatever the production mechanism turns out to be it must give a linear or near–linear dependence on Fe. It seems to me too strong a coincidence that a depletive fall from an initial value of ^{7}Li $>$ 3, a time–dependent process, should conspire to mimic a rise which exhibits a linear Fe–abundance dependence. Thus the "weakness" of the requirement for a galactic source for ^{7}Li, and hence for [^{7}Li/H]$_p$ $\sim$ 1.5x10^{-10} turns out to be a strength. Clearly we should attempt to use evolution models to account quantitatively for the disc rise in ^{7}Li. It is generally agreed that cosmic ray spallation can produce no more than 10% of the ^{7}Li observed. It is also of interest to note that if type II supernovae were responsible then ^{7}Li should track the Oxygen abundance rather than the Fe abundance. This is because recent [O/H] determinations in Population II (Abia and Rebolo, 1989) show that [O/Fe] rises steadily to values of between +0.5 and +1 below [Fe/H] $\leq$ 1, implying that oxygen production preceded iron production at these low metallicities. This is ascribable to the early presence of type II supernovae, either simply because more massive stars evolve more rapidly than the others, or because the IMF was less biased to lower mass stars at those epochs, or for both reasons. If the upper envelope of Fig 2 showed that ^{7}Li rises truly linearly ith Fe between [Fe/H] = -1.4 and 0, then a type II supernova origin for ^{7}Li would be ruled out. The data do not yet warrant this weight of interpretation, but do point somewhat in that direction, leaving novae and red giants as the more probable producers.

2.3. LITHIUM DEPLETION IN CLUSTERS AND IN THE FIELD

Before resuming the primordial abundance quest, I will take a short digression about depletion, to act as a reference framework. ^{7}Li is destroyed in stellar interiors at temperatures higher than 2.4x10^{6}K. In cool stars, with Teff$\leq$ 5200 K (spectral type G5) convection goes sufficiently deep to take all surface ^{7}Li to the combustion level where it is consumed. Simplistically we might expect to find main sequence stars with Teff$>$ 5500 K, say, showing surface abundances of ^{7}Li with which they were formed, and those of Teff$<$5200 K with no ^{7}Li at all (convective turnover times are of order days). In practice, observations of ^{7}Li $vs.$ Teff in a fairly young cluster, the Hyades, are as shown in Fig 3, compared with the same curve for stars with [Fe/H]$<$ -1.4, the oldest stars in the solar neighbourhood. For Teff lower than 6300 K (F8) there is a steadily increasing depletion to lower Teff implying that non–convective processes play a critical role. Although a number of authors have made substantial attempts to model these, and a number of depletion mechanisms are in contention (D'Antona and Mazzitelli 1984, Michaud et al. 1984, Pinsonneault et al. 1989), there is as yet no consensus on this matter. Varying the stellar rotation rate as a parameter is one possible method of elucidating the physics (Rebolo and Beckman, 1988) and incidentally enhancing our knowledge of stellar structure. For 7000K$>$Teff$>$6400K we find the "^{7}Li gap" of low Li abundances, first found by Boesgaard and Trippico (1986), systematically investigated by Boesgaard and her co–workers (1986b, 1987, 1988) and by Hobbs and Pilachowski (1986a,b, 1988), see Fig 4. Given the absence of major sub–photospheric convection in this range of Teff, mechanisms invoking different forms of diffusion have been proposed (Schatzman 1977; Michaud, 1986; Vauclair 1988; Delyannis et al., 1989). Here too, there is scope to use ^{7}Li depletion as a structure diagnostic. For Teff $>$ 7000 K, the ^{7}Li abundance reaches, in a given cluster, i.e. for a given age, a maximum asymptotic value. For clusters of solar age or younger, values for [^{7}Li/H] $>$ 10^{-9} are typically found in this upper temperature range. In this context, Hobbs and Pilachowski (1988) found that for the open cluster NGC 188, whose age had been assessed at some 10 Gyr using stellar isochrones by Vandenberg (1985), in the least depleted range [^{7}Li/H] lies very close to 10^{-9}. They inferred that this was too early in the disc lifetime for ^{7}Li to be produced by galactic processes, and hence that a primordial value of order 10^{-9} was preferred. A subsequent thorough re–examination of the age of NGC 188 by Twarog and Twarog (1989) puts it at $\sim$ 6 Gyrs, not much above that of the sun, and vitiates Hobbs and Pilachowski's conclusion. Furthermore, accurate spectroscopic estimates of the iron abundance in open clusters of different ages show that between the present epoch and that of M67, some 5 Gyrs ago, there has been little if any increase in Fe/H in the outer galaxy (Garcia Lopez *et al.* , 1989). The constancy of the undepleted ^{7}Li abundance over the same age range can thus readily be explained if ^{7}Li production tracks that of Fe. We need not assume a constant initial, universal, and primordial ^{7}Li abundance of $>10^{-9}$ to explain this.

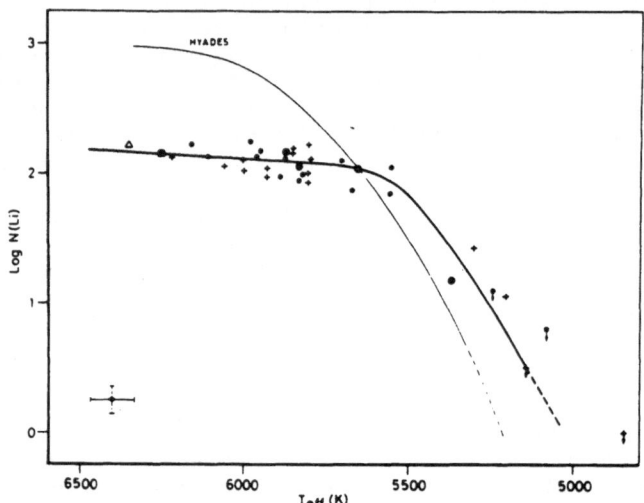

Figure 3. ^{7}Li abundance, logN(Li) *vs.* effective temperature, T_{eff}, for two groups of stars: (a) Extremely metal deficient (EMD) stars, with [Fe/H]≤ -1.4 (symbols plus thick curve), (b) The Hyades main sequence (least squares fit, fine curve). <u>Note:</u> (i) The EMD stars have an abundance "plateau", rising slightly to higher T_{eff}, for $T_{eff} >5800$K. (ii) The EMD stars for $T_{eff} <5600$K have higher ^{7}Li abundances than the Hyades stars indicating much lower depletion rates in the former.

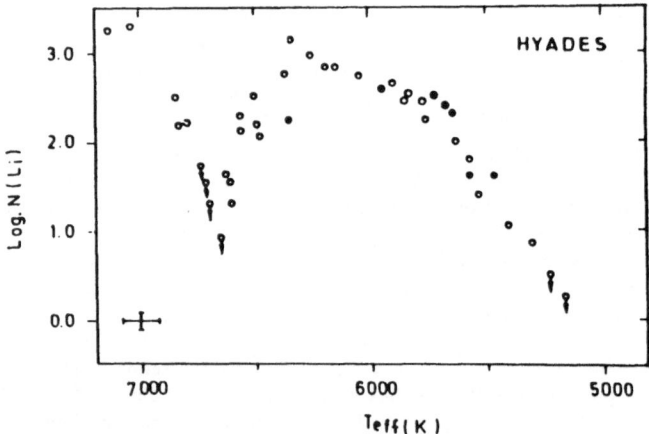

Figure 4. ^{7}Li abundance, logN(Li) *vs.* effective temperature, for the Hyades, illustrating the "Lithium gap" between $T_{eff}=6400$K and 7000K.

A further argument of those who are not convinced that the Population II ^{7}Li abundance is primordial is that in these very old stars, depletion mechanisms having surely operated over much longer periods than in the disc, we could not expect to see the initial abundance. A fully satisfactory answer here will be possible only when we have an adequate theoretical model for ^{7}Li depletion in the appropriate surface temperature range (6300 K > Teff > 5800 K). At present different theoretical studies give opposed answers. Recent examples are those of Vauclair (1988) who claims to be able to derive the "^{7}Li plateau", shown in Fig 3, using rotationally stimulated turbulent diffusion over periods of order 10 Gyrs and favours the 10^{-9} value for primordial ^{7}Li, whereas Delyannis et al. 1989, who fit their diffusion models to clusters of different ages, believe that the depletion rate is inhibited at low metallicities, due to the suppression of convection. Their models are consistent with $\sim 10^{-10}$ for primordial ^{7}Li. A significant observation shows that depletion can indeed be strongly suppressed at low metallicities. Comparing the run of ^{7}Li vs. Teff for stars with [Fe/H] ≤ -1.4 (age $\sim$ 10 Gyrs) and for the Hyades (age $\sim$x10^8 Gyrs), in Fig 3, we see that for Teff$\leq$ 5600 K, the absolute depletion for the low metallicity stars is less than that of the young cluster (even if the older stars had started life with ^{7}Li $\sim 10^{-9}$). Given the ratio of their ages, the depletion rate in the metal deficient stars has been at least twenty times less than that of their younger counterparts, at the same Teff. Whilst we should prefer to have acceptable physical models for all these cases, there is no reason to doubt that low metallicity stars can have their ^{7}Li, essentially undepleted without violating any physics, and that such evidence as we have does in fact point in that direction. This argument would apply a fortiori to the range close to 6300 K where the Population II abundance has been determined, and lends support to the $\sim 10^{-10}$ value for primordial ^{7}Li.

2.4. THE EVIDENCE FROM ^{6}Li, ^{9}BE AND ^{11}B

Our discussion so far has centred on the question of whether the primordial ^{7}Li abundance might be higher than the 2x10^{-10} Population II value, but might it in fact be less? Evidence for this was claimed by Sahu, Sahu and Pottasch (1988) who set an upper limit to interstellar Li/H in the Large Magellanic Cloud below 10^{-10} using the line of sight to SN 1987a. The conclusion is not however, definitive as a primordial abundance estimate, because it does not deal adequately with the grain depletion, which is often a problem in interstellar abundance studies, nor does it deal convincingly with the fraction of Li which could be in singly ionized form, and hence undetected. Furthermore, these authors appear to have chosen an extremely low value for the K depletion in the LMC which also weakens their analysis. Thus we cannot draw strong conclusions from this observations although it is probably hard to reconcile with a primordial value as high as 10^{-9}. A potential source of galactic pre–population–II ^{7}Li would have been cosmic ray spallation due to any initial strong burst of star formation in the galaxy. This has been ruled out by observations of ^{9}Be (Molaro and Beckman, 1984; Rebolo et al. , 1988b; Beckman et al., 1989) and ^{11}B (Molaro, 1987) in metal deficient stars. Since these nuclides would have been produced with their present galactic abundances in any burst capable of producing measurable ^{7}Li, the abundance upper limits for ^{9}Be and ^{11}B, ten times lower than their current values, which are found in stars with [Fe/H] ≤ -1.2, as shown in Fig 5 for ^{9}Be, demonstrate that little if any of the Population II ^{7}Li could have been produced in such a burst. We can state confidently that no more than 0.1 dex of the 2.2 dex Population II ^{7}Li could be spallogenic, and hence galactic in origin. A similar though less restrictive limit is placed by the failure, so far, to detect a ^{6}Li contribution to the ^{7}Li doublet at λ6708 Å in stars with Fe/H ≤ -1 (Maurice, Spite and Spite, 1984). Spallation will produce ^{7}Li/^{6}Li $\sim$2, and a failure to find as much as 20% ^{6}Li, in metal–deficient stars also limits the amount of ^{7}Li which could have been produced by this process, prior to the formation of Population II. The small sample of stars on which this conclusion is based, however, has led us to obtain a further sample of some dozen stars with [Fe/H] ≤ 1 in order to attack the problem more comprehensively. A quick look shows no obvious ^{6}Li, but any conclusions must be based on a rigorous analysis (Beckman et al., 1990).

Up to this point we have not been concerned with cosmological models, using the SBBN only as a guideline which has led to the importance of ^{7}Li being recognized. Even within that context, improvements in the nuclear physics (Kawano et al. ,) have led to some revisions of the cosmological inferences compared with the earlier models such as that of Wagoner (1973), notably the range of η corresponding to a measured [^{7}Li/H]. Thus if we adopt SBBN, the ^{7}Li at 1.5×10^{-10} with theoretical and observational uncertainties included, leads to a range for η of $1.5 \times 10^{-10} \leq \eta \leq 5 \times 10^{-10}$, and a corresponding range in Ω_b of $0.025 \leq \Omega_b \leq 0.08$ in other words it is very difficult to obtain a baryon density higher than 10% of the closure density.

Since inhomogeneous Big Bang models have been tailored to obtain the observed ^{7}Li abundance and those of the other three light nuclides it is at present not methodologically possible to use their measured values as direct tests, and in particular to distinguish the inhomogeneous from the homogeneous case. However, other nuclides are produced in potentially measurable quantities in the inhomogeneous models, which could offer observational targets. One such is ^{9}Be, which according to Boyd and Kajino (1989) could show up at a level of ^{9}Be/H $\sim 10^{-13}$ in stars with [Fe/H] ≤ -1.5. This is not too far below the present observational limits set by Rebolo, Molaro and Beckman (1989) in this metallicity range. Other limits, notably r-process nuclei have been predicted for inhomogeneous models to show up at low metallicities (Alcock et al. 1987; Applegate et al., 1988; Malaney and Fowler, 1988), in quantities not absurdly below present levels of detectability. There is some scope for observational advance here. Further advance can be made in pinning down the ^{7}Li/^{6}Li ratio in Population II. As has been shown by Audouze and Silk (1989) primordial light element synthesis via hadronic showers from exotic massive particle decays (Dimopoulos et al., 1988) already appear excluded by the present upper limit of $\sim 20\%$ for ^{6}Li/^{7}Li in Population II (Maurice et al., 1984). However detection of ^{6}Li in Population II at any level would be critical in proving that no significant Li depletion has occurred in these stars since their formation, since ^{6}Li is much more fragile than ^{7}Li. By the same token however, failure to detect ^{6}Li would not ipso facto imply significant depletion of ^{7}Li. Clearly the observers will need to look at ^{6}Li, ^{9}Be, and r-process elements in Population II.

I have shown the kinds of data and understanding of stellar physics needed to obtain a reliable estimate of primordial ^{7}Li/H and outline my reasons for believing that the best value we have for this abundance is close to 1.5×10^{-10}. It is clear that this value already places strong constraints on cosmological models, whether purely baryonic, with $\Omega_b \simeq 0.05$, or non-baryon dominated with $\Omega_b \approx 0.05$, $\Omega_o = 1$, or purely baryonic with $\Omega_b = \Omega_o = 1$. Neither ^{7}Li, nor ^{4}He, nor combination of these with D + ^{3}He, can, however, offer a definitive distinction between these three cases. We will need to look to other nuclides, and other tests. Of these, the cosmic microwave background fluctuations are amongst the most direct, and I devote the remainder of this article to a discussion of CMB fluctuations as constraints on baryonic cosmologies.

3. The importance of CMB fluctuations as constraints

I will confine my attention here to the two most sensitive measurements published to date of cosmic microwave background fluctuations, namely the arcminute scale observations of Readhead et al. (1988) and the degree scale observations of Davies et al. (1987), together with the more recent updating of the latter, in which the present author is a participant. Fluctuations in the radiation field of the CMB are predicted by adiabatic models to be in the range 10^{-5} to 10^{-4} for the measured amplitude, $\Delta T/T$, depending on the angular scale, the value of Ω, and the nature of the dominant particle species. Adiabatic models in which baryons dominate appeared to have been ruled out by the arcminute scale upper limit of Uson and Wilkinson (1984), a conclusion strengthened by the upper limit of 1.7×10^{-5} at 20 GHz reported by Readhead et al. (1989). This limit, the severest reported so far, might if taken in isolation, also appear to rule out low Ω (≤ 0.1) universes, even those where non-baryonic matter may dominate (Peebles, 1982; Bond and Efstathiou, 1984; Vittorio and Silk, 1984). However, as pointed out by Peebles (1981) reionization at a relatively low redshift (z = 3-10) which could occur during galaxy formation, might erase arcminute scale fluctuations, thus vitiating any sweeping conclusion about the original background spectrum. Progress may then be made along theoretical and observational lines. Theorists may explicitly calculate the quantitative effects of reionization as a function of angular scale on the sky, as has been carried out for a variety of models by Bardeen *et al.* (1987) and Efstathiou (1988). Observers may go to degree angular scales, which not only escape reionization smearing, but are larger than the horizon scale at recombination ($\Theta \sim 2°$), so that causal processes in the early universe cannot remove anisotropy on degree scales. It is also important to note that in an inflationary model which contains a first–order quark–hadron phase transition and leads to inhomogeneous nucleosynthesis, the scale of the fluctuation spectrum invoked to control the neutron enrichment and impoverishment which yields the observed light element abundances while allowing $\Omega_b = 1$, corresponds to mass scales of less than one solar mass. The fluctuation measurements on arcminute scales correspond to galactic masses, and on degree scales to galactic cluster masses. Thus the measurements of fluctuations can in principle test Ω_o, and even set useful constraints on the presence in significant quantity of non–baryonic matter. What they cannot do is to distinguish directly at the microscopic level between homogeneous and inhomogeneous Big Bang nucleosynthesis.

4. Degree scale observations

Davies *et al.* (1987) published a measurement at an 8° scale, on a frequency of 10.46 GHz, of sky fluctuations at an r.m.s. level $\Delta T/T = 3.7 \times 10^{-5}$, obtained on a declination strip at $\delta = +40°$. In that paper the authors claimed that extended galactic emission was not likely to be the major cause of these fluctuations, basing this conclusion on global comparisons with surveys at 408 MHz (Haslam *et al.* , 1982) and 1.4 GHz (Reich and Reich, 1986). They also suggested that galactic cirrus was unlikely to contribute significantly at these levels, and indeed a conservative calculation based on reported IRAS measurements shows that a generous upper limit to a cirrus contribution should be $\Delta T/T \simeq$ a few times 10^{-6} at this frequency. Convolved point sources cannot readily offer an explanation for the reported $\Delta T/T$, as we can infer using the most complete point source catalogues, such as that of Kuhr (1981), and extrapolating conservatively. In fact the beam dilution for the 8° individual beams used by Davies *et al.*, in their triple–beam configuration, is sufficiently strong that individual sources weaker than 0.5 Jy would not register above the r.m.s. noise, and it would take combinations of such weaker sources for which there is no evidence at other frequencies, to contribute at the observed levels. Thus Davies *et al.* claimed that there was no clear reason not to attribute the detected fluctuation levels to intrinsic CMB fluctuations, while acknowledging that other contributions, notably extended galactic emission, needed further study.

Recently the same group, using an augmented set of data from the same experiment (Watson *et al.* , 1988), claimed that a specific "bump" at $\alpha = 15$ hrs, $\delta = +40$, in the Bootes void region, is present in the scan. This bump, which has a peak height of $\Delta T/T \simeq 10^{-4}$, if abstracted from their data, leaves an underlying level of 3×10^{-5}, still above their quoted r.m.s. noise level.

I can now comment on the results of a follow–on experiment using the same equipment except for a modified pair of antenna horns, lengthened to narrow the sky beams to a 5.6° half-power width, but retaining the 8.1° interbeam throw. The experiment was performed over a similar two–year time scale to the previous one, during which time not only was a very deep scan at $\delta = +40°$ obtained, but strip scans at levels below $\Delta T/T \simeq 10^{-4}$ were measured at neighbouring declinations in $2 1/2°$ intervals between $+35°$ and $+45°$, from which the sky map in Fig 5 was derived. In addition, a deep scan at 0° declination was obtained which, like that from the previous experiment, showed a clean detection of 3C273, as well as very high signal to noise ratio galactic plane crossings. In Fig 6 we show the $\delta = +40°$ scan from the 5° beam experiment compared with the same scan from the previously reported 8° experiment (Davies *et al.*, 1987). The new results confirm the previous ones in both the level of fluctuations detected, $\Delta T/T \simeq 4 \times 10^{-5}$, and in the presence of a bump at $\alpha = 15$ hrs, $\delta = +40°$. The bump does not appear to have a spread significantly larger than a point source diluted to the 5° scale.

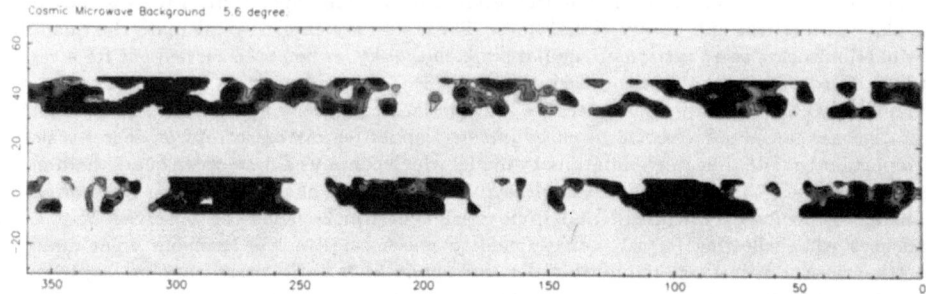

Figure 5. Gray scale representation of the two strip maps on the sky at $\delta = +40°$ and $\delta = 0°$ obtained with the Tenerife CMB radiometer. These maps are only illustrative, showing anisotropies of order 0.3mK in amplitude, obtained by deconvolving sets of scans such as that in Fig. 6 (5.6 degree beamwidth).

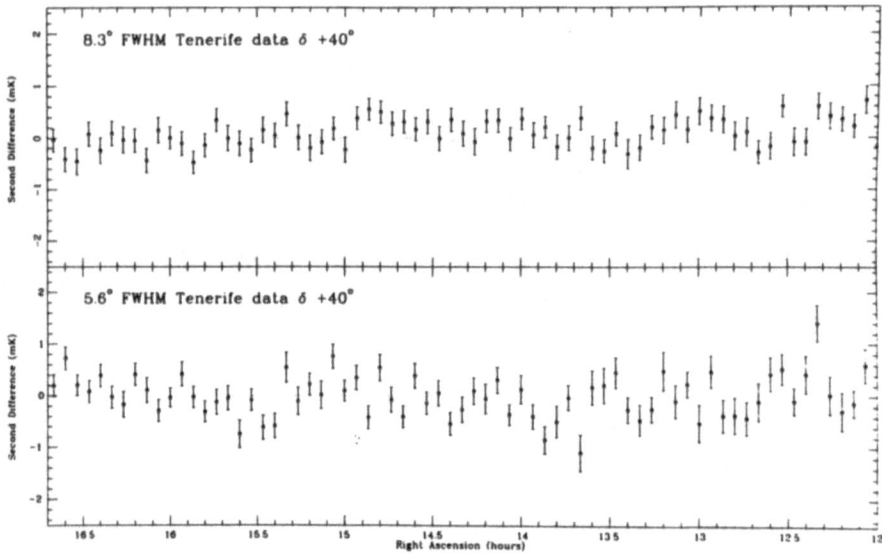

Figure 6. One–dimensional second difference scans with the Tenerife CMB anisotropy radiometer, over the same piece of sky, at $\delta = +40°$, from R.A. 12^h to 16^h45^m; 8.3° and 5.6° experiments are compared. Scans are sums of all available data, and show comparable S:N ratios; $\Delta T/T \approx 4 \times 10^{-5}$ for both.

I should say that this summary of our results is a considerable simplification, giving no descriptions of the intense degree of care taken to eliminate spurious effects in the mainbeam and sidelobes of the twin–horn triple–beam pattern on the sky. For brief details the reader is referred to Watson *et al.* (1989), and for greater detail to Lasenby *et al.* (1990). To summarize we minimized atmospheric noise by rejecting scans when the sky showed any cloud, or when noise levels either of whole scans or of single data points were higher than a threshold determined by the system noise level. We also did not use data when either sun or moon was in the first sidelobe of the beam patterns, (and a fortiori when either was in the main lobe!). The site of the experiment, Izaña, Tenerife, has consistently low precipitable water above it, so that a minority of days had to be rejected for bad weather. Space also forbids a detailed description of the method used to estimate the probability and level of a detection, for which the reader is referred to Lasenby *et al.* (1990). It comprised the construction of a likelihood function, predicated on gaussian statistics and using the measured system noise, in a two–dimensional parameter space with dimensions of angular scale and fluctuation amplitude. Where the value of the likelihood function exceeded a threshold of 20, a detection was deemed to have been made. This condition was satisfied for the $\delta = +40°$ experiment, on a range of angular scales between 3° and 10°, with a nearly constant amplitude of 4×10^{-5} for the detections. Outside this angular range the effective sensitivity did not permit us to claim detection. Again in this new experiment, subtraction of the "bump" at $\alpha = 15$ hrs, as if dealing with a point source, reduces the residual fluctuations level to $\sim3\times10^{-5}$ for $\Delta T/T$.

5. Consequences of a detection for baryonic cosmologies

If we believe that the levels of $\Delta T/T$ reported above in fact correspond to fluctuations in the CMB, the consequences for baryonic cosmologies would of course be significant. They are well summarized in Fig 7, which has been adapted from Efstathiou (1989), and is a plot in which the abscissa is baryon density (assuming no other major contributor to omega), and the ordinate is the index n of a hypothetical power law representing amplitude against wavenumber for the background fluctuation spectrum. The blank area in the top right and bottom left of the diagram are those parts of parameter space permitted by the Readhead *et al.* (1989) upper limits on arcminute scales; the shaded region between them is excluded. The models implied entail the assumption of complete reionization of the universe at intermediate z; partial reionization would further restrict the permitted parameter space, and in a model with no reionization the lower left hand permitted zone would diasappear completely. Thus the Readhead *et al.* result alone already eliminates standard Big Bang, purely baryonic, models without reionization. In the diagram our degree scale fluctuation result is reproduced by the locus which arises from top right to bottom left.

Taken as a detection, it gives us two possibly viable ranges for baryonic cosmologies. One has Ω_b between 0.8 and 1 with a somewhat unexpected fluctuation spectral index close to -2; this solution range would be valid with or without reionization. The other had $\Omega_b \leq 0.08$, a spectral index close to -1, and needs complete reionization. If our fluctuations represent an upper limit to the true CMB value, that would not exclude either range of Ω_b, but would compress the permitted parameter space, putting pressure on baryonic models.

Comparing the CMB inferences with those from the light element abundances, we can draw some interesting conclusions. In the "low Ω_b" range, the presently permitted upper limit for Ω_b, 0.08, appears to coincide for both techniques. Here there is no room for more density; the inflationary value is ruled out, and so are some inhomogeneous Big Bang nucleosynthesis models (Reeves, 1989). If our degree scale fluctuations were a true measurement of the CMB, it would not appear to be possible to invoke non–baryonic matter to make up the inflationary value, i.e. $\Omega = 1$ ($\Omega_b \leq 0.08$) or even up to $\Omega = 0.2$ as required by some virial estimates of the masses in galaxy clusters. This can be inferred by comparing the tabulated predictions of a set of such models in Bardeen *et al.* (1987) with both Readhead *et al.* upper limit, and our value, taken as a detection. Neither cold dark matter dominated, nor neutrino–dominated models, with $\Omega = 1$, or with $\Omega = 0.2$, satisfy both observational requirements. This conclusion would be strengthened if the "bump" at $\alpha = 15$ hrs, $\delta = +40°$ were a CMB feature. Using the technique of Martinez Gonzalez and Sanz (1984) we can show that a bump of this amplitude is highly unlikely in an isocurvature CDM cosmology, with a 15σ probability that it would not be produced. Although many fine details, especially about the statistical nature of the fluctuations, need to be considered, the general tenor of this conclusion is unlikely to be changed for different CDM models.

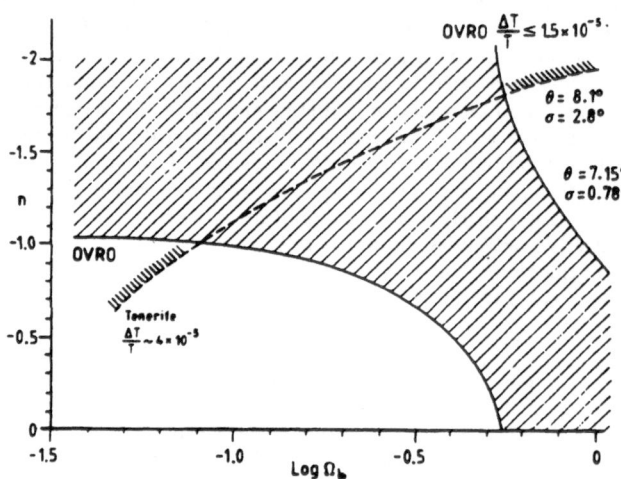

Figure 7. Ω_b the universal baryon density, in units of the closure density, against n, the slope of the power spectrum of primordial fluctuations, for isocurvature baryonic models with post–recombination reheating (after Efstathiou 1988). The shaded area is excluded by the OVRO data (Readhead *et al.* 1989). The dashed line represents a detection of true fluctuations by the Tenerife experiment. If the detected fluctuations are not cosmological, the excluded parameter space is to the shaded side of the line.

In the "high Ω_b" range we would be able, given the postulate of an inflationary value of unity for Ω, to specify the fluctuation spectral index of the background, and use this as an input for galaxy formation theories. The result $\Omega_b = \Omega = 1$ is naturally welcome to advocates of inflation and inhomogeneous primordial nucleosynthesis, but leaves open the question of where the 90% of non-luminous galactic matter may be.

With so much at stake, it is clearly of the highest importance to establish whether or not we have indeed observed CMB fluctuations on degree scales. In order to check whether or not our measurements may be due to diffuse galactic emission, we will carry out two new experiments on similar lines to those reported here, but at higher radio frequencies. One at 15 GHz, is just starting up (September 1989), at Izaña, Tenerife, after trials at Jodrell Bank, and should integrate to reach sensitivities high enough to distinguish between $\Delta T/T$'s which fall to higher frequencies (galactic) from those which are essentially frequency independent in this range (CMB), in a time of order one year. The second, at 33 GHz, will be tested at Izaña, before placing it at an even drier, antarctic site, within two years. This is an exciting time for such experiments, and the results of these, and similar enterprises by other groups will be keenly awaited.

6. Conclusions

The net is closing on baryonic cosmologies, thanks to a combination of light element abundance measurements (^{4}He and ^{7}Li) and CMB fluctuation searches on intermediate (arcminute and degree) angular scales. Already we can rule out purely baryonic models with $0.1 \leq \Omega_b \leq 0.8$. In the lower permitted range the universe would have been completely or largely reionized at intermediate z (20 $> z > 3$), and in the upper range we would need to find a hiding place for some 90% of the universe's baryons. More pressure on these models can be applied via searches for ^{9}Be or r-process elements in extremely metal deficient stars, and by pressing forward with CMB fluctuation observations to higher sensitivities and higher radio-frequencies, and the cosmology-particle physics connection will be further tightened soon when the Z_0 width will allow us to fix the number of massless neutrino flavours as an input to the primordial nucleosynthesis scheme.

7. References

Abia, C.; Rebolo, R.: 1990, *Ap.J.* (In press).
Alcock, C.; Fuller, G.M.; Matthews, G.J.: 1987, *Ap.J.*, **320**, 489.
Applegate, J.H.; Hogan, C.J.; Scherrer, R.J.: 1988, *Ap.J.*, **329**, 592.
Bardeen, J.M.; Bond, J.R.; Efstathiou, G.: 1987, *Ap.J.*, **321**, 28.
Beckman, J.E.; Abia, C.; Rebolo, R.: 1989, *Astrophys.Spa.Sci.* (In press).
Beckman, J.E.; Garcia Lopez, R.; Rebolo, R.; McKeith, C.: 1990 (In preparation).
Boesgaard, A.M.: 1985, *P.A.S.P.*, **97**, 784.
Boesgaard, A.M.: 1987, *Ap.J.*, **321**, 967.
Boesgaard, A.M.; Trippico, M.J.: 1986a, *Ap.J.*, **302**, L49.
Boesgaard, A.M.; Trippico, M.J.: 1986b, *Ap.J.*, **303**, 724.
Boesgaard, A.M.; Budge, K.G.; Ramsay, M.E.: 1988, *Ap.J.*, **327**, 389
Bond, J.R.; Efstathiou, G.: 1984, *Ap.J.*, **285**, L45.
Boyd, R.N.; Kajino, T.: 1989, *Ap.J.*, **336**, L55.
D'Antona, F.; Mazzitelli, I.: 1984, *Astron.Astrophys.*, **138**, 431.
Davies, R.D.; Lasenby, A.N.; Watson, R.W.; Daintree, E.J.; Hopkins, J.; Beckman, J.E.; Sánchez-Almeida, J.; Rebolo, R.: 1987, *Nature*, **326**, 462.
Dearborn, S.P.; Schramm, D.N.; Steigman, G.; Truran, J.R.: 1990, *Ap.J.* (Submitted).
Delyannis, C.; Demarque, P.; Kawaler, S.D.: 1990, *Ap.J.* (In press).
Dimopoulos, S.; Ezmailzadeh, A.; Hall, J.; Starkman, G.J.: 1988, *Ap.J.*, **330**, 545.
Efstathiou, G.: 1988, *Proc. Vatican Conference on Cosmology* (In press).
Garcia Lopez, R.; Rebolo, R.; Beckman, J.E.: 1988, *P.A.S.P.*, **100**, 1489.
Haslam, C.G.T.; Salter, C.J.; Stoffel, H.; Wilson, W.E.: 1982, *Astron.Astrophys.Supp.Ser.*, **47**, 1.
Hobbs, L.M.; Pilachowski, C.: 1986a, *Ap.J.*, **309**, L17.
Hobbs, L.M.; Pilachowski, C.: 1986b, *Ap.J.*, **311**, L37.
Hobbs, L.M.; Pilachowski, C.: 1988, *Ap.J.*, **334**, 734.
Kawano, L.; Schramm, D.N.; Steigmann, G.: 1988, *Ap.J.*, **327**, 750.
Kuhr, H.; Witzel, A.; Pauling Toth, I.I.K.; Nauker, U.: 1981, *Astron.Astrophys.Supp.Ser.*, **45**, 367.
Lasenby, A.N.; Davies R.D.; Watson, R.W.; Daintree, E.J.; Hopkins, J.; Beckman, J.E.; Rebolo, R.: 1990, *Mon.Not.R.Astr.Soc.* (In preparation).
Malaney, R.A.; Fowler, W.A.: 1988, *Ap.J.*, **333**, 14.
Maurice, E.; Spite, F.; Spite, M.: 1984, *Astron.Astrophys.*, **132**, 378.
Michaud, G.; Fontaine, G.; Beaudet, G.: 1984, *Ap.J.*, **282**, 206.
Molaro, P.: 1987, *Astron.Astrophys.*, **183**, 241.
Molaro, P.; Beckman, J.E.: 1984, *Astron.Astrophys.*, **139**, 394.
Peebles, P.J.E.: 1981, *Ap.J.*, **263**, L1.

278

Pinsonneault, M.H.; Kawaler, S.D.; Sofia, S.; Demarque, P.: 1989, *Ap.J.*, **338**, 424.

Readhead, A.C.S.; Lawrence, C.R.; Myers, S.T.; Sargent, W.L.W.; Hardebeck, H.E.; Moffet, A.J.: 1989, *Ap.J.* (In press).

Rebolo, R.; Beckman, J.E.: 1988, *Astron.Astrophys.*, **201**, 267.

Rebolo, R.; Beckman, J.E.; Molaro, P.: 1987, *Astron.Astrophys.*, **172**, L17.

Rebolo, R.; Molaro, P.; Beckman, J.E.: 1988, *Astron.Astrophys.*, **192**, 192.

Rebolo, R.; Molaro, P.; Abia, C.; Beckman, J.E.: 1988, *Astron.Astrophys.*, **193**, 193.

Reich, P.; Reich, W.: 1986, *Astron.Astrophys.Supp.Ser.*, **63**, 205.

Sahu, K.; Sahu, M.; Pottash, S.R.: 1988, *Astron.Astrophys.*, **207**, L1.

Schatzmann, E.: 1977, *Astron.Astrophys.*, **56**, 211.

Spite, F.; Spite, M.: 1982, *Astron.Astrophys.*, **115**, 357.

Twarog, B.; Twarog, B.A.: 1989, *Astron.J.*, **97**, 759.

Uson, J.; Wilkinson, D.T.: 1984, *Ap.J.*, **283**, 471.

Vandenberg, D.A.: 1985, *Ap.J.Supp.Ser.*, **58**, 511.

Vauclair, S.: 1988, *Ap.J.*, **335**, 971.

Vittorio, N.; Silk, J.: 1984, *Ap.J.*, **285**, L39.

Wagoner, R.V.: 1973, *Ap.J.*, **179**, 343.

Wagoner, R.V.; Fowler, W.A.; Hoyle, F.: 1967, *Ap.J.*, **148**, 3.

Watson, R.A.; Rebolo, R.; Beckman, J.E.; Davies, R.D.; Lasenby, A.: 1989, *Proc. IAU Symp. "Large scale structures and Motions in the Universe"*, Trieste (Eds. C. Mezzetti *et al.*). Kluwer. 133–137.

Yang, J.; Turner, M.S.; Steigman, G.; Schramm, D.M.; Olive, K.A.: 1984, *Ap.J.*, **281**, 493.

CAN HALOS CONSIST OF COMPACT STELLAR REMNANTS?

Joseph Silk
Departments of Astronomy and Physics
and
Center for Particle Astrophysics
University of California, Berkeley
Berkeley, CA 94720
USA

ABSTRACT. The hypothesis that galactic halos consist of compact stellar remnants, such as white dwarfs, neutron stars or black holes is explored. I focus on star formation considerations, including the nature of the initial stellar mass function both at present and in the past, and the rate of star formation in various types of galaxies. A protogalaxy model is presented, and I describe alternative schemes for forming halos of baryonic dark matter in the form of stellar relics. I conclude by describing possible manifestations of the enhanced primordial star formation rate that must be invoked for this hypothesis to be viable, and I suggest a possible technique for directly observing halo relics in our galaxy.

1. Introduction

Baryonic dark matter (BDM) is the one form of dark matter whose existence is undisputed. In units of the mean cosmological density, expressed relative to the critical value for closure of the universe, observations of galactic rotation curves require that $0.03 \lesssim \Omega_{halos} \lesssim 0.1$. This range is essentially identical with the baryonic dark matter content inferred from canonical primordial nucleosynthesis explanations of the abundances of ^{4}He, ^{3}He, ^{2}H and ^{7}Li, according to which $\Omega_{BDM}h^2 \sim 0.03$. Here h is the Hubble constant expressed in units of 100 km s^{-1} Mpc^{-1}. Of course, this coincidence does not *require* halos to consist of BDM. However, the luminous content of galaxies, identifiable with stars in the mass range 0.1-100 $\mathcal{M}_\odot$ stellar remnants and interstellar matter within the luminous confines of galaxies, amounts to $\Omega_{lum} \sim 0.007$.

By way of contrast, the dark matter content on large scales, $\gtrsim 10$Mpc, is inferred to be $\Omega \sim 0.2$ from dynamical measurements of supercluster infall, and a similar value is inferred from galaxy cluster mass-to-light ratios, if these are assumed to also be representative of field galaxies and their environments. This value is marginally compatible with the primordial nucleosynthesis constraint if all the dark matter is baryonic. However, the extreme value $\Omega = 1 + \mathcal{O}(10^{-4})$ inferred from inflationary cosmology requires non-baryonic dark matter

D. Lynden-Bell and G. Gilmore (eds.), Baryonic Dark Matter, 279–294.
© 1990 *Kluwer Academic Publishers.*

unless the inhomogeneous nucleosynthesis models can reproduce the observed light element abundances. Fine-tuning of quark-hadron phase transition parameters may even allow $\Omega_{BDM} = 1$.

Any physically motivated discussion of baryonic dark matter must inevitably commence with considerations of star formation. Stellar remnants, such as neutron stars, white dwarfs and black holes, and failed stars such as brown dwarfs, are the only BDM candidates which are known to exist and could conceivably be abundant enough to provide the halo dark matter. Dust grains or diffuse gas are not viable as halo dark matter, although there are more esoteric possibilities for BDM that include dense nuggets of strange matter.

I will, in this review, focus on stellar relics as a possible form for halo dark matter. This will necessitate a discussion of star formation considerations, including the nature of the initial stellar mass function, both today and in the past, and the rate of star formation. I will present a model for protogalaxies, and prescribe recipes for halos of BDM. I will conclude with a discussion of constraints on the primordial IMF and of the prospects of actually observing the halo relics. While some theoretical motivation will be given, I shall emphasize the following empirical argument: at least *some* primordial stars were in the 1-100 $\mathcal{M}_\odot$ range if the halo formed from stellar relics, and the BDM hypothesis is therefore testable by searching for traces of these primordial stars.

2. The Present-Day Stellar Mass Function

The mass range of stars is an important ingredient of star formation theory. Long-lived, low mass stars lock up mass, whereas the short-lived massive stars dominate the luminosity of a star-forming system, and, via mass ejection in the late stages of stellar evolution, are responsible for the synthesis of the heavy elements. The initial mass function represents the theoretician's attempt to take the observed stellar luminosity function and convert it into a more useful form. Two important corrections are made: stellar luminosity is converted to mass via the mass-luminosity relation for main-sequence stars, and one allows for the formation of massive stars that have already died over the age of the stellar system being studied. Only the low mass stars are survivors from throughout the star-forming phase.

The initial mass function (IMF) has been extensively studied in the solar neighborhood, where it is defined as the total number of stars ever formed in a cylinder of cross-section 1pc^2 normal to the galactic plane. For many purposes, one can approximate the IMF by a power-law in mass, characterized by three parameters: a low-mass-cut-off m_L, a high mass cut-off m_U, and a slope x, $dN/d\ln m \propto m^{-x}$. In fact, x is observed to increase slightly with increasing mass, from ~ 1.5 over $1 - 10\,\mathcal{M}_\odot$, to ~ 2 from 10 up to $m_U \sim 100\,\mathcal{M}_\odot$. The observed IMF peaks at $m_L \sim 0.2\,\mathcal{M}_\odot$, below which the uncertainties become very large, in part, because of the poorly known mass-luminosity relation for low mass stars (Scalo 1986).

Away from the solar neighborhood, one has no direct information on the least massive stars. In star-forming regions, the massive stars dominate the light, and the evidence is generally consistent with a universal slope to the IMF of $x \approx 1.5$ ($\pm$ 0.5). There is some evidence, however, that m_L and m_U may vary. For example, in the inner Milky Way, a mass-to-light ratio increase relative to the solar neighborhood is needed in simple models for the galactic rotation curve (Larson 1986). Enhanced m_L both results in a higher fraction of stellar remnants, and simultaneously accounts for the observed diffuse far-infrared and Lyman-continuum radiation that is produced by more massive stars. The nucleosynthetic yield is also enhanced in the inner galaxy, thereby providing an explanation for the observed metallicity gradient. This explanation is not unique, however. For example, the brown dwarf fraction may vary, as may the slope of the IMF. One has to look elsewhere for stronger evidence that m_L is variable.

One such manifestation of enhanced m_L may occur in regions of greatly elevated star formation rate, or starbursts. The far-infrared luminosity monitors the energy output by massive stars, and the blue, or visual, light that emanates from the low mass stars. This ratio is enhanced by a factor of 10 or even 100 in extreme starburst galaxies. One also has a crude measure of the gas content from CO observations, provided that a suitable calibration of CO brightness temperature to H_2 column density is adopted. One finds that for a star formation burst to last for $\sim 10^8$y, given a known gas supply and a known total stellar mass, the IMF must be truncated in many cases. Further constraints on the IMF are obtained from spectroscopic signatures: for example, the $H\beta$ line strength probes the intermediate stellar masses. These constraints restrict m_L to be in the range 2-5 $\mathcal{M}_\odot$ for such objects as Arp 220, ESO 338-IG04, NGC 253, Markarian 171, IC 2153, IZw 36, and NGC 6240 (Scalo 1986). However, in a complete sample of nearby (15-40 Mpc) IRAS galaxies, the ~ 10 percent that are undergoing starbursts are identified with barred spirals and do *not* require any IMF truncation provided that a substantial fraction (~ 30 percent) of the galactic bulge (the inner $\sim 10^9 \mathcal{M}_\odot$) formed in the current starburst (Devereux 1989).

Of course, a sufficiently flat IMF ($x < 1$) would accomplish a similar objective by reducing the amount of mass locked up in low-mass stars. There is no evidence, however, from direct measurements of the IMF that favors such a flat power-law rather than a truncated IMF in star-forming regions. For many purposes, it is not necessary to distinguish between these possibilities: it is only the characteristic stellar mass that matters.

Evidence that m_U may be variable is sparse. One could cite T- associations of exclusively low mass stars, but these are generally adjacent to regions of massive star formation. Perhaps they are destined to form massive stars, and indeed, the regions which contain the massive stars may once have undergone an exclusively low mass star forming phase. Early indications that the typical stellar mass in a star-forming region increases with time have been found to be ambiguous. Massive star forming regions have been revealed by 2μ imaging to contain many low mass stars, but whether the entire observed IMF in these regions forms contemporaneously is not known.

Another site that is frequently identified with exclusively low mass star formation is that of the galaxy cluster cooling flows. Here, the evidence is very indirect. Modelling of the x-ray surface brightness and temperature reveals that the intracluster gas does not appear to be in hydrostatic equilibrium, if one assumes that energy input to the gas is negligible. The central cooling time-scale in up to 50 percent of known x-ray clusters is shorter than the Hubble time-scale (Forman 1988). One infers that there is a net inward drift of cooling gas to the cluster core, at a rate that is of order 100 $\mathcal{M}_\odot yr^{-1}$ in luminous x-ray clusters such as Perseus. The fate of the gas challenges astronomers' ingenuity: evidently it does not survive as cool gas, nor does it form stars with a solar neighborhood IMF. A popular resolution of this difficulty is to assert that the cooling flow gas, almost exclusively, forms low mass stars, and possibly brown dwarfs. Direct observations of the nearest cluster cooling flow suggest that $m_U \lesssim 1 \mathcal{M}_\odot$ (Fabian 1990).

3. Primordial Star Formation

Given our incomplete knowledge of present-day star formation, it would be surprising indeed if one could unambiguously identify the primordial IMF. In fact, one can make a series of plausible assertions, in which one has varying degrees of confidence. For example, there were primordial stars that spanned the mass range 0.5-50 $\mathcal{M}_\odot$. At the low mass end, one measures metallicities relative to solar that are as low as [Fe/H] = -4.5 in CD $-38°$ 245 (Bessel and Norris 1984) and –5.6 in G 77-61 (Dearborn *et al.* 1986). These are low mass stars: in the latter case, there is even a direct mass determination since the star is a binary that has evidently undergone mass transfer from an evolved companion. At the massive end, one can examine the abundance patterns in the oldest stars. One finds relative abundance patterns that are suggestive of massive star precursors: there is both an [O/Fe] enhancement and an r-process signature in extremely metal-poor stars (Wheeler *et al.* 1989).

The remaining arguments are more speculative. The paucity of low metallicity stars in the disk is sometimes attributed to a primordial IMF with $m_L \gtrsim 2\mathcal{M}_\odot$: one simply did not form many low mass stars during the low metallicity phase. This explanation is not unique: in fact, pre-enrichment of disk matter via infall from the thick disk and halo is an alternative. There is an analogous problem for population II stars: their metallicity distribution is consistent with a solar neighborhood IMF. Gas longevity in spiral disks is another contentious issue. Is it a coincidence that with the standard IMF ($m_L \sim 0.2\mathcal{M}_\odot$), most luminous spirals are just beginning to exhaust their gas supply, via lock-up in low mass stars and stellar remnants, at the present epoch? One solution is to enhance m_L, thereby reducing the lock-up fraction (Sandage 1986). Massive stars recycle most of their mass back into the interstellar medium, the surviving relics mostly being white dwarfs or neutron stars of mass below 2/msun.

Isochrone fitting of old disk stars suggests that rapid chemical evolution occurred over the first billion years of galactic history. Although one can more readily achieve this with enhanced m_L, there is sufficient spread between different fits to the observational data that one cannot make a definitive case. The fact that the intracluster medium appears, at least for the Virgo Cluster, to be enhanced in [O/Fe] is suggestive of a massive star-enhanced primordial IMF. The observed intracluster Fe abundance is so large, [Fe/H] $\approx$ -0.3, that, given current levels of mass ejection from evolving stars, one can only appeal to primordial, or at least, protogalactic, star formation rates to provide the required nucleosynthetic yield.

From observed colors of galaxies, one can infer the current rate of global star formation, with appropriate assumptions about the IMF (Gallagher *et al.* 1984). Past star formation rates are found to have been higher in spiral galaxies, although not by more than a factor of 2 or 3. In late-type galaxies, past star formation rates were actually lower than at present. Population synthesis of the spectra of ellipticals reveals that their lack of intermediate mass stars and red colors require a highly elevated initial rate of star formation, up to a factor of 100 larger per unit mass than is observed in the Milky Way. Equivalently, the bulk of the stars in spheroidal components were formed over between one and ten percent of the age of the disk in a typical spiral.

This enhanced star formation in proto-ellipticals is reminiscent of extreme starbursts. There is a further circumstantial connection. Theoretical arguments suggest that ellipticals may have formed by mergers of gas-rich protodisk galaxies, and indeed mergers most likely were common at early epochs. Extreme starbursts also generally show evidence that a merger or strong interaction is occurring (Telesco et al. 1988).

Evidence that the upper mass limit of the IMF may have been reduced in primordial star formation is indirect, and inferred from observations of low metallicity globular star clusters. The ejecta from a small number of supernovae suffices to produce the observed heavy elements, yet the narrowness of the giant branch indicates an intrinsic metallicity dispersion that is less than 10 percent in $\delta Z/Z$. If the clusters are self-enriched, one would expect considerable inhomogeneity. One resolution of this paradox is that the globulars formed stars very efficiently over a time-scale much more rapid than a dynamical time, thereby allowing the possibility of self-enrichment and homogenization of ejecta from massive stars (Murray and Lin 1989). It is rather more likely that the globular clusters formed out of halo gas that was pre-enriched by prior generations of massive star formation, and that very few, if any, massive stars formed during the initial star formation phase of the protoglobular cluster. The discovery of pulsars in globular clusters indicates that rapidly spinning neutron stars are present. However, supernova theory cannot easily be reconciled with the required number of neutron stars, since most neutron stars are born with a velocity that exceeds the globular cluster escape velocity. The abundance of neutron stars required in a globular cluster is estimated from the binary capture rate of neutron stars which accounts for the millisecond pulsars that are spun-up by mass transfer, and its explanation *may* require a preexisting component of neutron stars (Kulkarni et al. 1989). Again, a pre-enrichment phase in which the protoglobular cluster matter already consisted of enriched gas *and* a stellar component including neutron stars would account for the required neutron star abundance.

4. Speculations About the IMF

If Eddington's precepts are followed, the characteristic mass of a star may be inferred from the appropriate combination of fundamental constants,

$$N \sim (hc^2/Gm_p^2)^{3/2} \sim 10^{57} \text{protons}.$$

Within dimensionless constants, this scale is either the maximum mass of a stable white dwarf, the maximum mass of a gravitationally stable star, or the minimum mass of a non-degenerate star. The latter is presumably the most relevant interpretation, since the solar neighborhood IMF is observed to peak near 0.1–$0.2\,\mathcal{M}_\odot$, somewhat above the minimum hydrogen-burning scale of $0.08\,\mathcal{M}_\odot$. However, any such interpretation begs the question of how the parent molecular cloud was able to subdivide into stellar masses. Molecular cloud masses range from $10^3 - 10^6\;\mathcal{M}_\odot$, and one has to understand their evolution in order to develop a theory for the IMF.

Molecular clouds form because the dissipative galactic gas component forms a cold disk that is gravitationally unstable. Spiral density waves impart non-circular motions that enhance the cloud collision rate, and a spectrum of cloud sizes develops over a rotation period. Individual clouds become unstable, either by accretion of mass due to coalescence with other clouds or by an enhancement in external pressure due to interaction with an HII region or stellar outflow. To what extent fragmentation occurs during the ensuing collapse is highly uncertain. What seems most likely is that the collapse is extremely irregular, and generically forms a dense sheet or filament that is itself unstable to further fragmentation. The most rapidly growing mode results in formation of sub-fragments, whose characteristic wavelength is given by linear theory, and corresponds, for example, to mass $\sim c_s^4/G^2\mu$, where c_s is the sound speed and μ is the surface density of an infinitesimally thin, unstable self-gravitating sheet (Larson 1985). Provided the external pressure exceeds $\sim G\mu^2$ (or the central pressure in the fragment), as expected because of enhanced cooling, the fragments collapse. The masses of the fragments themselves are well defined, and depend primarily on the gas temperature. One finds that $\mu \approx 150$ g cm^{-2} over a wide range of clouds in the interstellar medium, and the variation in T, from $\sim 10K$ in generally small, cold clouds to $\sim 50K$ in massive warm clouds, determines characteristic fragment masses. The reason for this is that the mean pressure in the interstellar medium apparently regulates the gravitational instabilities that determine molecular cloud masses, although whether the dominant effect is by thermal or magnetic pressure is uncertain.

The final outcome must be to form protostars, but there remain many uncertainties that make it difficult to relate the final stellar masses to that of the fragments. The fragments are unstable to further subfragmentation, and subfragments may coalesce or undergo accretion. The relation of the eventual protostellar mass to its parent fragment is further complicated by the necessity for the final protostar to have shed most of its initial specific angular momentum and magnetic flux. The associated processes must presumably have involved mass loss. Rather than attempt to relate stellar masses to the parameters of the parent cloud, it is more fruitful to consider the physics of protostar formation, motivated by phenomenological considerations of protostellar outflows. It is apparent that while there is as yet no definitive mechanism for bipolar flows, such energetic outflows are almost certainly responsible for limiting the accretion of molecular cloud gas onto the forming protostar, and thereby play an important role in determining its final mass.

The following simple model helps demonstrate how protostellar masses may be constrained. The final stages of cloud collapse are highly non-homologous. The dense core rapidly develops and undergoes quasi-static Kelvin-Helmoltz contraction while still surrounded by an accreting envelope of molecular gas. The infall rate is regulated by the molecular cloud environment. In the simplest case of spherical accretion, the rate is $(\Delta v)^3 G^{-1}$ g s^{-1}, where Δv is the effective sound speed (including contributions from turbulence and magneto-acoustic waves). Even in the more likely situation that sufficient angular momentum remains for a massive protostellar disk to form, the accretion rate outside the centrifugal radius should be quasi-spherical, and the preceding estimate should still apply at the outer edge of the disk. The continuing infall results in a standing shock that heats the surface of the protostar, at a rate

$$\sim (\Delta v)^5 G^{-1} \text{ erg s}^{-1}. \tag{1}$$

One may compare this energy input with the energy released within the protostellar core by the Kelvin-Helmholtz contraction, which, if dominant, results in a highly convective protostar. It is the interaction between this convection and the differential rotation and magnetic flux of the core that provide a possible mechanism for tapping the gravitational energy reservoir of the protostar to drive a vigorous outflow. The observed outflows are known to *not* be radiatively driven, rather a mechanical or magnetohydrodynamic energy source must be sought (Pudritz 1988). However, even in the case of a magnetically driven wind, involving winding of field lines and Alfven wave generation that transfers angular momentum outward as a disk develops, support of the protostellar disk requires magnetic and gravitational forces to be in approximate balance, and the ultimate source of energy for the outflow is gravitational energy release. This emanates from the bottom of the gravitational potential well, and in particular from the inner edge of the accretion disk and the surface of the central protostar.

While there is no consensus on a detailed mechanism for driving bipolar flows, one can make semi-quantitative progress with the following argument. I have argued elsewhere (Silk 1989) that as long as convection is suppressed by the infall, the outflow mechanism remains inoperative, and the protostellar mass increases by accretion until eventually the Kelvin-Helmholtz contraction energy release becomes dominant. That this situation *must* eventually prevail may be inferred because the energy released by gravitational contraction increases rapidly with protostellar mass m: for example, the Kelvin-Helmholtz luminosity

$$\sim L/(\kappa\rho R) \sim \sigma m^3 \kappa^{-1} (Gm_p/k)^4, \tag{2}$$

where κ is the Rosseland mean opacity in the core. Comparison of (1) and (2) leads to a constraint on protostellar mass that is imposed by the molecular cloud environment, namely

$$m < m_{crit}, \text{ where}$$

$$m_{crit} \approx 2(\Delta V/2km \text{ s}^{-1})^{5/3} \mathcal{M}_\odot, \tag{3}$$

ensures that convection is suppressed. This allows infall to continue, so that m_{crit} is a lower bound on the protostellar mass. In fact, m_{crit} is also a characteristic stellar mass if the IMF is sufficiently steep, as is observed. Note that in cold clouds, $m_{crit} \lesssim 0.1 \mathcal{M}_\odot$, whereas in warm clouds, $m_{crit} > (1-2) \mathcal{M}_\odot$.

This model is only suggestive, neglecting, for example, any deviations from spherically symmetric infall. However, it does point to an intriguing interpretation of the observation of apparently bimodal star formation, whereby molecular cloud characteristics determine the lower mass cut-off of the IMF. If m_L is identified with m_{crit}, it is evident that m_L would not be a sharp cut-off when averaged over, say, a starburst galaxy or a protogalaxy that contains a broad spectrum of molecular clouds.

5. Star Formation Rate

The rate of star formation is another critical ingredient that must be extrapolated to the protogalactic environment in order to meaningfully address the possibility of baryonic dark halos. While one is far from an understanding of the star formation rate in a forming galactic spheroid, progress has been made toward modelling star formation rates in galactic disks. The following semi-empirical rule works remarkably well:

$$\text{SFR} = \epsilon(1 - R)^{-1}(\mu_{\text{gas}} - \mu_{\text{thr}})\,\Omega(r), \tag{4}$$

where R is the returned gas mass fraction per cycle of star formation, μ_{gas} is the surface density and Ω (r) is the local rotation rate at radius r. The parameter ϵ is equivalent to the star formation efficiency. Equation (4) is motivated by the requirement that a self-gravitating disk of gas and stars should be marginally unstable to non-axisymmetric instabilities which govern the molecular cloud and star formation rates. For example, marginal instability of a thin one-component disk requires

$$1 \lesssim Q \lesssim 2, \text{ where } Q \equiv \sigma K/\pi G\mu, \tag{5}$$

K ($\approx 2\Omega$ if $\Omega \propto r^{-1}$) is the epicyclic frequency, and σ is the velocity dispersion. Local perturbations are swing-amplified in this regime (Toomre 1981), developing into density waves which drive formation of molecular clouds. If dissipation maintains Q in the required range, the instability growth rate is approximately equal to Ω, and results in a molecular cloud, and presumably star formation, rate proportional to $\mu_{\text{gas}}\Omega$ (r) at radius r. Condition (5) sets a threshold value for the total gas surface density in terms of measured $\sigma(\sim 5 - 10 \text{kms}^{-1})$ and Ω (r), below which the disk is stabilized and star formation is suppressed, $\mu_{\text{thr}} \equiv \sigma K/\pi GQ \sim 10\ \mathcal{M}_\odot \text{pc}^{-2}$, for characteristic values of $\sigma(6\text{kms}^{-1})$ and K (10^{-15}s^{-1}). This prescription then yields a detailed fit to observed radial gradients in the star formation rate in disk galaxies (Kennicutt 1989). It also provides a global model for the variations in star formation rate observed along the Hubble sequence, as well as for the production of metallicity gradients in disks (Wyse and Silk 1989).

Self-regulated disk instability motivates a star formation law that is also capable of accounting for the elevated star formation rates observed in starburst galaxies. Enhancement of both the shear rate, parameterized by $\Omega(r)$, and the gas density by an order of magnitude is likely to occur in a merger between galaxies. A strong tidal interaction or presence of a bar will lead to milder effects, but should still stimulate cloud aggregation and star formation. Simulations suggest that the inelasticity of the gaseous component produces considerable angular momentum transfer and results in the gas being concentrated into a dense central ring over the inner several hundred parsecs of the disk. Conditions are also optimal for increasing m_L under such conditions, and the gas return fraction, equal to about 30 percent for a solar neighborhood IMF, increases to R $\sim$ 0.9 if $m_L \gtrsim 2\mathcal{M}_\odot$. The star formation rate can therefore plausibly be enhanced by up to three orders of magnitude in strongly interacting gas-rich disks.

It is intriguing to note that this model is in good accord with IRAS observations which probe star formation rates over a wide range in disk galaxies. The data suggest that the ratio of far-infrared to blue luminosity, a monitor of specific star formation rate, is systematically enhanced by factors of $\sim$ 2, $\sim$ 10, $\sim$ 100 in barred spirals, tidally interacting galaxies, and merging galaxies, respectively. Strongly interacting and merging galaxies are also found to have higher far-infrared luminosities by about an order of magnitude at a given CO luminosity, which is considered to be approximately equivalent to a specified mass in molecular gas (Solomon and Sage 1989).

6. Protogalaxy Model

My next objective is to describe a model for protogalaxy formation, motivated by the success of a collisional trigger for inducing starbursts. Empirical evidence that supports such a hypothesis is that galaxy morphology is strongly correlated with local density. The spiral fraction drops and the elliptical fraction rises steeply with increasing density, above $\sim 100\text{Mpc}^{-3}$ or about 10^4 times the field number density of luminous galaxies. The S0 abundance is intermediate between these extremes. It is difficult to account for this phenomenon unless environment has affected the galaxy formation process. The following schematic model relates environment to galaxy formation.

Suppose that a protogalactic cloud remains predominantly gaseous until it interacts strongly with another protogalaxy. The stronger the interaction, the more extreme will be the amount of induced star formation. In a protocluster, mergers between galaxies of comparable mass, or of a massive galaxy with several smaller galaxies, are expected to occur. In the field, however, the typical encounter will be between a massive galaxy and a much smaller system, and such encounters will be relatively rare. In order to drive star formation, and this will generally apply in the field, in galaxy groups, and in the initial protocluster collapse phase, relative encounter velocities must typically be low enough (less than several hundred km s^{-1}) to guarantee strong dynamical interactions and eventual mergers.

The resulting disruption is especially marked for a satellite on a prograde orbit. Such orbits should be common, given the general tendency of angular momentum acquisition on all non-linear scales by tidal torquing. But more significantly, the preponderance of gas that our hypothesis requires for the early interactions will ensure strong dissipation and prolific star formation provided the encounter velocities are not too large. Mergers between massive gaseous systems will result in considerable heat input to the gas. Little small-scale substructure should survive. The vigorous stirring and mixing of the gas will enhance coalescence of interstellar clouds, and giant cloud complexes would be expected to develop. Instead of the turbulent velocity of v $\lesssim$ 10kms^{-1} associated with giant molecular clouds in our galaxy, one would expect gas random motions up to an order of magnitude larger. Since typical cloud masses scale as $\sim$ v^4, cloud complexes of $\gtrsim 10^9$ $M_\odot$ should form.

An independent argument supports this conclusion. If the primordial fluctuations from which galaxies developed were generated by random gaussian fields, then the rare, high-threshold, peaks which are more strongly clustered than low threshold peaks, are the plausible predecessors of elliptical galaxies . These objects also have low primordial spin or specific angular momentum, and one therefore anticipates that coalescence will play a greater role for formation of galactic-scale structures. Orbits are predominantly radial and this favors collisions. The net effect of enhanced coalescence is that substructure in galaxies should consist of relatively massive clouds.

The ensuing evolution of a gas-rich, star-forming protogalaxy will be very different depending on whether it is initially warm (massive substructures or cloud complexes predominate) or cloud (low mass clouds are the norm). There are two distinct physical effects to be considered. Massive clouds form stars much more efficiently than low mass clouds, the latter being easily disrupted by energy input from massive stars. In fact, v $\sim$ 100km s^{-1} is the critical galaxy escape velocity above which even supernova ejecta are not capable of causing much disruption and interrupting the star formation process. Another consequence of the presence of massive star-forming clouds is that during protogalactic collapse, the stellar cluster cores are likely to be more efficient transporters of angular momentum than predominantly gaseous fragments (Zurek *et al.* 1988). As a consequence both of dynamical friction and gas drag, the more massive and dense cloud-star complexes should spiral in towards the inner protogalaxy, where their internal star formation rate is likely to be further enhanced due to stronger tidal interactions with the local gravitational field and to collisions with other clouds.

This scheme lends support to the observational inference that ellipticals must have undergone highly efficient star formation over $\lesssim$ 2Gyr, whereas spirals are continuously forming stars at a more or less constant rate. Environmental processes provide a natural discriminant between formation of the different morphological types. In dense regions, warm initial conditions within protogalaxies arise as a consequence of collisions and tidal interactions that enhance coalescence. Protogalaxies contain massive cloud complexes that lead to efficient star formation over a galaxy collision time (typical $\sim$ 1 Gyr) before the development of a hot intergalactic gas in virialized rich clusters strips or overpressures the interstellar gas in the protogalaxies. In low density regions, collisions are relatively rare. The "natural" scale for cloud masses is associated with thermal instability (Fall and Rees 1985), and of order $(T_{cloud}/T_g)^2 M_g$, where $T_{cloud} \sim 10^4$K and $T_g \sim (1-3) \times 10^6$K. These "cold" initial conditions result in low mass ($\sim 10^6 \mathcal{M}_\odot$) clouds that form stars inefficiently and are easily disrupted.

Consequently, galaxies form slowly in the field and remain gas-rich, there being no environmental effect capable of disrupting their gas reservoir apart from a rare merger with another galaxy of comparable mass. Spiral and elliptical galaxies accordingly dominate in the field and in clusters, and form stars inefficiently and efficiently, respectively. S0 galaxies form an intermediate population that a more detailed theory of environmental impacts should be able to describe: S0's are not simply stripped spirals, but may be protospirals that have merged with smaller companions, retaining much of their initial disks but boosting the relative sizes of the spheroidal components. A satellite merger is capable of concentrating the gas reservoir of the larger galaxy, producing a transient active galaxy that finally may resemble a gas-poor S0. Perhaps also ram pressure stripping of protospirals would result in formation of S0's, thereby providing a population of galaxies that would form over a wider range of galaxy cluster environments than ellipticals, for which we require mergers between protospirals of comparable mass, as well as in the field via satellite mergers.

7. Dwarf Galaxies

Dwarf galaxies play an important role in galaxy formation theory. For example, dE's did not presumably form by mergers, yet are indistinguishable in several respects from luminous ellipticals. dE's generally appear to cluster with luminous galaxies (Binggeli *et al.* 1985). The luminosity function of field dE's appear not to continue rising to low luminosities:

rather dIrr's are more numerous (Binggeli *et al.* 1989). This is the converse of what is seen in clusters where samples are more complete; here the extreme low luminosity dwarf population consists exclusively of dE's.

Several explanations have been advanced to explain the dwarf galaxy populations. One is that the dE's are failed galaxies, unable to retain their gas reservoir once star formation was initiated because of their shallow potential wells (Larson 1974). In the cold dark matter model, natural candidates for forming such common, low mass objects are the low threshold fluctuations, which are weakly clustered and provide a potential hiding place for the unseen dark matter postulated by the CDM model (Dekel and Silk 1986). The large M/L values reported for dwarf spheroidals support such a model, but the lack of segregation of dE's argues against it. Another possibility is that dE's are overstimulated, burnt-out relics. The low metallicities reported for dE's argue against this possibility, as does the intuitive expectation that star formation should be highly inefficient in such shallow potential wells.

It is likely that dIrr's provide initial clues to the nature of dwarf galaxies. It has been argued that dIrr's are resuscitated dE's which have undergone accretion of gas (Silk *et al.* 1987). This would favor the presence of dIrr's in groups but not in clusters of galaxies. Gas stripping of these dIrr's would produce nucleated dE's, which must therefore have formed during the initial cluster formation phase and would be more frequent at cluster peripheries than in cluster cores relative to the unnucleated dE's that have not undergone episodes of gas infall.

An alternative possibility for dIrr's is suggested by the environmental impact hypothesis and does not appeal to gas infall. Galaxies remain gas-rich until they suffer a strong interaction or merger. This happens in dense regions; the result is a star formation burst and production of dE's at an early epoch, so that dE's cluster along with luminous ellipticals. Wind-driven stripping ensures that dE's remain metal-poor. Near spirals, however, dIrr's remain quiescent until their orbits carry them within tidal shocking range, when vigorous star formation is initiated. This may be a cyclical affair, as gas infall is also possible once the burst of star formation has ceased. Finally, in isolated regions, dwarfs, as well as giants with high primordial specific angular momentum, form stars at an exceedingly slow rate, being below the critical surface density for self-regulated star formation. Intergalactic gas clouds and extremely low surface brightness disks are nearby examples of such nascent galaxies. Most dIrr's would then be forming stars at a high rate for the first time in their history; any old star population should be diffuse and characterize a very low but essentially constant star formation rate over a Hubble time. One would expect dE's near spirals and in the peripheries of clusters to be considerably younger than dE's in rich cluster cores.

The semistellar nuclei in many dE's might then simply be a central massive star cluster that is a relic of star formation that occurred at an intermediate epoch, perhaps 1-2 Gyr ago. This would require the nucleated dE's to be those in regions of lower density where environmental impact would have been delayed relative to the dense cluster cores. Other morphological characteristics of dE's are readily understandable in the context of the supernova-driven wind ejection model. The energy input from massive stars, both via stellar winds as well as supernova remnants, will puff up these protogalaxies before star formation is terminated. The result is an object that is morphologically distinct from a dIrr, in which the bulk of star formation occurs in a disk and the energy input from massive star formation has not yet accumulated to a disruptive level. dE's should, therefore, be round but axially symmetric and rotationally supported. Luminous ellipticals, which form by mergers, are expected to also be round, but generally triaxial and pressure-supported. Merger simulations find that the low angular momentum cores of the merging systems produce a low angular momentum core in the final system. Consequently, while merger products tend to retain preexisting metallicity gradients, the oldest stars forming on the outside, and younger stars as well as the preexisting metal-rich nuclei dominating the inner protogalaxy, one would expect dE's to show a distinctly different pattern. The younger stars form later in the nucleus and in the thick disk/spheroid than in the thin disk according to the above model, and one might expect much weaker or even inverted metallicity-gradient kinematics correlations to those ordinarily encountered in massive spheroids.

There are interesting implications at high redshift. At $z \sim 2$-3, the damped Lyman-alpha systems seen in absorption against distant quasars have properties reminiscent of disk galaxies, including line width, column density and heavy element abundance patterns (Wolfe 1989). The abundances are low, characteristic of dIrr's or protogalactic disks rather than large spirals. The sizes are not measured directly, although a 21cm absorption observation of one system suggests a minimum extent of $\sim$ 10-20 kpc. The gas content of these systems at $z = 2$-3 amounts to $\Omega \sim 0.005$, comparable to that of nearby luminous disks. The effective cross-section of the damped Lyman systems is equivalent to $\sim 10^{-3}$ $(R/100kpc)^2$ Mpc^{-3} in terms of the comoving number density of absorbers. Lyman emission has been detected from one system, and upper limits set on several others at values far lower than expected for luminous spirals (Hunstead and Pettini 1989). The centering of the detected Lyman-alpha emission line on the damped Lyman-alpha absorption line implies that this absorbing system is small, of diameter $\lesssim$ 20 kpc to avoid a large offset due to galactic rotation. If the cloud diameters are this small, they must be as numerous as dwarf galaxies (comoving number density $\sim 0.1 Mpc^{-3}$). The damped Lyman-alpha systems could therefore be protodwarfs, which are still gas-rich and would both exist in sufficient number and be large enough in size compared to nearby dwarfs to give the required effective cross-section. The expected surface density is of order $10^9 \mathcal{M}_\odot/\pi(10kpc)^2$ or $\sim 3 \mathcal{M}_\odot pc^{-2}$, equivalent to a column density $N_H \sim 3 \times 10^{20} cm^{-2}$, similar to that in the damped Lyman-alpha systems.

These objects have no low redshift counterparts: local dwarfs are much smaller systems. Perhaps the explanation is that by $z \sim 1$, the gas-rich dwarfs should be undergoing bursts of star formation. They may be sufficiently numerous to account for the large number of faint blue fuzzy objects, some $\sim 10^5$ per square degree, seen in deep galaxy counts below a J magnitude of ~ 26 (Tyson 1988). At high redshift ($z \gtrsim 5$), dwarf galaxies may already have formed a sufficient number of massive stars to contribute to the metagalactic ionizing flux. The requirement imposed by the Gunn-Peterson constraint, if the intergalactic medium were photo-ionized, is that approximately 100 ionizing photons per baryon are needed, and this requires, for a solar neighborhood IMF, an ionizing photon flux equivalent to that produced in the course of synthesizing a metal abundance of order solar metallicity in galactic disks (Shapiro and Giroux 1989). One would need at least as much baryonic matter in high z dwarfs as is seen in damped Lyman-alpha systems, and perhaps as much as $\Omega \sim 0.1$, most of which would have formed stars and metamorphosed into gas-poor dE's and intergalactic gas by $z \sim 3$. Since intracluster gas has a metallicity of about 1/3 solar, one infers that if it were present throughout the universe in a similar ratio to the stellar mass in clusters of galaxies, a sufficient number of ionizing photons would have been generated provided the intracluster iron originated at $z \gtrsim 4 - 5$. The apparent excess of oxygen $[O/Fe] \sim 0.5$ in intracluster gas parallels that seen in old halo stars, and is suggestive of a very early origin, before most of the Fe production had occurred. Dwarf galaxies would not have retained the massive star ejecta and thereby could account for the O/Fe enhancement via early supernova-driven winds.

Finally, and very speculatively, the submillimeter excess reported in the cosmic microwave background has been interpreted as being due to dust emission from protogalaxies at large redshift (Carr 1989). The stringent limits on fluctuations set near 1mm require an enormous comoving number density of sources, $\gtrsim 10\mathrm{Mpc}^{-3}$, to be present at very high redshift (z > 30) in order for there to be sufficient dust optical depth. Dwarf galaxies could therefore play an even more exotic role in the early universe: the required abundance is $\Omega \sim 0.1$ in stars at $z \sim 30$ to provide the requisite energy source that shows up as an excess today amounting to ~ 20 percent of the CMB energy density.

8. Recipes for a Halo of Stellar Remnants

Two alternatives may be envisaged for producing halos of stellar remnants in the form of white dwarfs, neutron stars and black holes. At very high redshift, $z \gtrsim 100$, star formation in dwarf galaxies with $m_L \gtrsim 2\text{-}3\ \mathcal{M}_\odot$ would have inefficiently processed the gas. The prominent role of mergers at such high z, when galaxies are practically overlapping, motivates the large value of m_L via the stirring mechanism proposed above. Supernova-driven winds drive out modestly enriched gas, up to $[Z] \sim -1$. With $\Omega \sim 0.1$ in dwarfs, one would produce $\Omega \sim 0.07\text{-}0.09$ in hot, enriched intergalactic gas and $\Omega \sim 0.001\text{-}0.003$ in dark remnants embedded within loosely bound dwarfs. Compton drag would effectively couple the gas to the microwave background until $z < 100$, leaving the baryonic remnants to cluster and eventually form massive dark halos. Gas infall at low redshift, once the dark potential wells were in place, could plausibly supply the 10 percent contamination of baryons that is responsible for galactic disks and spheroids. According to this prescription, 90 percent of the baryonic matter would be in the enriched intergalactic medium. Because of the inefficiency of the initial star formation mode, it is difficult to envisage this scenario producing much more baryonic dark matter than is required for dark halos.

The second option is that a massive ($\sim 10^{12}\ \mathcal{M}_\odot$) protogalactic cloud collapses and forms stars almost exclusively with $m_L \gtrsim 2\ \mathcal{M}_\odot$. The stellar debris is contained in the protogalaxy and recycled, star formation in this massive mode continuing until $\gtrsim 90$ percent of the initial gas mass is locked up in stellar remnants. The remaining gas is highly enriched if the IMF extends to $\sim 100\ \mathcal{M}_\odot$ and the yields are similar to those in conventional star formation. This creates a difficulty, since the initial enrichment of the disk was about 10 percent of the solar metal abundance. There are two possible resolutions: either the mixing was very inefficient, and one can tolerate at most one or two percent leakage of the ejecta into pristine protodisk gas, or the yields are greatly suppressed below conventional values, as would occur if the primordial IMF is truncated above 10-15 $\mathcal{M}_\odot$. Even if the IMF were truncated at high masses, one still has the concern that intermediate mass stars synthesize and eject helium, so that one again has to impose a mixing limit that only allows up to $\sim$ 10 percent of the halo ejecta to contaminate the disk gas (Ryu *et al.* 1989). Carbon is also ejected, but may be less of a problem for primordial stars which may not undergo helium flashes and carbon dredge-up. The mixing efficiency is difficult to estimate, depending on the turbulent diffusion coefficient D and mixing length scale, which one may estimate to be of order $(Dt)^{1/2} \equiv \{1/3 \times 100\text{km/s} \times 1\text{kpc} \times 3\,\text{Gyr}\}^{1/2} \sim 10$ kpc over several protohalo gas-e-folding time-scales ($\lesssim$ several Gyr if $m_L > 2\,\mathcal{M}_\odot$). Further uncertainties arise with the orbit distribution of the halo gas, radial orbits mixing efficiently, and the relative sizes of protodisk and protohalo.

One could tolerate $\Omega_{\text{BDM}} \gtrsim 0.1$ according to this recipe, since the gas ejecta are effectively recycled. Of course, only $\Omega_{\text{BDM}} \lesssim 0.03$ is in the form of dark halos that are associated with luminous galaxies. The bulk of the efficiently processed baryonic dark matter would be in dark halos that have failed to trap a dense enough gas cloud to form stars at a rate large enough to result in a luminous galactic disk. Failed giants are an inevitable consequence in this case, and objects such as Malin 1, namely low surface brightness but gas-rich giant disks, are possible prototypes of such objects.

Another implication is that the efficient processing of baryons results in considerable stellar luminosity being produced at large redshift. This would show up as a contribution to the diffuse background light. For an effect comparable to existing limits on diffuse extragalactic light ($\lesssim 0.001\Omega_{\text{CMB}}$), one only requires $\Omega_{\text{BDM}} \equiv 0.01[(1+z)/10](100\Delta\Omega_{\text{EGL}}/\Omega_{\text{CMB}})(0.5/h)^2$ to have been processed to 10 percent of solar metallicity at $z \sim 10$.

9. Conclusions

The radical notion that halos consist of dark stellar remnants is of little interest unless it is subject to experimental test. Relic white dwarfs are likely to be an inevitable constituent of such halos. The discover that the white dwarf luminosity function declines rapidly below $\sim 10^{-4.5}\mathcal{L}_\odot$ has led to the realization that the fall-off is sufficiently steep that it must be attributed to formation of the disk (Winget *et al.* 1987). The disk age inferred is about 8 Gyr, and leads to the expectation that an older population of white dwarfs should still be observable. Modeling of white dwarf formation rates shows that a survey to $\sim$ 21st magnitude should be capable of detecting a halo population that contains white dwarfs up to 6 Gyr older than the disk population (Tamanaha *et al.* 1989). Moreover, the halo component should show up in deep proper motion surveys as having large tangential motions: indeed, the first population II white dwarfs have recently been identified in a survey that is complete only to 18th magnitude.

The $\sim r^{-2}$ density profiles usually associated with massive halos are usually considered to rise dynamically, from infall of cold dark matter or possibly via mergers. Dissipational collapse, inevitable if the dark halos consist of compact stellar remnants, is also capable of producing $\sim r^{-2}$ halos, as may be inferred from the similarity solution for isothermal collapse of a uniform spherical gas cloud.

In summary, baryonic dark matter is an attractive option for galaxy halos, since baryons are the observed form of matter. Compact stellar remnants are natural candidates for the dark matter particles, since such objects are known to form today, and to have formed at the earliest epochs of galaxy formation. The only uncertainty is whether they have formed efficiently and in sufficient numbers. I have argued that enhancement of m_L, as inferred to occur in starbursts, can be explained as a consequence of the vigorous gas stirring that occurs in galaxy mergers, which are known to be starburst triggers and which occurred prolifically in the early universe according to several cosmic evolution schemes.

I presented two alternatives for the sites of primordial star formation. An inefficient option appealed to a vast population of dwarf galaxies at high redshift that resulted in an hot enriched intergalactic medium with $\Omega_{IGM} \sim 0.1$ and a "cold" component of dark remnants with $\Omega_{BDM} \sim 0.03$ that later condensed into galaxy halos in which only a few percent of the intergalactic gas was trapped. The efficient option postulated that giant galaxies formed stars prolifically with $m_L > 2\,\mathcal{M}_\odot$, and recycled the stellar ejecta until most of the gas was locked up into compact remnants. The former possibility pre–enriches the intergalactic gas to a level inferred for protodisks, and the high redshift dwarfs may already be observable from damped Lyman-alpha absorption systems and in deep galaxy counts. The latter predicts that isolated, dim gas-rich giants, such as Malin 1, may be abundant today, with the gas forming a low density disk that is stabilized by the surrounding, massive halo. Both possibilities face difficulties in accounting for the pristine nature of the oldest halo stars and of the Lyman-alpha forest absorption line systems. Perhaps both dwarfs and giants formed dark matter with differing efficiency, and the legacy would be $\Omega_{BDM} \gtrsim 0.1$. The main objection to so high a value of Ω_{BDM} is the diffuse radiation that is produced by the primordial stars: this can be effectively hidden if the star formation redshift is sufficiently large.

ACKNOWLEDGEMENTS

I am indebted to my collaborators C. Lacey, R. Schaeffer, and especially R. F. G. Wyse for many discussions during the course of this work. This research has been supported in part by grants from NASA and NSF.

References

Bessel, M.S. and Norris, J. 1984, *Ap. J.*, **285**, 622.

Binggeli, B., Sandage, A. and Tammann, G. A. 1985, *Ap. J.*, **89**, 66.

Binggeli, B., Tarenghi, M. and Sandage, A. 1989, *Astr. Ap.*, (in press).

Carr, B. J. 1989, in *The Epoch of Galaxy Formation*, ed. C. S. Frenk *et al.*, (Dordrecht: Kluwer Academic), 227.

Dearborn, D., *et al.*, 1986, *Ap. J.*, **300**, 314.

Dekel, A. and Silk, J. 1986, *Ap. J.*, **303**, 39.

Devereux, N. A. 1989, *Ap. J.*, (in press).

Fabian, A.C. 1990, in *Baryonic Dark Matter*, eds D. Lynden-Bell & G. Gilmore (Dordrecht: Kluwer), 195.

Fall, S. M. and Rees, M. J. 1985, *Ap. J.*, **298**, 18.

Forman W. 1988, in *Cooling Flows in Clusters and Galaxies*, ed. A. C. Fabian (Dordrecht: Kluwer Academic), 17.

Gallagher, J., Hunter, D. and Tutukov, A. 1984, *Ap. J.*, **284**, 544.

Hunstead, R. W. and Pettini, M. 1989, in *The Epoch of Galaxy Formation*, ed. C. S. Frenk *et al.* (Dordrecht: Kluwer), 115.

Kennicutt, R. C. 1989, *Ap.J.*, **344**, 685.

Kulkarni, S. R., Narayan, R. and Romani, R. W. 1989, *Ap. J.*, (in press).

Larson, R. B. 1974, *M.N.R.A.S.*, **169**, 224.
Larson, R. B. 1985, *M.N.R.A.S.*, **214**, 379.
Larson, R. B. 1986, *M.N.R.A.S.*, **218**, 409.
Murray, S. and Lin, D. C. 1989, *Ap. J.*, **339**, 933.
Pudritz, R. E. 1988, in *Galactic and Extragalactic Star Formation*, ed. R. E. Pudritz and M. Fich (Dordrecht: Kluwer Academic),135
Ryu, D., Olive, K.A. and Silk, J. 1989, *Ap. J.*, (in press).
Sandage, A. 1986, *Astr. Ap.*, **161**, 89.
Scalo, J. 1986, *Fund. Cosmic Phys.*, **11**, 1.
Shapiro, P. and Giroux, M. 1989, in *The Epoch of Galaxy Formation*, ed. C.S. Frenk, *et al.*, (Dordrecht: Kluwer Academic), 153.
Silk, J., Wyse, R. F.G. and Shields, G.A. 1987, *Ap. J. Letters*, **322**, L59.
Silk, J. 1989, in *The Physics and Chemistry of Interstellar Molecular Clouds*, ed. G.Winnewiss and T. Armstrong (Berlin: Springer-Verlag), 285.
Solomon, P. and Sage, L. 1988, *Ap.J.*, **344**, 613.
Tamanaha, F.*et al* 1989, *Ap. J.*, (submitted).
Telesco, C. M., Wolstencroft, R. D. and Done, C. 1988, *Ap. J.*, **329**, 174.
Toomre, A. 1981, in *The Structure and Evolution of Normal Galaxies*, ed. S. M. Fall and D. Lynden Bell (Cambridge: Cambridge University Press), 111.
Tyson, N. 1988, *Ap. J. Letters*, **329**, L57.
Wheeler, J. C., Sneden, C. and Truran, J.W. 1989, *Ann. Revs. Astr. Ap.*, **27**, 279.
Winget, D. *et al.* 1987, *Ap. J. Lett.*, **315**, L77.
Wolfe, A. M. 1989, in *The Epoch of Galaxy Formation*, ed. C. S. Frenk *et al.* (Dordrecht: Kluwer Academic), 101.
Wyse, R.F.G. and Silk, J. 1989, *Ap.J.*, **339**, 700.
Zurek, W. H., Quinn, P. J. and Salmon, J.K. 1988, *Ap. J.*, **330**, 519.

SUBJECT INDEX

LIST OF ACRONYMS

AGB	Asymptotic Giant Branch.
AGN	Active Galactic Nucleus.
AXAF	Advanced X-ray Astrophysics Facility.
BBF	Binary Birth–rate Function.
BBNS	Big Bang Nucleo-Synthesis.
BDM	Baryonic Dark Matter.
BIM	Binary Ion Mixture.
CCD	Charge-Coupled Device.
CDM	Cold Dark Matter.
CMB	Cosmic Microwave Background.
COBE	COsmic Background Explorer.
DIRBE	Diffuse Infra-Red Background Experiment.
EMD	Extremely Metal Deficient.
EXOSAT	European X-ray Observatory SATellite.
FPCS	Focal–Plane Crystal Spectrometer.
FIDS	Filtered Index catalogue of visual Double Stars.
GCE	Galactic Chemical Evolution.
GMC	Giant Molecular Cloud.
GREG	Giant Red-Envelope Galaxy.
IBBN	Inhomogeneous Big Bang Nucleosynthesis.
ICM	Intra–Cluster Medium.
IDS	Index catalogue of visual Double Stars.
IGM	Inter Galactic Medium.
IMF	Initial Mass Function.
IRAS	Infra-Red Astronomy Satellite.
IUE	International Ultra-violet Explorer.
KS	Kolmogorov-Smirnov.
LF	Luminosity Function.
LMS	Lower Main Sequence.
ML	Maximum Likelihood.
MS	Main Sequence.
NGP	North Galactic Pole.
OCP	One Component Plasma.
QCD	Quantum Chromo–Dynamics.
QSO	Quasi Stellar Object, Quasar.
SBBN	Standard Big Bang Nucleosynthesis.
SFR	Star Formation Rate.
SGP	South Galactic Pole.
SIRTF	Space Infra-Red Telescope Facility.
SN	Super-Nova.
WD	White Dwarf.
YBS	Yale Bright Star catalogue.
ZAMS	Zero Age Main Sequence.